# Machining of Complex Sculptured Surfaces

J. Paulo Davim
Editor

# Machining of Complex Sculptured Surfaces

Springer

J. Paulo Davim
Department of Mechanical Engineering
University of Aveiro
Campus Santiago
3810-193 Aveiro
Portugal
e-mail: pdavim@ua.pt

ISBN 978-1-4471-2355-2 e-ISBN 978-1-4471-2356-9
DOI 10.1007/978-1-4471-2356-9
Springer London Dordrecht Heidelberg New York

British Library Cataloguing in Publication Data
A catalogue record for this book is available from the British Library

Library of Congress Control Number: 2011943797



*Cover design:* eStudio Calamar S.L.

Printed on acid-free paper

Springer is part of Springer Science+Business Media (www.springer.com)

# Preface

The machining of complex sculptured surfaces is an important technological topic in modern manufacturing, namely in the molds and dies sector. Today, this sector, with great importance to automotive, aircraft and others advanced industries, is placed in all industrialized or emerging countries. In the recent past, the traditional technology employed in molds and dies manufacture was a combination of conventional milling and electro-discharge machining (EDM) or electrochemical machining (ECM). Nowadays, high-speed milling (HSM) is used in roughing, semi-finishing and finishing of molds and dies with great success. This technology required modern CAM systems and process planning for 3 and 5-axis machining. HSM presents several advantages when compared with the traditional technology in terms of workpiece precision and roughness as well as in manual polishing after the machining operations.

Chapter 1 of this book provides the flank milling of complex surfaces. Chapter 2 is dedicated to 5-axis flank milling of sculptured surfaces. Chapter 3 described high performance 5-axis milling of complex sculptured surfaces. Chapter 4 contains information on milling tool-path generation in adequacy with machining equipment capabilities and behavior and Chap. 5 is dedicated of intelligent optimization of 3-axis sculptured surface machining on existing CAM systems. Chapter 6 contains process planning for 5-axis milling of sculptured surfaces based on cutters accessibility analysis. Finally, Chap. 7 is dedicated to manufacturing of sculptured surfaces using EDM and ECM processes.

The present book can be used as a research book for final undergraduate engineering courses or as a topic on manufacturing at the postgraduate level. Also, this book can serve as a useful reference for academics, manufacturing researchers, manufacturing, industrial and mechanical engineers, professional in machining and related industries. The interest of scientific in this book is evident for many important centers of the research, laboratories and universities as well as industry. Therefore, it is hoped this book will inspire and enthuse other researches for this field of the machining of complex sculptured surfaces.

The Editor acknowledges Springer for this opportunity and for their enthusiastic and professional support. Finally, I would like to thank all the chapter authors for their availability for this work.

Portugal, January 2012 J. Paulo Davim

# Contents

# Contributors

**Dr. P. G. Bernardos** Department of Manufacturing Technology, School of Mechanical Engineering, National Technical University of Athens, Heroon Polytehneiou 9, 15780 Athens, Greece

**Dr. Yaman Boz** Manufacturing and Automation Research Center, Koc University, Sariyer, 34450 Istanbul, Turkey

**Dr. A. Calleja** Department of Mechanical Engineering, University of the Basque Country, Alameda de Urquijo s/n, 48013 Bilbao, Spain

**Prof. F. Campa** Department of Mechanical Engineering, University of the Basque Country, Alameda de Urquijo s/n, 48013 Bilbao, Spain

**Dr. S. K. Ehsan Layegh** Manufacturing and Automation Research Center, Koc University, Sariyer, 34450 Istanbul, Turkey

**Dr. Huseyin Erdim** Mitsubishi Electric Research Laboratories, Cambridge, MA 02139, USA

**Dr. L. Geng** Department of Mechanical Engineering, National University of Singapore, 9 Engineering Drive 1, Singapore, Singapore

**Prof. Wit Grzesik** Department of Manufacturing Engineering and Production Automation, Opole University of Technology, P.O. Box 321, 45-271 Opole, Poland, e-mail: w.grzesik@po.opole.pl

**Prof. Jean-Yves Hascoet** Institut de Recherche en Communications et Cybernetique de Nantes (IRCCyN), UMR CNRS 6597, 1 rue de la Noe, BP92101, 44321 Nantes Cedex 03, France, e-mail: jean-yves.hascoet@irccyn.ec-nantes.fr

**Dr. A. Kimpenis** Department of Manufacturing Technology, School of Mechanical Engineering, National Technical University of Athens, Heroon Polytehneiou 9, 15780 Athens, Greece

**Prof. A. Lamikiz** Department of Mechanical Engineering, University of the Basque Country, Alameda de Urquijo s/n, 48013 Bilbao, Spain

**Prof. Ismail Lazoglu** Manufacturing and Automation Research Center, Koc University, Sariyer, 34450 Istanbul, Turkey, e-mail: ilazoglu@ku.edu.edu

**Prof. L. N. López de Lacalle** Department of Mechanical Engineering, University of the Basque Country, Alameda de Urquijo s/n, 48013 Bilbao, Spain, e-mail: norberto.lzlacalle@ehu.es

**Prof. Frédéric Moniès** Institut Clément Ader, Université Paul Sabatier, 118 route de Narbonne, 31062 Toulouse Cedex 09, France

**Dr. D. Olvera** Department of Mechanical Engineering, University of the Basque Country, Alameda de Urquijo s/n, 48013 Bilbao, Spain

**Dr. Mattieu Rauch** Institut de Recherche en Communications et Cybernetique de Nantes (IRCCyN), UMR CNRS 6597, 1 rue de la Noe, BP92101, 44321 Nantes Cedex 03, France

**Prof. Walter Rubio** Institut Clément Ader, Université Paul Sabatier, 118 route de Narbonne, 31062 Toulouse Cedex 09, France, e-mail: rubio@cict.fr

**Prof. Adam Ruszaj** Faculty of Mechanical Engineering, Cracow University of Technology, al. Jana Pawla II 37, 31-864 Cracow, Poland

**Dr. Johanna Senatore** Institut Clément Ader, Université Paul Sabatier, 118 route de Narbonne, 31062 Toulouse Cedex 09, France

**Prof. G.-C. Vosniakos** Department of Manufacturing Technology, School of Mechanical Engineering, National Technical University of Athens, Heroon Polytehneiou 9, 15780 Athens, Greece, e-mail: vosniak@central.ntua.gr

**Prof. Y. F. Zhang** Department of Mechanical Engineering, National University of Singapore, 9 Engineering Drive 1, Singapore, Singapore, e-mail: mpezyf@nus.edu.sg

# Chapter 1
# Flank Milling of Complex Surfaces

**D. Olvera, A. Calleja, L. N. López de Lacalle, F. Campa and A. Lamikiz**

In this chapter the main methods, machining strategies and possible problems when flank milling complex surfaces, are deeply explained. Flank milling is an operation defined by using large axial depth of cut with end milling tools, high cutting speed and relatively small radial depths of cut. This process is especially recommended for ruled surfaces machining, whose tangential contact of the involving cylinder with the cutting tool body is the key factor to define the tool paths. Due to the complexity of these kinds of surfaces, 5-axis milling is required taking special care of the geometrical interferences between the tool and the complex geometry of the pieces in order to avoid collisions. Finally, a new model for the prediction of roughness and dimensional accuracy on thin-walled component is presented, along with examples of parts with surfaces which need the flank milling operations due to their complexity.

## 1.1 Complex Surfaces and Milling

The book now in the reader's hand is focused on machining technologies for complex surfaces production regarding different applications. High speed ball-end milling is the most spread technology currently used for free form or sculptured surfaces machining [1]. The main industries using the process are mould and die making. However, there are other complex surfaces that can be included into the general category of warped surfaces i.e., a surface generated by a straight line

D. Olvera · A. Calleja · L. N. López de Lacalle (✉)
F. Campa · A. Lamikiz
Department of Mechanical Engineering, University of the Basque Country,
Escuela Técnica Superior de Ingeniería Industrial,
c/Alameda de Urquijo s/n, 48013 Bilbao, Spain
e-mail: norberto.lzlacalle@ehu.es

J. P. Davim (ed.), *Machining of Complex Sculptured Surfaces*,
DOI: 10.1007/978-1-4471-2356-9_1, 

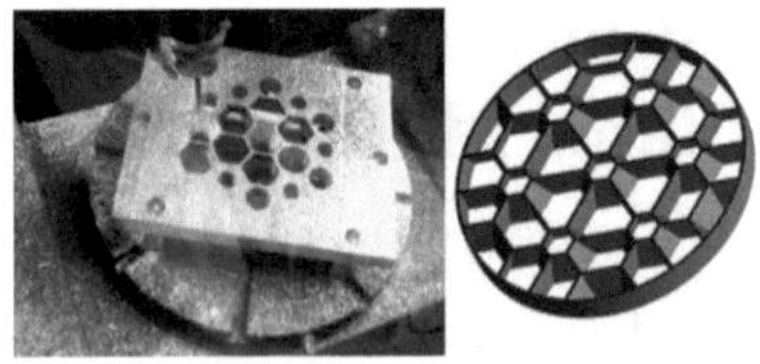

Fig. 1.1 A honeycomb structure for a satellite, continuous 5-axis milling

movement so that no two of its consecutive positions shall be in the same plane; these are also known as ruled surfaces [2]. This chapter is devoted to describe the milling and production of these kinds of surfaces.

In this field the development of cutting tools, machining strategies, CAM software and machine's programming are pieces of the same solution [3–5].

### 1.1.1 Ruled Surfaces and Applications

In algebraic geometry, ruled surfaces were originally defined as projective surfaces in projective space containing a straight line through any given point. This immediately implies that there is a projective line on the surface through any given point, and this condition is now often used as the definition of a ruled surface: ruled surfaces are defined to be abstract projective surfaces satisfying the condition that there is a projective line though any point. This is the key to an easy way of milling, the so-called flank milling: to keep tangential contact of the cylindrical envelope to end milling tool along this surface straight line, and applying as long axial depth of cut as allowed by part geometry and spindle power [6].

One example is shown in Fig. 1.1, where a small honeycomb structure in Al7075-T6 is produced by simultaneous movement of the five axes of a high speed milling machine. In this case only one end mill of 8 mm diameter was used for all milling operations. There were no differences between roughing and finishing operations because the spindle was kept at a constant rotational speed of 18,000 rpm. The total thickness of the plate was 30 mm therefore, this value was defined as the axial depth of cut. The total machining time was 54 min for a $400 \times 400$ plate. In this case the inclination for the walls was 20°.

### 1.1.2 Thin Featured Parts

The previous case presented is an example of aerospace parts in which ruled surfaces are a common feature, often included in part designs without special interest from designers: a wall twisted or inclined along a boundary is actually a ruled surface but a consequence of the design requirements. This is the usual case in industrial applications, not to build a revolution hyperboloid, which are common in architecture and obviously not obtained by machining.

Besides the geometrical shape of surfaces, usually there is another reiterative geometrical factor, the little thickness of walls which defines the so-called thin-

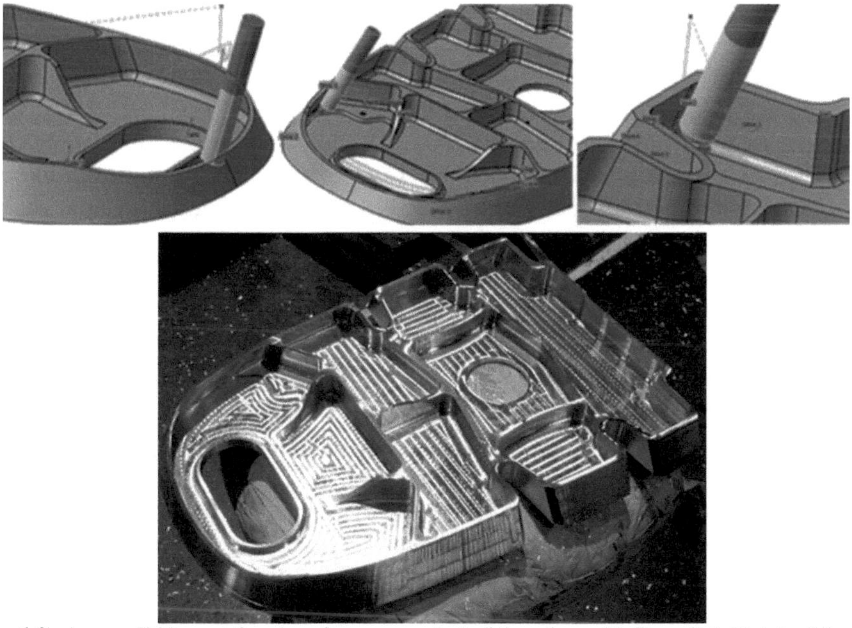

Fig. 1.2 A monolithic part for aircraft components (colaboration with Tecnalia). Detail of the CAM programming (*Up*). Actual piece (*Down*)

walled parts. The pieces shown in Figs. 1.1 and 1.2 are good examples of it. In Fig. 1.2 a light component for aircraft structure in its CAM stage is shown. In this case the 5-axis flank milling is illustrated showing a detail when using the flank of the tool to machine the wall surface. The final manufactured part is also shown after removing the 95% of material from the starting raw block.

At present, airframes are mainly composed of monolithic components, instead of small parts joined using welding or riveting. Inside this category, ribs, stringers, spars and bulkheads can be mentioned. After milling they are assembled and joined to the aircraft skins, the latter being milled as well. The aim of these components is to obtain a good strength-to-weight ratio based on their homogeneity.

The milling of a monolithic structural part implies removing up to 95% of the weight from the raw block material. Therefore, to achieve a removal rate as high as possible is the main objective. However, at high removal rate conditions (high feed, large depth of cut) milling implies high cutting forces inducing over the part deflection or vibration in low stiffness zones such as thin walls and/or floors. These static and dynamic problems often lead to geometry inaccuracy, poor surface quality and in the worst cases damage of the machine tool's spindle.

When manufacturing thin walled components, the spindle speed must be the maximum permitted by current spindle technology, based on asynchronous motors introduced in spindles and supported by hybrid bearings (steel races with ceramic ball bearings); a value between 18,000 and 25,000 rpm is a usual maximum speed for the current machine tools. This milling type is usually known as HPC (High

Performance Cutting); the main difference in comparison with classical approach of High Speed Milling (HSM) is the depth of cut, several millimetres in the former and only some hundredths of a millimetre in the latter [7, 8].

Nowadays, HPC is quite extended in aeronautical production. However, some of the problems derived from this process usually lead to non-conforming parts and as consequence to a considerable waste of time and money regarding the added value of airframe parts because of the price regarding the material and the value of previous machining operations [9].

## 1.2 5-Axis Milling

The multi-axis machining advantages can be divided into two main groups. First, the industrial advantages can be referred [10], involving the capability of five axes machining process to improve productivity and precision by using machine additional axes. The two additional orientation axes allow the machining of very complex parts, which cannot be machined using three axes machines. For example, in automotive sector all part faces must be machined, so different set-ups and fixturing are avoided with a 5-axes machine. This improves both productivity and precision by reducing set-up idle times and errors occurred between different set-ups. Additionally, more suitable tools for each operation can be used in order to increase productivity just by positioning the tool and the workpiece. Finally, the tool length necessarily large when deep cavities are machined is reduced. Therefore, the tool stiffness is higher which increases the machining precision and reduces the risk of tool breakage [11].

Some of these advantages are shown in Fig. 1.3 (upper) the total machining of a complex part in only one fixture and also the use of shorter and stiffer tools. As shown in Fig. 1.3 (lower) tool stiffness [12, 13] is directly related with the L3 factor, hence a reduction in tool length dramatically reduces tool deflection and the lack of precision due to this effect. In the past years during the EMO fairs and other national exhibitions, a lot of 5-axis milling centres were exhibited machining in a $3 + 2$ operation mode [14], orienting tool axis with respect to a target surface and machining only with interpolation of the three cartesian axes. As example, in Fig. 1.3 (upper) a test polyhedral part is presented; this aluminium part was made in only 3 min with a three inserts face milling plate.

On the other hand, some technological advantages can be highlighted. The tool orientation can be used to increase productivity by changing both the type of tool (using a stiffer or more productive one) and the machining strategy [15]. Three examples can be illustrative of this:

- In the finishing operation of ruled surfaces, the flank milling strategy can be used [16], machining with the cylindrical part of the tool with a big depth of cut. This strategy can reduce machining time and improve surface finish.
- Another example is the machining of inclined planes: using the correct tool-axis orientation face milling operation can be carried out instead of a ball end

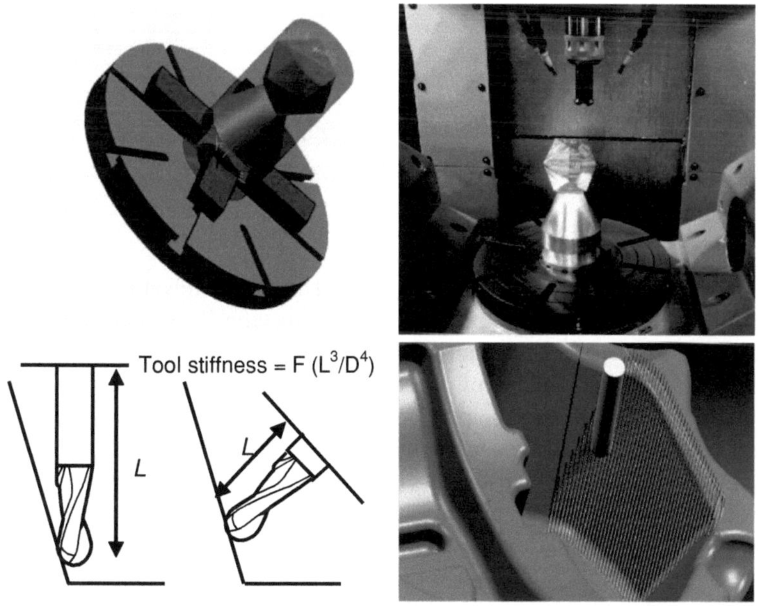


**Fig. 1.3** Advantages of 5-axis milling

sculptured milling operation. Machining time reduction and surface quality improvement is also obtained.

- The use of ball end mills can be substantially improved slightly changing the orientation of the tool axis. In this way it can be avoided to cut with the tool tip. As shown in Fig. 1.4 there is a low cutting speed area in the tool tip of a ball end mill, therefore the cutting process is very unfavourable at this point. Thus, using a better workpiece-tool orientation by means the numeric control the cutting speed and the whole process can be improved. Moreover, this ability to change the tool orientation allows the use of high performance ceramics or PCBN tools; the main snag for these tools is the need for continuous high cutting speed. A lack of continuity is the reason for typical failures of the tool tip due to the inherent brittle behaviour of ceramic materials.

In Fig. 1.3 (down), the continuous interpolation of the 5-axes machine avoids the tool tip cutting. The machine tool manufacturer Starrag gives another example; the Sturz (P-milling ©) machining strategy uses bull nose tools for the milling of freeform surfaces, in which an optimum tool axis orientation reduces time by a factor of three.

Five-axis machining in finishing operations requires special attention to the toolpath generation. Tool positioning is done based on local geometrical characteristics of the surface, but not on the interferences of tool body with other part zones, which can lead to severe collisions during machining. In different papers [17–21]

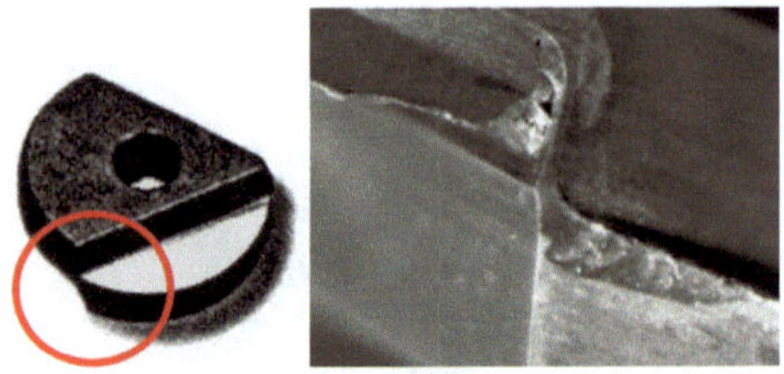

**Fig. 1.4** Advantages of 5-axis milling to reduce the tool tip failure

other kinds of positioning methods are proposed, but they are yet to be implemented in any commercial CAM software. There is also specific CAM software focused on different part geometries, especially for impeller and other turbo machinery components [22]. The high number of papers about this geometrical problem shows the difficulty for a correct 5-axes toolpath generation. However, these papers are focused on the algorithms for CAM calculations, and not in the structure and work methodology at the CAM stage which should consider CAM as the whole planning process.

However, there are two main problems in developing the 5-axes machining process. On the one hand those related to the CAM and toolpath generation, on the other hand the possible interferences during the process, collisions of tool against part and fixtures and even between different parts of the machine. The geometrical calculation of the position of the tool centre point (TCP) and orientation of the tool axis is directly calculated for all commercial software with 5-axis capabilities (Unigraphics®, Catia®, Openmind®, GibbsCam® and others), and these are not a problem for a skilled CAM user. The algorithms implemented inside these systems are explained in its theoretical manuals and there is abundant technical information about them [23]. Therefore, this matter is out of the scope of the CAM users.

The main concern during toolpath generation appears in the postprocessing step, when the toolpath generated by CAM is translated into CNC code. There are many different configurations of five axes machines, and the postprocessors have to be adapted for each of them. For example, using a machine with two rotary additional axes in bed is absolutely different from those with two orientation angles (twist and tilt angles) in the tool head. The same part and even the same APT code obtained from CAM, drives to very different CNC codes.

Another real problem is the possibility of collisions during the machining process. Collisions can damage the high speed spindle hybrid bearings (steel races with ceramic balls bearings), which involve high repair costs and long off-production times. Even if the machine is not damaged, 5-axes process is usually applied in complex and high added value parts, such as impellers made on a titanium and/or superalloys, or near to net shape precision cast parts; machining errors can damage the workpiece wasting a lot of previous machining time and the expensive base material.

In this regard as explained in [24] for three axes machining of complex surfaces, in 5-axes a new approach to the CAM stage can be applied improving reliability of the whole process. Definition of reliability for a machining process is "achieving a good productivity with a low risk of wasted parts due to be out of tolerances or with irrecoverable errors". In 5-axes milling production, the CAM and the CAM user are

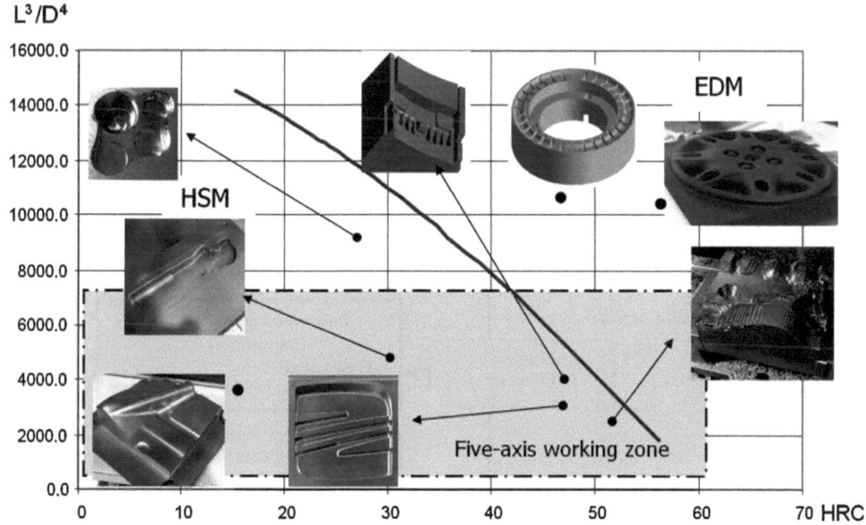


**Fig. 1.5** The borderline between high speed milling and electro discharge machining

the centre of gravity of the planning process, because workshop workers can only change the actual values of cutting speed and feed rate making use of the machine dials (which modify the actual feed and spindle rotation speed with respect to that programmed in the NC code), being impossible changes of the complex toolpath directly in the CNC interface. A new intelligent CAM procedure is presented and some interesting examples are described. That production scheme include a real knowledge approach based on a scientific model to evaluate the cutting force, showing that new CAM planning process trends to include the process knowledge obtained from the complex modelling of the machining process.

### *1.2.1 5-Axis Milling Against EDM*

In Fig. 1.5 the borderline between high speed milling and electrodischarge machining is shown, along with several examples of hard parts made in the last 10 years by the University of the Basque Country. The X-axis is the hardness of the part to be machined, whereas the Y-axis is the tool overhang regarding the basic deflection relation Eq. 1.1:

$$\delta = \frac{64F}{3\pi E}\frac{L^3}{D^4} \tag{1.1}$$

It may be seen in Eq. 1.1 that tool deflection in the static model is a function of the following parameters: E = Young's modulus for the tool material, L3/D4 – Tool slenderness parameter, D equivalent tool diameter, L overhang length and finally F, the cutting force perpendicular to the tool axis [25].

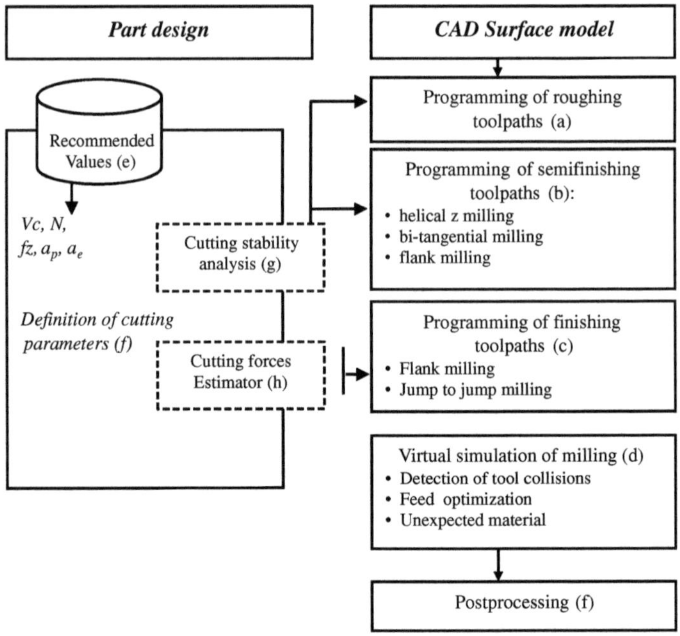


**Fig. 1.6** A reliable CAM approach for flank milling

Five-axis milling allows reducing the tool overhand and therefore the tool deflection, enlarging the application area of high speed milling in decrement of the slow EDM [23].

### *1.2.2 The Virtual Machining for a Reliable Process*

As may be seen in Fig. 1.6, there are three stages in the generation of CAM cutting paths, according to the type of operation: roughing (a), semifinishing (b) and finishing (c).

Roughing (a) is of critical importance in HSM. The aim is to achieve not only productivity but also a highly uniform stock allowance, which will be removed in the course of finishing. In roughing there are three possibilities, each very much related to the size and hardness of the workpiece.

Semifinishing (b) mainly intended to remove the uneven material and keep even part stock allowance for the subsequent finishing operations.

The object in finishing (c) is to achieve a roughness and tolerance specified by the client for each surface. The traditional strategy is zigzag, but in this case the main drawback is that it intercalates two different cutting types, downmilling (also called climb milling) and upmilling. A solution may be to cut along one direction (zig), either downmilling (the most commonly used) or upmilling, but not along both of them. The best option is the use of milling strategies more closely adapted to each of the part zones and its geometry, depending on factors like control of the cusp, and as a consequence the maximum roughness $R_t$, by varying the radial depth of cut in accordance with the workpiece slope.

At the definition of cutting parameters, two developed utilities (g, h) are newly available to assist in the selection of toolpaths and cutting conditions. These utilities are applied after the selection of the recommended conditions given by toolmakers, directly obtained from links to the commercial databases (e) of these companies, usually calculated for suffering low tool wear [26].

A check stage (d) is included for the NC programs, using an ad hoc software utility such as Vericut®, NC-Verify®, NC-Simul®. This software allows a virtual simulation previous to actual machining, in which different problems can be effectively detected, as collisions, machining outside of the machine workspace and problems due to tool gauging into the workpiece. The virtual simulation is a fundamental step in the multi-axis machining toolpath generation.

The simulator software allows the user to perform a virtual simulation of the process previous to actual machining; different problems can be effectively detected and corrected:

- Collisions and interferences between tool and part, toolholder and part, or even between spindle head and machine bed.
- Machining outside of the machine work volume.
- Problems due to tool gauging into the workpiece. This is an important aspect in the machining of corners.

The movements during machining are not predictable from a visual toolpath analysis due to the complex kinematical characteristics of the machine. The same toolpath that lead to small movements in a machine can lead big ones in other machine with a different configuration. From a kinematical point of view, axes interpolation performed by the numerical control in 5-axes is much more complex than that done in three axes interpolation. Therefore it is not easy to predict the existence of collisions in these kinds of machines. This fact involves the need to use specialized software of machining simulation such as Vericut®, Virtual Machine®, Nc-Verify® and Nc-Simul®. At present, some general CAM systems integrate such kinds of virtual machining environments (Fig. 1.7).

A virtual machining environment can be divided into four stages

- First, the machine representation with its kinematical configuration and motions. Usually, kinematics is defined as a tree structure, where the local matrix transformations between the coordinate reference systems associated to each element are related. The mathematical formulation is 'transparent' to the software user, who only describes the tree chains of both the tool movement and the part movement. But internally, a formulation similar to the well-known Denavit and Hartenberg [27] in spatial mechanisms is used.

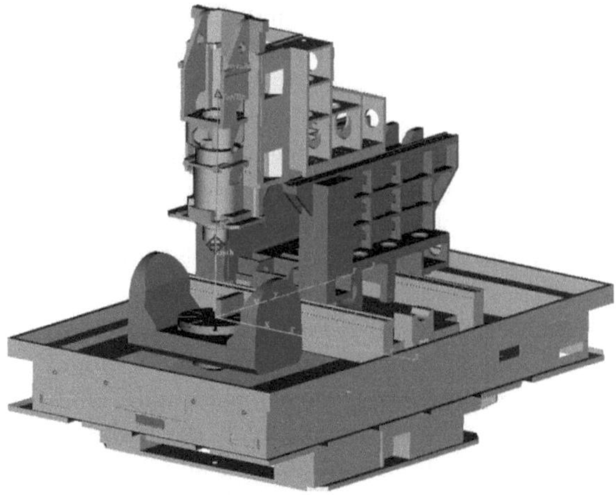

**Fig. 1.7** 5-axis milling centre kinematics modelled under verification software

- Second, the numerical control of the machine must be included in order to correctly interpret the machine code and the axes movements. Here the G and M codes syntaxes, the maximum transverse course for each machine axes and the positive sense of machine axes, are described.
- Third, the simulation must include tool and toolholder solid models and the raw part geometry.
- Fourth and last, the CNC code. The better case is the use of the real ISO or other type of CNC code directly obtained after postprocessing. Thus, the same program that will be performed by the CNC is checked.

However, the result is far away from getting a true reliable toolpath. Virtual simulation takes into account only geometric collisions, so problems due to the cutting process itself are not revealed. The virtual simulation guarantees that there will be no collisions during machining but it does not mean that it will be a trouble-free machining. In spite of its limitations, the virtual simulation is a powerful tool in order to achieve a good machining process, which allows very fast error detection during the machining process. As conclusion, in three axes machining virtual simulation was recommended but in five axes it is essential.

At present, due to the useful information provided by the verification systems, the trend is to introduce the virtual simulation in the CAM software or include a direct link to other partner software.

## 1.3 Milling of Thin-Walled Components

Several 5-axis milled complex surfaces exhibit features such as the so-called thin walls or/and thin floors. Therefore, is important to mention some of the main characteristics related to their milling process.

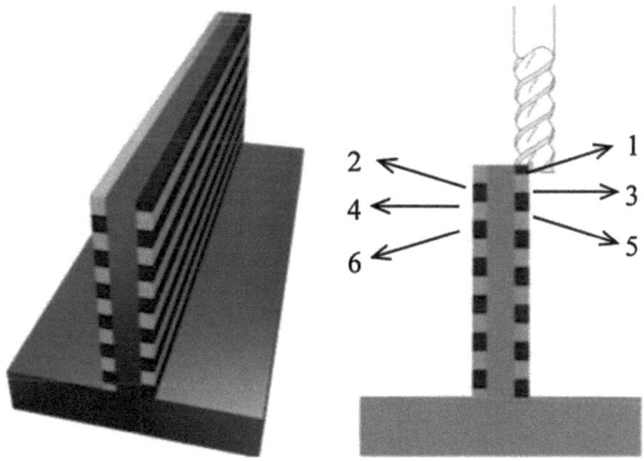


**Fig. 1.8** Jump to jump milling for thin wall

Once the roughing operations have been carried out, achieving a workpiece near to the net shape, the subsequent finishing of walls and floors of the pockets along the airframe component must be performed.

The case of thin wall can be regarded (and simplified) as a shell clamped at its bottom border, being excited by a force applied at the tool contact point. Bending of the part will be the highest at the starting passes, applied at the top level (next to the wall free border). This part deflection causes a lower radial depth of cut, and as consequence a thicker section at the part top, even more than a tenth of millimetre for 1 mm thick walls.

Some solutions to minimise this part bending caused by milling are:

- Machining with 'jump to jump' toolpaths. Following this method alternative tool passes on both sides of the wall to be machined are used and a higher local rigidity at the cutting point is constantly achieved. During every toolpath, the section down the tool contact point is the stiffest possible. This technique is shown in Fig. 1.8, in which consecutive toolpaths are numbered.
- To select an optimal stock offset to be left on the wall just before the finishing step. This offset will be the radial depth of cut of the last finishing toolpaths. If the thickness of this offset was very small, the jump to jump effect would not be high enough; on the contrary, a thicker offset would greatly increase cutting forces causing more deflection. Therefore, a compromise value for this parameter taking into account these two points must be chosen.
- To use the upmilling mode. In the opposite case, downmilling, the deflection component of the cutting force (generally that perpendicular to the wall) prevails over the tangent component to the wall. This component pushes and separates the wall from the milling tool, cutting with a low depth of cut and then causing overthickness. Otherwise when upmilling is used, the force tends to engage tool into part when the tool tooth enters in the material. Thus, the actual radial depth of cut is affected, being higher at the top of thin walls in upmilling than in downmilling, giving a different part thickness. Therefore, precision is better in the downmilling case.

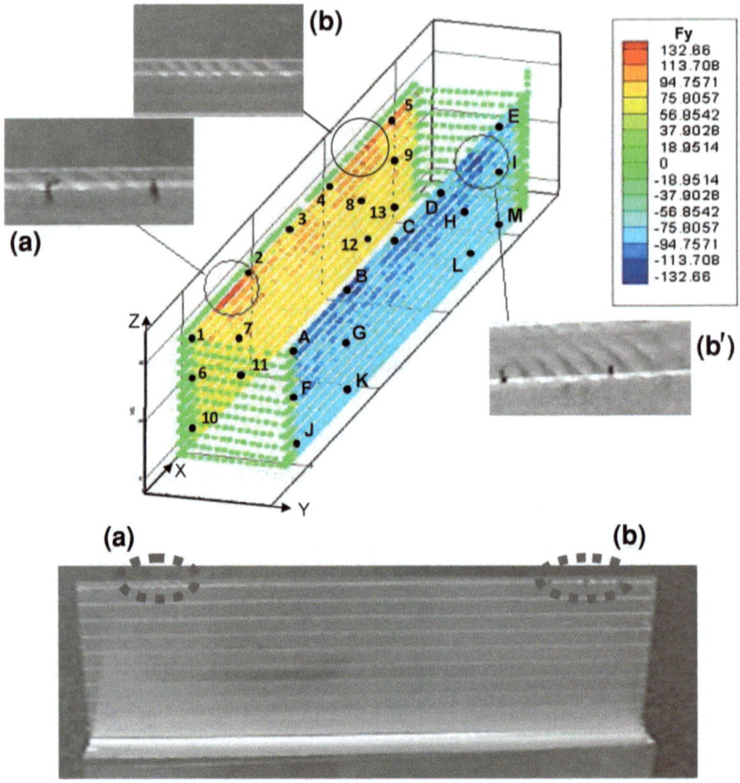


Fig. 1.9 Continuous force monitoring of a thin wall machining process. **a** and **b** are the worst finished points

- Tool corner radius. The corner radius at the tool tip has a strong influence on the cutting force components. If high, it reduces the normal component (to the wall) and increases the component along the tool axis, which is positive for reduction of wall deformation. However, in the case of thin floors, this feature will act in just the opposite way

Even using the above recommended suggestions and jump to jump strategy, the thin wall milling is still a complex task. Figure 1.9 shows the cutting forces recorded along a 0.43 mm thick wall, applying high speed milling with a 16 mm diameter end mill with relieved shank for preventing wall from tool recutting, ap 5 mm, ae 2 mm, fz 0.07 mm, $v_c$ 201 m/min, f 560 mm/min, N 4,000 rpm. The technique of recording real forces along geometry is itself an innovative technique [28]. As shown in Fig. 1.9, several marks on surface were produced, and oversize thickness happened. In Table 1.1 values of cutting force components on several surface points and values of thickness in those wall points are shown, taking into account that the programmed part was 0.43 mm thick.

**Table 1.1** Cutting forces components and final wall thickness for part of Fig. 1.9

| 1 side | A | B | C | D | E | F | G | H | I | J | K | L | M |
|---|---|---|---|---|---|---|---|---|---|---|---|---|---|
| Fx(N) | 101.07 | 98.87 | 106.2 | 108.4 | 96.68 | 101.07 | 90.087 | 90.09 | 89.35 | 84.96 | 86.43 | 86.43 | 83.5 |
| Fy(N) | 82.76 | 116.45 | 100.3 | 150.88 | 75.44 | 137.7 | 115.7 | 11.8 | 120.12 | 109.86 | 97.41 | 114.26 | 124.5 |
| 2 side | 1 | 2 | 3 | 4 | 5 | 6 | 7 | 8 | 9 | 10 | 11 | 12 | 13 |
| Fx(N) | 91.55 | 93.75 | 94.48 | 108.39 | 95.21 | 88.62 | 91.55 | 87.89 | 87.76 | 82.76 | 81.3 | 86.43 | 85.69 |
| Fy(N) | 85.69 | 73.97 | 98.88 | 112.8 | 131.1 | 123.78 | 85.69 | 108.4 | 87.89 | 131.1 | 109.13 | 101.8 | 121.58 |
| e(mm) | 0.886 | 0.686 | 0.622 | 0.680 | 0.889 | 0.591 | 0.482 | 0.487 | 0.598 | 0.473 | 0.413 | 0.408 | 0.44 |

## 1.4 Vibrations in the Flank Milling of Thin Walls

Vibrations in the flank milling of thin-walled parts are always present due to the low dynamic stiffness, stiffness plus damping, inherent to these parts and the high frequency of the tooth impacts, frequently near the modes of the wall. This problem results in parts with a poor surface roughness that does not meet the geometrical requirements. Hence, special care is needed to avoid manual finishing or part rejection. Usually, machinists face the problem with an experimental approach. The milling strategy can avoid problems, as it has been shown with the jump-to-jump strategy [29]. The use of fixturing to stiff or damp the wall during the machining is very common in an industrial context, by means of vacuum fixtures, materials with high damping properties as rubber or foams, low melting point materials such as wax and other alloys or simply by using modular fixtures [30]. Active damping techniques have also been tested, but their use in an industrial context is not as feasible as the previous methods [31]. Nevertheless, the theoretical approach to the problem of milling thin walls has brought new solutions. From this point of view, two kinds of vibrations must be taken into account, the forced vibrations due to the tooth impacts on the wall, and the self-excited vibrations due to the relation between the wall displacements and the cutting forces.

The forced vibrations due to the periodical impacts of the tooth against the part are always present in the milling process. Leaving apart solutions based on fixtures, one solution is distributing the excitation against the wall on a broader range of frequencies varying the spindle speed online, or using tool with variable pitch or variable helix angle [32]. Under this situation, instead of a high excitation at the same frequency, the excitation is made at several frequencies with lower amplitude. In Fig. 1.10 an example is shown. The pattern of forces on a wall and the corresponding Fourier Transform (FFT) are shown for the milling of an aluminium 7075T6 wall with an end mill of 4 flutes and 12 mm of diameter. The radial depth of cut is 0.1 mm and the feed per tooth is 0.05 mm. The forces have been calculated by means of a mechanistic model of the milling forces, which can represent accurately the real cutting forces [33, 34]. First, the patterns are shown for a constant pitch angle between edges. Then, the same is shown for a tool with a variable pitch angle. Comparing the FFTs, although there is a broader frequency content, the amplitude of the peaks for the variable pitch tool is almost half of those peaks for the regular tool in the same operation.

Another option is to customize the tool. In milling, when the depth of cut is equal or a multiple of the helix height divided by the number of cutting edges, the cutting forces become constant, as if they were static forces instead of dynamic ones. Under these conditions, the tool teeth are permanently engaged on the wall so the impacts of the tooth against the wall no longer exist. Therefore, it is possible to design a custom tool with a helix height and a number of flutes that matches the height of the thin wall. The immediate result is a more continuous machining, with lower vibration. If the height of the wall is variable, it is possible to design a custom tool with a very large helix angle. The smaller helix height, makes it easier

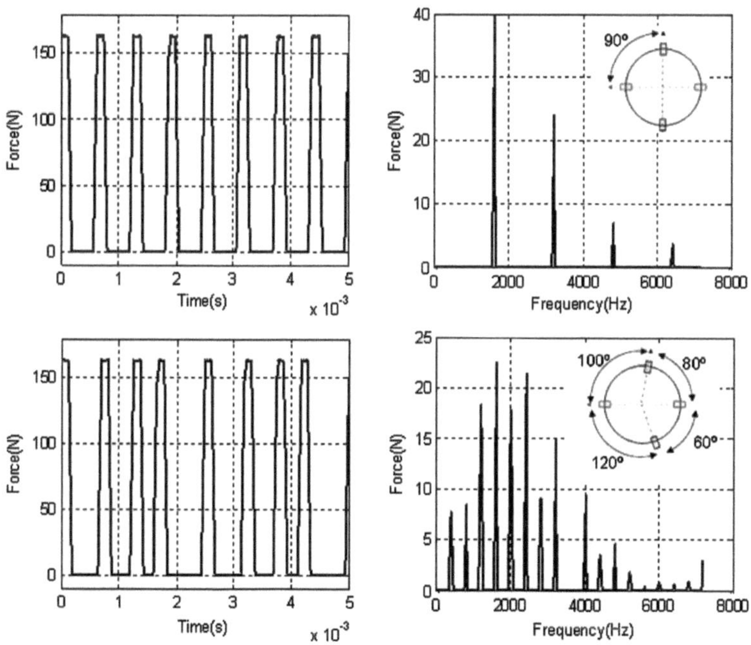


**Fig. 1.10** Cutting force in the normal direction to a thin wall and the FFT using a tool with constant pitch (*up*) and a tool with variable pitch (*down*)

to find depths of cut that provide constant static-like forces. In Fig. 1.11, the pattern of the cutting forces on a wall is again calculated with a model of the process. It is shown how the cutting force becomes constant for the key values of the depth of cut, which are 12.56 mm and its multiples for an end mill of 16 mm of diameter, 4 flutes and a 45° helix angle.

The self-excited vibrations during the milling of thin walls are the most damaging vibration, see Fig. 1.12. Regenerative chatter and period doubling chatter are the most common ones. The first is due to the regeneration of the chip thickness when a surface that has been previously machined by other tooth is being cut. Depending on the phase between the undulations left in the surface and the vibration of the tooth in cut, the chip thickness will vary and so the cutting forces, exciting as a result the modes of the wall. The excitation leads to a higher vibration, a higher chip thickness, and again to a higher excitation. This is the regenerative mechanism which results in a growing vibration [35, 36]. On the other hand, period doubling chatter occurs when ratio between the natural frequency of the wall and the tooth passing frequency is equal to (2n−1)/2 where n = 1, 2, 3,... Under these conditions, the tooth impacts alternatively provide more kinetic energy to the wall than what is being subtracted.

The appearance of self-excited vibrations is always due to the lack of dynamic stiffness from the wall, in these cases the use of fixturing to damp or stiff the wall is very positive to avoid the vibration problem. The use of variable spindle speed or

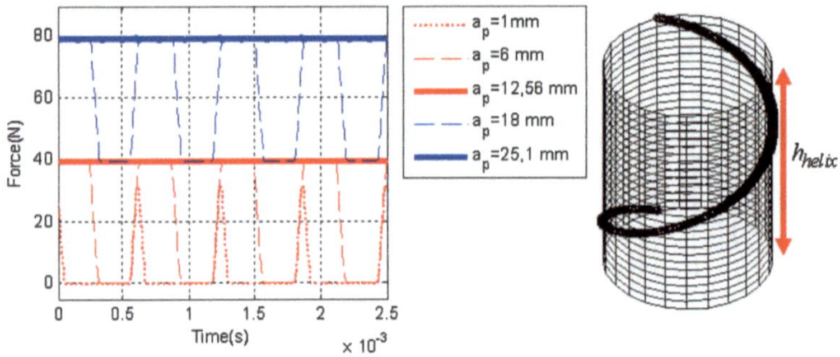


**Fig. 1.11** Influence of the helix angle on the cutting forces profile

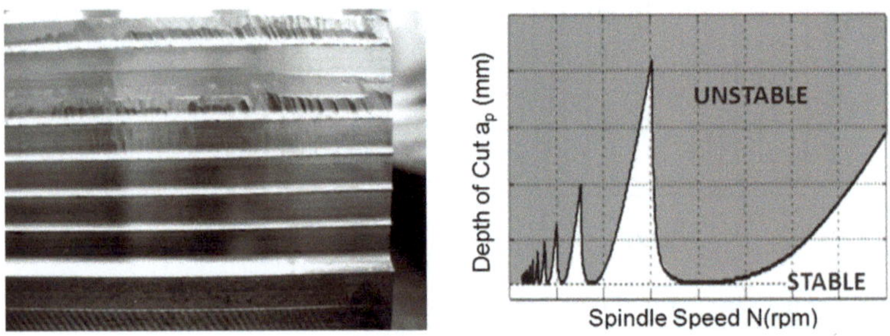


**Fig. 1.12** Chatter vibration patterns in the surface of a thin wall (*left*). Stability lobes diagram (*right*)

variable pitch tools is also an alternative solution. But once again, the theoretical approach, based on the modelling of the milling process and the dynamic behaviour of the wall, can provide a solution to this problem by means of the stability lobes diagram for the milling of a thin wall. This diagram indicates which combinations of the spindle speed, the radial and axial depths of cut can be used in the machining to avoid chatter [37].

The shape of the stability lobes diagram depends on the modal parameters of the system, the tool geometry, the material and the radial immersion of the tool. From all of them, the prediction of the modal parameters is of crucial importance, since an error in that prediction will result in a proportional error in the location of the stable speeds in the diagram. However, the modal parameters of the thin wall are variable along the wall height and also during the machining due to the mass removal. What is more, the tool position along a given mode determines whether that mode is highly excited or not. As a conclusion, the stability lobes of the milling of a thin wall vary along the machining, and a three-dimensional diagram along the tool-path must be arranged, see Fig. 1.13 [38–40]. This theoretical approach is certainly complex to apply in an industrial context, and it can be justified only when there is a thin wall geometry that is repetitively machined, which is the case of the blades manufactured for the aeronautical industry.

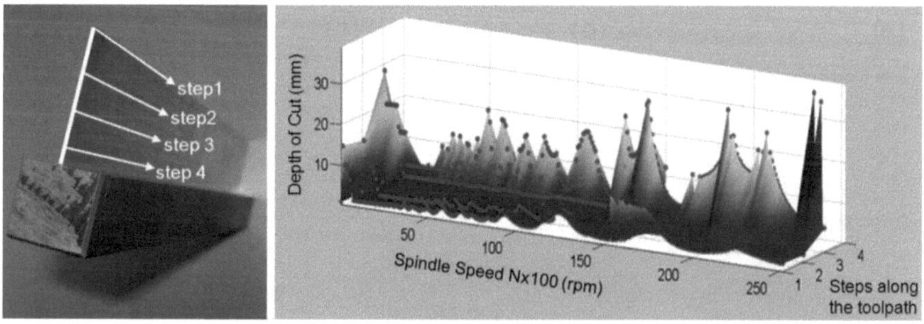


**Fig. 1.13** Three-dimensional stability diagram for the milling of a thin wall

To sum up, the avoidance of vibrations is one of the challenges that machinists must face during the optimization of the flank milling of a thin wall. The main efforts of the machinist must be focused always on providing the greatest dynamic stiffness to the wall and the selection of a tool with a proper geometry in terms of helix angle and pitch. It is also very important the selection of the spindle speeds as the chatter theory indicates that there are better, more stable, spindle speeds than others.

## 1.5 End Milling Surface Topography Prediction

The current technological advances in the area of computation and modeling allow the development of new applications in fields apart from the original purposes for which they have been developed. It is possible to take advantage of the powerful solid substraction engine contained in the CAD software for other objectives. The modelling of the peripheral milling process to obtain the machined surface topography is one of these alternative applications succesfully proven already.

### *1.5.1 The Run-out and Its Geometrical Definition*

The run-out is a well-known set of parameters considered for the milling process modelling. It is a consequence of the imperfect clamping of the tool inside the tool holder and can also be a consequense of the deviations between the tool holder and the spindle rotation axes [24]. Currently the machine tool industry has minimized the uncertainty of this issue with the development of more accurate tool holders, the heat shrink tool holders are a good example of improvement, reaching values of concentricity within 0–5 μm, but in so many applications of the machining industry the mechanical clamping is still used or required with its inherent consequences.

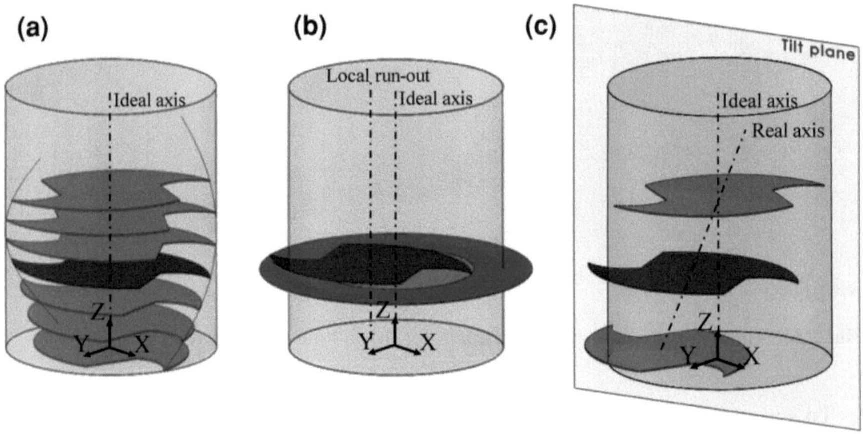


**Fig. 1.14** Schematic decomposition of the run-out. **a** Ideal tool, **b** local run-out, **c** real tool

Thus, taking the language into the analytical field in order to analyze the problem, the final effect of the deviations chain can be considered as the sum of two main effects: first, a parallel local eccentricity of the tool centre with respect to the spindle axis and second, an angular deviation between the tool and the spindle axis, as shown in Fig. 1.14. It is important to highlight that the real tool axis could or could not intersect the ideal axis. This effect will be explained in detail below.

Idealistically, the milling process mechanics is up to now well understood. The cutting force modelling is based on the assumption of the flutes trajectories describing trochoids [33, 41], the Fig. 1.15 (left) illustrates an example for a three-fluted end mill tool. The overlapping of such trajectories defines the localized instantaneous chip thickness if a discretized slide of the tool is considered along the tool axis in a plane parallel to XY. Thus, the cutting forces are directly related to the instantaneous chip thickness and slight variations of it have a significant effect on the cutting forces which provoke quasi-static effects such as tool or work part deflections either even dynamic effects such as forced vibration or regenerative vibration (chatter).

## *1.5.2 The Run-out Effect Over the Chip Thickness*

In order to achieve a more realistic approximation of the cutting forces over the milling process it is necessary to consider two different effects, the run-out in first place and the deflections as a subsequent step. In this section will be described the run-out effect in the trajectory followed by the flutes of the tool, and also the geometric approximation followed to obtain a more realistic chip thickness during the cutting process.

As shown in Fig. 1.14b if we considered a differential slide of the tool along its axis, the XY section view shows the kinematics of the tool, Fig. 1.16. The

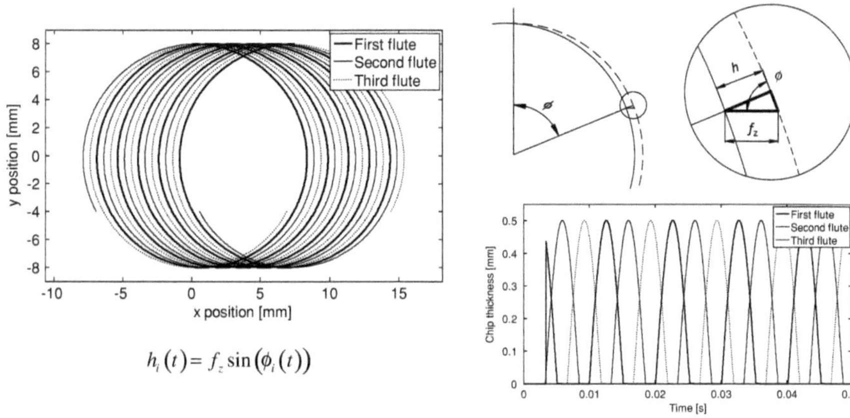


**Fig. 1.15** Chip thickness for a three-fluted mill 16 mm diameter

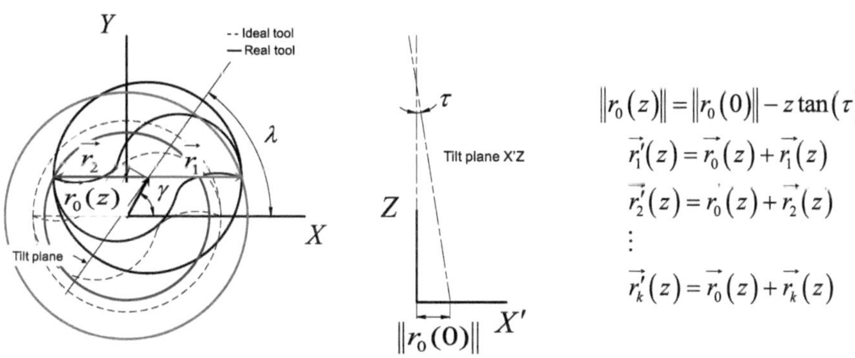


**Fig. 1.16** Real radius of each flute considering the tilt angle and the run-out zero position magnitude

eccentricity of the tool at Z position generates variation in the radius for each single flute. Such variations for each instantaneous flute position are easily described with respect to a reference flute once the local run-out orientation $\gamma$ and magnitude $r_0(z)$ are known. The analysis can be extrapolated to the successive section views, due to the variation of the run-out magnitude within the tilt plane as the Z position increases.

Therefore, the consideration of the radius variation affects directly the chip thickness for each single flute. This is clearly observed in Fig. 1.17 which represents the aforementioned example, a three-fluted mill, but here with a run-out of $r_0(z) = 100$ μm at $\gamma = 0°$. The chip thickness is different for each single flute, and the result consists in overloads and unloads over the flutes with more or less chip load. This can cause either breakage of the flutes or uneven tool wear.

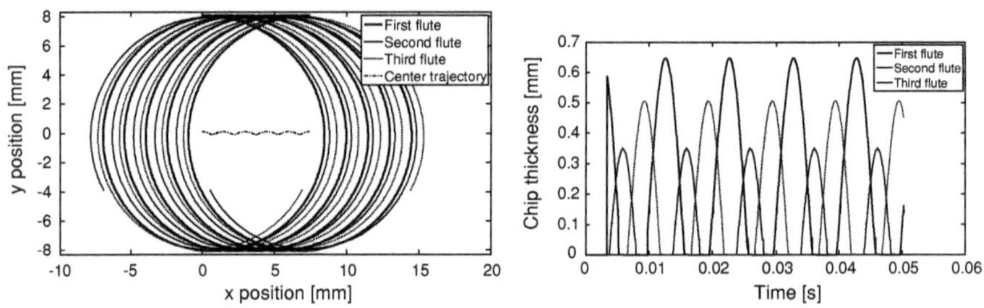


**Fig. 1.17** Chip thickness for a three-fluted mill 16 mm diameter considering run-out for a single z-section view

## *1.5.3 Topography Prediction for High Stiffness*

If only the geometrical error of the tool is considered and the asumption of high stiffness is taken into account for both the tool and the machined workpiece, the eccentricity in combination with the tilt angle generates different roughness paths along the axial depth of cut [42].

If a two-fluted end mill is considered as example (for its simplicity), in some cases such is the combination of eccentricity and tilt that the surface finishing is left for only one of the flutes, Fig. 1.18 left. This is due to one of the flute trajectories having larger diameter than the other along the whole depth of cut. And even if the flute with lower diameter trajectory cuts material, the flute with larger diameter trajectory removes the effect left by the previous one.

In some other combinations of excentricity and tilt, both flutes can generate alternatively the surface finishing at different Z positions and the interference bands phenomena takes place. An interference band is a transition zone between the marks left by different flutes. In the example shown in Fig. 1.18 right, flute 1 is responsible for the surface marks for depths of cut from 0 to 10 mm, from 10 to 1 mm there is a clear transition in which flute 1 gradually loses importance at the time flute 2 is increasing its importance up to the end where only flute 2 is responsible for the roughness path. Clearly the middle point can be found at 12 mm of depth, where both flutes generate the same roughness path [43].

As previously mentioned the CAD software can be used for different purposes than the original ones. The flank milling surface prediction is one of them and it takes advantage of the powerful solid subtraction engine content as the core of such software. The modelling consists in the incorporation of the process kinematics and the solid representing the tool with the run-out geometrical properties. Hence, as the tool turns and moves forward, it is removing the interference volume between the flutes and the workpiece; in this way a virtual machining process is carried out, Fig. 1.19. The resulting surface is later numerically processed to obtain a topography represented as gradient. Figure 1.18 illustrates the results of this kind of prediction for the same milling operation under different clamping

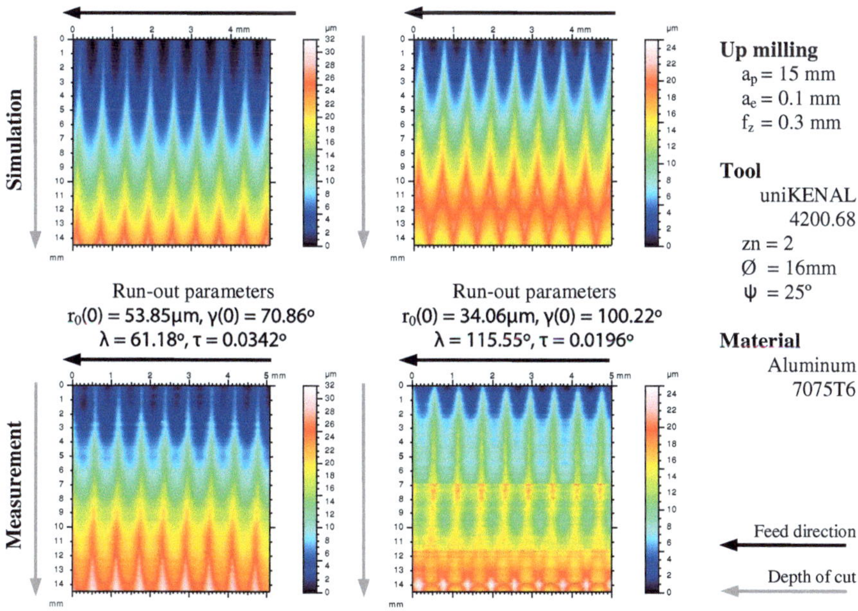


**Fig. 1.18** Simulation of the run-out effects over the machined surface

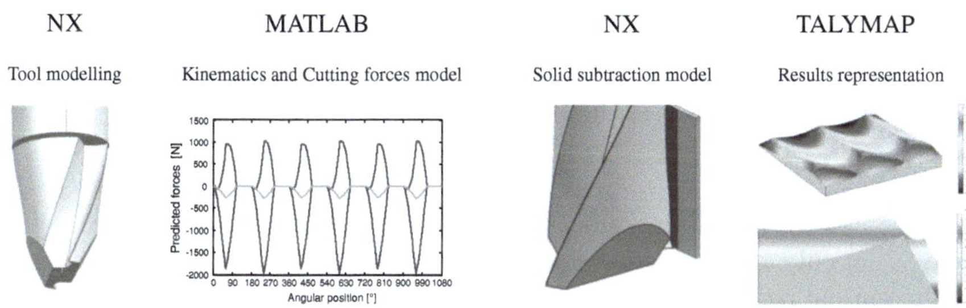


**Fig. 1.19** Solid subtraction model scheme

errors over the tool. The prediction corresponds accurately with the real machined surface measured using a high precision profilometer.

### *1.5.4 The Workpiece as Flexible Element: Thin Walls*

The reality does not always allow to model the elements of the milling process as infinitely rigid. Deformable bodies in big or small scale always take place and the resulting deformation due to the cutting forces need to be considered if accuracy is to be achieved in the surface topography modelling.

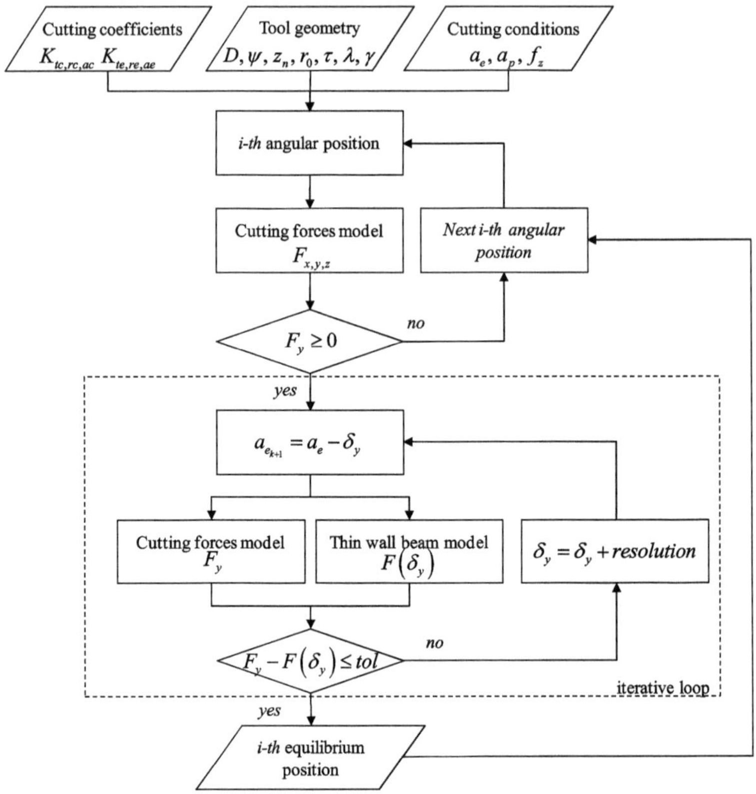


**Fig. 1.20** Thin wall displacements algorithm

In the case of the milling process of thin-walled elements, the surface topography is not only a consequence of the starting inputs for the process. Those inputs are the run-out parameters, the static deformation and the cutting conditions. The surface depends on the real-time interaction between them for the following reasons:

- The cutting conditions and the run-out geometrical parameters define theoretical cutting forces generated during the chip removal process.
- The cutting forces, due to the flexibility of the work-part, cause quasi-static deformations which reduce the engagement conditions (radial immersion).
- The variation on the radial immersion modifies the theoretical chip thickness and the cutting forces as well. Thus, a recalculation of the whole process is needed until the equilibrium between the cutting forces and the static deformation response is reached for each single angular position of the flute.

The aforementioned process is better clarified in the flowchart, Fig. 1.20.

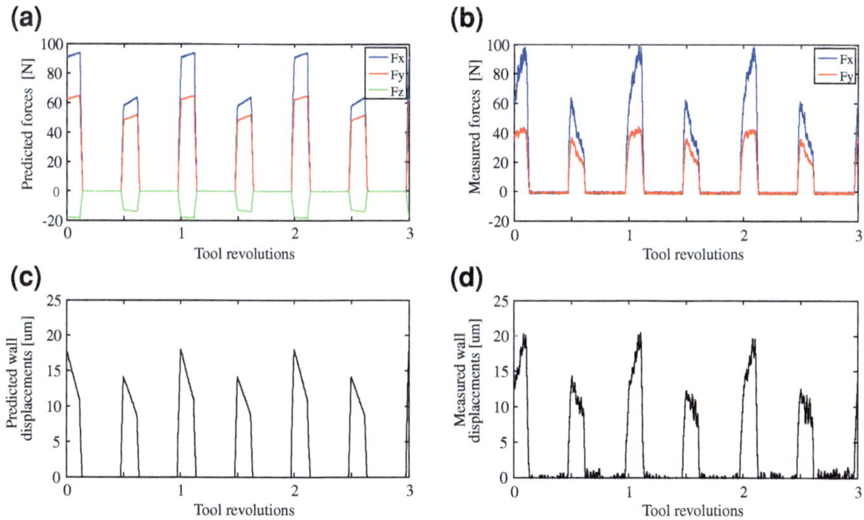


**Fig. 1.21** Comparison predicted and measured peripheral milling variables. **a** Predicted forces, **b** measured forces, **c** predicted displacements, **d** measured displacements

The algorithm has as results the time domain behaviour of the cutting forces and the displacements of the work-part, Fig. 1.21. The latter define part of the kinematics of the new process which involves a continuous change of position between the flute of the tool and the work-part. Evidently any modification of the relative displacement has important consequences on the machined surface and the solid modelling of the milling process helps to understand the main effects of such consequences.

The most remarkable effect of the flexibility over the surface takes place in the tool axial direction. Instantaneous changes in the radial depth of cut caused by part deflection reveal modifications not only on the shape error but also in the surface error. This means the roughness at different z-coordinates is sensitive to the part flexibility.

Figure 1.22 illustrates the comparison between a surface generated under non deformable bodies' assumption versus flexible elements, thin walls made of aluminum 7075-T6 with 50 and 70 mm height respectively and 6 mm thick. These topography predictions were validated experimentally with certainty less than 5% in the shape error comparison and less than 2% in the surface comparison. Some other cases were evaluated obtaining results as precise as the shown in the illustration.

The increase in the shape error can be observed as the flexibility increases in the zones of the depth of cut where the chip load is higher. For such reasons, the zones with the lower chip load, as the tool entrance at the tip and the flute exit at the wall border, exhibit a deeper axial immersion; meanwhile at the medium area where the chip load is greater the displacement of the wall reduces the radial immersion, less material is cut and for that reason the shape error reaches its maximum values.

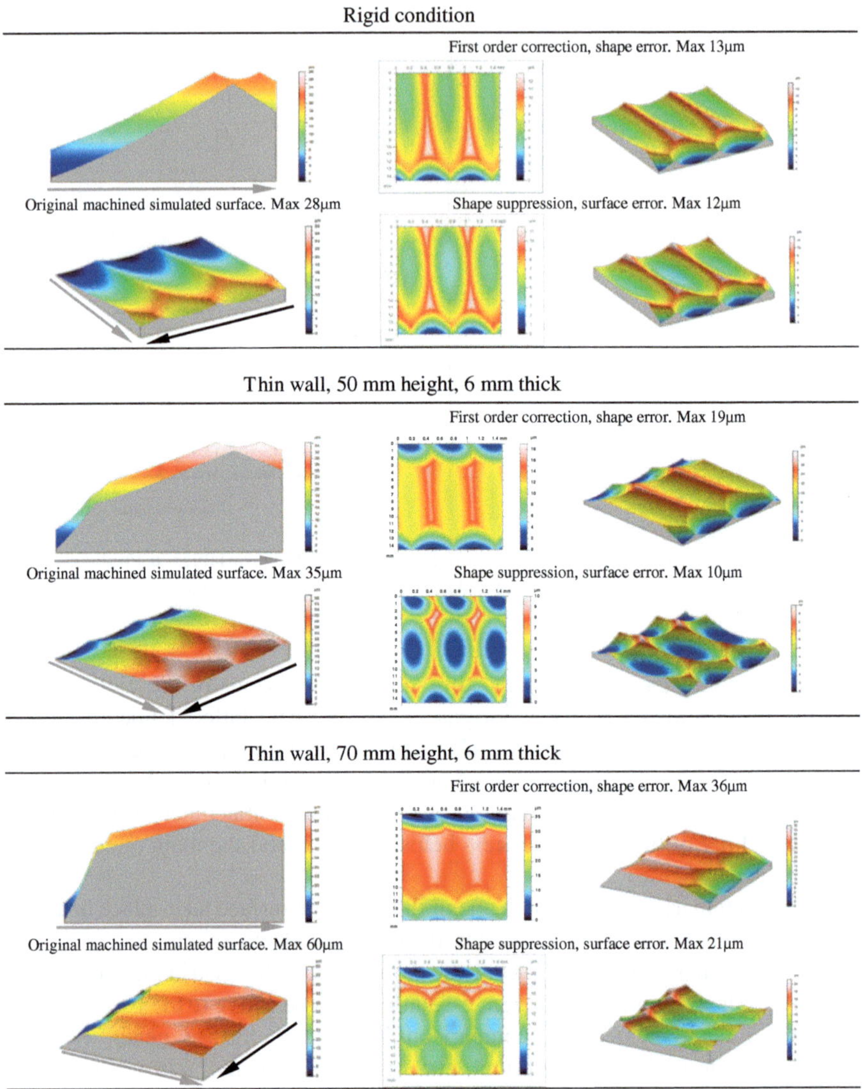


**Fig. 1.22** Comparison between rigid condition and a flexible condition

This recently developed application of the CAM engine could be incorporated in the designing software as a decision tool for the CAM programer. The theoretical prediction of the surface finishing would offer the opportunity to improve cutting conditions in order to keep the surface quality under a pre-established shape error and surface error tolerance. The possible drawback for its suitability, consists in the need for the run-out real geometry definition and the flexibility of the system machine-tool-tool holder.

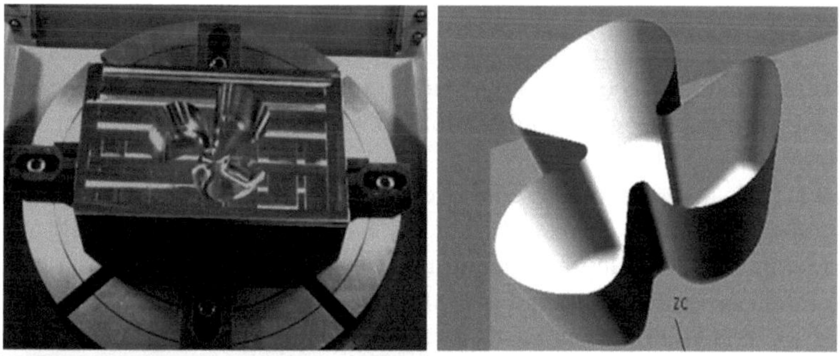


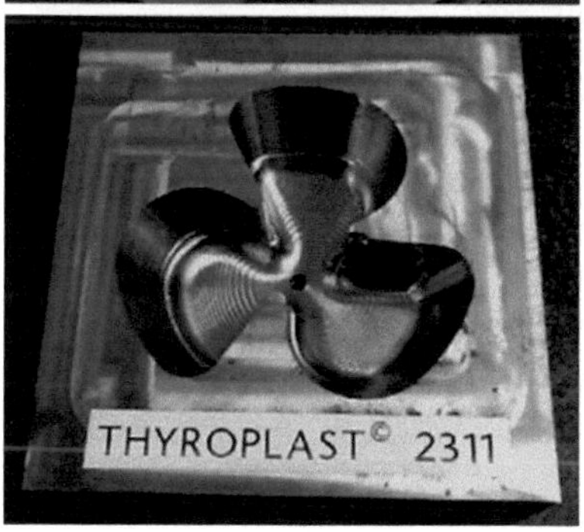


**Tool:** Bull Nose Ø12 r2

| Operation | $V_c$ [m/min] | $f_z$ [mm/toth] | S [rpm] | Time [min] |
|---|---|---|---|---|
| Roughing | 234,4 | 0,20 | 6200 | 43 |
| Roughing | 234,4 | 0,12 | 6200 | 11 |
| Flanks finishing | 377,0 | 0,10 | 10000 | 6 |
| Floor finishing | 377,0 | 0,10 | 10000 | 8 |

**Fig. 1.23** Hardened steel mould greometry and process conditions

## 1.6 Examples of Complex Surfaces

### *1.6.1 A Plastic Injection Mould for a toy*

The first example is a plastic injection mould made on steel hardened up to 35 HRC. The simple mould geometry and the possibility of using different cutting strategies with the same tool make them a representative example of complex surface milling, see Fig. 1.23.

The experimental design is composed as follows:

- A 5-axis machine Ibarmia ZV25U, a transverse-column milling machine with two rotational axis in the table and with a spindle suitable for 18,000 rpm. The acceleration and the maximum speed are not the same in the different axis because of the machine configuration
- Unigraphics CAD/CAM software, it was used for the toolpath generation. Virtual environment and machining simulation were performed using the well-known Vericut software.
- The workpiece, whose material was a 35HRC Thyroplast 2311 (AISI P20 steel).
- A 4-flute bull nose end mill, 12 mm diameter and 2 mm nose radius was used.

Although it is possible to machine this part in a 3-axis milling centre, with a 5-axis machining centre an improvement in productivity and a reduced machining time were obtained, keeping an accurate surface finishing. The methodology displayed in Fig. 1.6 was applied to toolpath generation in order to avoid collisions.

Figure 1.23 shows geometry, cutting conditions and the bull nose tool's diameter and number of cutting edges. Part flanks had been machined using a Z-level simultaneous 5-axis strategy, finishing them in only 6 min, and with a maximum cutting force perpendicular to tool axis of 35 N. This cutting mode is to maintain the flank of the end milling tangent to a closed line, with toolpaths defined by tool pivoting around the centre of a cone. Axial depth of cut is applied in the direction of the directrix lines of the cone. A spiral strategy in downmilling was used to machine the floor. The cutting force did not exceed the value of 27 N during this operation. The total machining time is 68 min, instead of the more than 2.5 h needed in a 3-axis machining centre. The main reason for this time saving is based on the fact that in a 3-axis milling centre the lateral side of the tapered star form must be sculptured with a ball end milling tool (step between tool passes in z-level 0.2 mm), instead of the 5 mm axial depth of cut used in the five axis z-level with an end milling tool.

### *1.6.2 A Test Part for 5-Axis Milling Machining Centres*

The second example includes 'test parts' to check the behaviour of milling centres during machining operations. Nowadays these tests are only focused on 3-axis milling centres. The NAS workpiece (ISO 10791-7:1998) is a commonly known test developed in 1969 for testing machining centres working with aluminium in aeronautical production. However, the NAS test does not include complex surfaces. That is why different test parts have been designed in the past years, such as Mercedes or NCG parts (available at NC-Gesellschaft association, www.ncg.de), which have been widely spread out in order to satisfy customer demands [26]. However, none of them are included in the ISO standard normative. Furthermore, there are no specific test parts for testing 5-axis machining centres, forcing customers to develop their own to check the capabilities of the machine they are interested to acquire.

The geometry of the workpiece shown in Fig. 1.24 is an example of both a test part that could be carried out in a 5-axis milling centre and a complex geometry workpiece that needs flank milling operations. The geometry of the inside walls corresponds to ruled surfaces whose finishing operations are machined using with flank milling strategies.

In conclusion, 5-axis machine builders and users have to design their own tests, which is time-consuming and is only focused on the particular demand of one specific customer. Figure 1.24 shows the result of a 5-axis test for a milling centre developed from a collaboration between the University of the Basque Country UPV/EHU and the machine builder company Ibarmia S.A.

Fig. 1.24 Test developed for a 5-axis milling center test

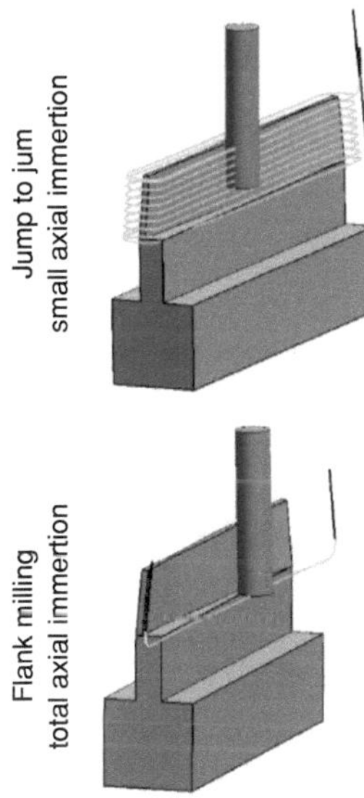


Maximum deflection controlled for each step along depth of cut. Kept under 300 μm

| Stage | ae [mm] | Cutting Force [N] |
|---|---|---|
| 1 | | 450 |
| 2 | | 575 |
| 3 | 5.25 max | 675 |
| 4 | | 800 |
| 5 | | 825 |
| 6 | | 825 |

Maximum deflection Kept under 300 μm

Variable radial immertion:

| Stage | ae [mm] | Cutting Force [N] |
|---|---|---|
| 1 | 4 | 900 |
| 2 | 1 | 270 |
| 3 | 1 | 90 |

Fig. 1.25 Strategies for the machining of a thin wall ruled surface

## *1.6.3 Thin Wall Ruled Surface*

Thin wall ruled surfaces are one of the most characteristic examples of flank milling complex surfaces. In this case a wall whose thinnest depth consists of 1mm is machined following two different strategies that represent two ways of tackling the problem. These kinds of surfaces tend to suffer from deformation and chatter problems. Predicting an admissible deformation for the wall, it is possible to calculate the cutting forces that are going to be tolerated during the machining process, defining the adequate process parameters.

As shown in Fig. 1.25 one of the strategies is based on small axial depth machining and the other is based on flank tool cutting with a small radial depth immersion. The admissible deformation calculated for the wall is 300 μm and theoretical forces predicted can be seen in the table in Fig. 1.25.

The cutting forces were measured by machining monitoring during the machining processes for the flank milling strategy. In the case of the last stage it

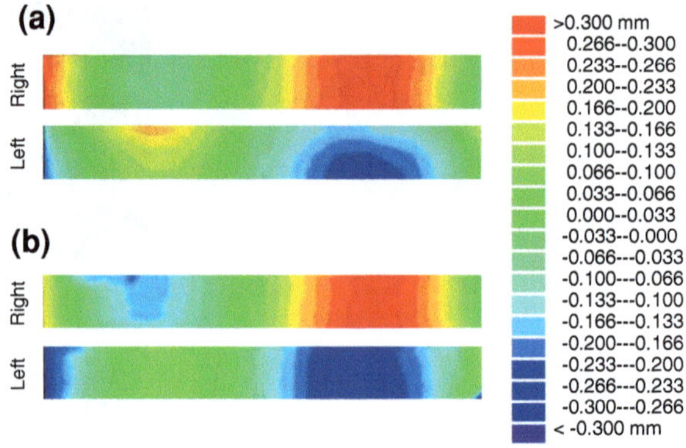


**Fig. 1.26** CMM machine measures. **a** Flank milling strategy (total axial immersion), **b** jump to jump strategy (small axial immersion)

was corresponding to the finish passes for both left and right faces for which the radial immersion was 1 mm. The predicted forces were 90N max to keep the deformation under established limits. Thus, the monitored values were 81N y83N for the right and left side of the walls respectively.

A numeric controlled CMM ZEISS 850 machine was used to validate the geometrical accuracy of the workpiece on control points over the ruled surface equally spaced each 2 mm. The digitalized data file of the geometrical dimensions of the surface can be obtained and used to compare the CAD model with the real one. Figure 1.26 illustrates the surface errors according to the nominal values.

Although the behaviour of both cutting strategies seem to follow similar patterns, there had been improvements when using a flank milling with total axial immersion strategy. On the one hand machining time using flank milling strategy is 77% of the machining time of the jump to jump strategy. On the other hand, as can be seen in Fig. 1.26, the wall surface errors already considered are closer in the flank milling strategy.

### *1.6.4 A Compressor Disk for Helicopter Turbines*

The last example is a compressor disk for helicopter turbines. Compressor disks are widely used in many aeronautical applications. The complex geometry of an impeller is susceptible to collisions between the tool and the blade. Therefore it is necessary to use 5-axis CNC machining centres instead of the traditional 3-axis CNC centres, being necessary to control the tool axis vector to avoid collision and adjust the cutter's twisting angle to a proper location to finish surface cutting [44].

The compressor disk was made in Aluminium (Al 7075-T6). The design and the toolpath generation were made with Unigraphics CAD/CAM software, Fig. 1.27. There are several steps that must be followed in the tool path planning for a

**Material:**
Al 7075-T6
**Machining times:**
- Total 177’
- Rough 67’
- Finish 110’

**Tools:**
- Rough
  Ø40mm
  nose radius 6mm
  Ø12mm
  nose radius 1mm
- Finish
  Ø12mm
  nose radius 1mm

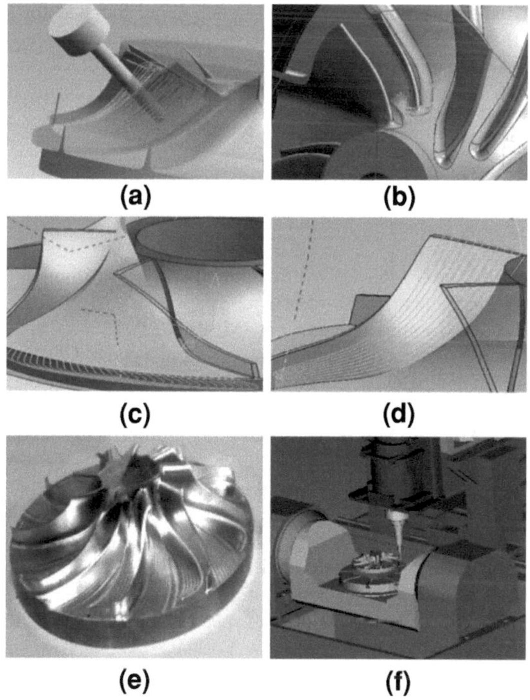


**Fig. 1.27** A compressor disk detail of the virtual simulation and CAM strategies in a 5-axis centre and machining conditions. **a** Rough milling, **b** rest milling, **c** hub finishing, **d** blade finishing, **e** final piece, **f** virtual verification

compressor disk machining. They are rough milling operations, blade finish operations and hub finish operations. The rough milling is adopted to roughly cut out the shape of the impeller from initial dimensions of the stock material. It cuts out the material between the pressure surface and the suction surface. At this stage the efficiency of the removal rate is the main factor to be considered. In this case rough operations take 67 min and were carried out with an end mill Ø 40 and 6 mm nose radius and an end mill Ø 12 and 1 mm nose radius.

Generally, the blade and hub surfaces are machined using swarf milling, which provides a line contact between the tool and the workpiece. The periphery of the tool perfectly touches the workpiece geometry. Hence, very high surface quality in the direction of feed and the direction perpendicular to the feed is obtained. Head milling with a flat end mill, with appropriate lead angle, is highly recommended in this case because of technological and geometrical advantages of using a flat end mill [45]. Finish operations took 110 min and the tool used is an end mill Ø 12mm and 1mm nose radius.

The software used for compressor disk virtual verification is Vericut. Taking into account the high probabilities of collision, virtual verification before real machining is of vital importance to prevent from part and machining collisions.

**Acknowledgments** Our thanks to Miguel Angel Salgado for the work performed in his Ph.D. Special thanks to Maestro Eduardo Sasia for his time dedicated to discuss several aspects of this chapter. Also special thanks are addressed to the ETORTEK pro-Future project. Special thanks to IBARMIA Innovatec and Tecnalia for their assistance during the test part development.

## References

1. Choi BK, Jerard RB (1998) Sculptured surface machining: theory and applications. Kluwer academic Publishers, Dordrecht
2. Senatore J, Monies F, Redonnet JM, Rubio W (2005) Analysis of improved posi-tioning in 5-axis ruled surface milling using envelop surface. Comput Aided Design 37:989–998
3. Tönshoff HK, Gey C, Rackow N (2001) Flank milling optimization—the flamingo project. Air Space Eur 3:60–63
4. Bedi S, Mann S, Menzel C (2003) Flank milling with flat end milling cutters. Comput Aided Design 35:293–300
5. Li C, Bedi S, Mann S (2006) Flank milling of ruled surface with conical tools-an optimization approach. Int J Adv Manuf Technol 29:1115–1124
6. Lartigue C, Duc E, Affouard A (2003) Tool path deformation in 5-axis flank milling using envelop surface. Comput Aided Design 35:375–382
7. Arnone M (1998) High performance machining. Hanser gardner publications, Cincinnati
8. King R (1985) Handbook of high speed machining technology. Chapman and hall advanced industrial technology series, New York
9. Lopez de Lacalle LN, Sanchez JA, Lamikiz A (2004) High performance machining (in Spanish) Ed Tec Izaro
10. Campshure KJ (1997) Mapping your way to 5-axis machining, modern machine shop. www.mmsonline.com
11. Tlusty G (2000) Manufacturing processes and equipment. Prentice Hall, New Jersey
12. López de Lacalle LN, Lamikiz A, Sánchez JA, Salgado MA (2004) Effects of tool deflection in the high speed milling of inclined surfaces. Int J Adv Manuf Technol 24:621–631
13. Kang MC, Kim KK, Lee DW, Kim JS, Kim NK (2001) Characterisation of inclided planes according to de the variations of cutting direction in high speed ball-end milling. Int J Adv Manuf Technol 17:323–329
14. Salgado MA (2003) Verification of CNC programs. Report project UE/96, ETSI of Bilbao, Spain
15. López de Lacalle LN, Lamikiz A, Sánchez JA, Salgado MA (2007) Toolpath selection based on the minimum deflection cutting forces in the programming of complex surfaces milling. Int J Mach Tools Manuf 47:388–400
16. Lee JJ, Suh SH (1998) Interference-free tool-path planning for flank milling of twisted ruled surfaces. Int J Adv Manuf Technol 14:795–805
17. Warkentin A, Ismail F, Bedi S (2000) Multi point tool position strategy for 5-axis machining of sculptured surfaces. Comput Aided Geom Des 17(1):83–100
18. Lee JJ, Suh SH (1998) Interference-free tool-path planning for flank milling of twisted ruled surfaces. Int J Adv Manuf Technol 14:795–805
19. Yown JW, Jun Y, Park S (2003) Interference-free tool path generation in 5-axis machining of a marine propeller. Int J Prod Res 41:4383–4402
20. Lauwers B, Kiswanto G, Kruth JP, Leuven KU (2003) Development of a 5-axis milling tool path generation algorithm based on Faceted models. Ann CIRP 52:85–89
21. Xu XJ, Bradley C, Zhang YF, Loh HT, Wong YS (2002) Tool-path generation for 5-axis machining of free-form surfaces based on accessibility analysis. Int J Prod Res 40:3253–3274
22. Tsay DM, Yan WF, Ho HC (2001) Generation of 5-axis cutter paths for turbomachinery components. J Manuf Sci Eng Transact ASME 123:731–738

23. Krajnik P, Kopac J (2004) Modern machining of die and mold tools. J Mater Process Technol 157–158:543–552
24. Arizmendi M, Fernández J, López de Lacalle LN, Lamikiz A, Gil A, Sánchez JA, Campa FJ, Veiga F (2008) Model development for the prediction of surface topography generated by ball-end mills taking into account the tool parallel axis offset: experimental validation. CIRP Ann Manuf Technol 57:101–104
25. Salgado M, López de Lacalle LN, Lamikiz A, Muñoa M, Sánchez JA (2005) Evaluation of the stiffness chain on the deflection of end-mills under cutting forces. Int J Mach Tools Manuf 45:727–739
26. López de Lacalle LN, Lamikiz A, Salgado MA, Herranz S, Rivero A (2002) Process planning for reliable high speed machining of moulds. Int J Prod Res 40:2789–2809
27. Denavit J, Hartenberg RS (1995) A kinematic notation for lower-pair mechanisms based on matrices. Transact ASME J Appl Mech 23:215–221
28. López de Lacalle LN, Lamikiz A, Sánches JA, Fernándes de Bustos I (2005) Recording of real cutting forces along the milling of complex parts. Mechatronics 16:21–32
29. Smith S, Dvorak D (1998) Tool path strategies for high speed milling aluminum workpieces with thin webs. Mechatronics 8:291–300
30. Rong Y, Tao R, Tang X (2000) Flexible fixturing with phase-change materials. Part 1: experimental study on magneto-rheological fluids. Int J Adv Manuf Technol 16:822–829
31. Sims ND, Zhang Y (2004) Piezoelectric active control for workpiece chatter reduction during milling, smart structures and materials. In: Alison B (ed) Smart structures and integrated systems. Proc SPIE, 5390, 335–346
32. Budak E (2000) Improving productivity and part quality in milling of titanium based impellers by chatter suppression and force control. Ann CIRP 49(1):31–64
33. Engin S, Altintas Y (2001) Mechanics and dynamics of general milling cutter. Part I: helical end mills. Int J Mach Tools Manuf 41:2195–2212
34. Lamikiz A, López de Lacalle LN, Sánchez JA, Salgado MA (2004) Cutting force estimation in sculptured surface milling. Int J Mach Tools Manuf 44(14):1511–1526
35. Koenigsberger F, Tlusty J (1971) Structures of Machine Tools. Pergamon Press, Oxford
36. Merrit H (1965) Theory of self-excited machine tool chatter. J Eng Ind 87:447–454
37. Budak E (1994) Mechanics and dynamics of milling thin walled structures. Ph.D. thesis, University of British Columbia
38. Bravo U, Altuzarra O, López de Lacalle LN, Sanchez JA, Campa FJ (2005) Stability limits of milling considering the flexibility of the workpiece and the machine. Int J Mach Tools Manuf 45:1669–1680
39. Thevenot V, Arnaud L, Dessein G, Cazenave-Larroche G (2006) Integration of dynamic behaviour in stability lobes method: 3D lobes construction and application to thin walled structure milling. Int J Adv Manuf Technol 27:638–644
40. Campa FJ, Lopez de Lacalle LN, Celaya A (2011) Chatter avoidance in the milling of thin floors with bull-nose end mills: model and stability diagrams. Int J Mach Tools Manuf 51: 43–53
41. Altintas Y (2000) Manufacturing automation. Cambridge University Press, Cambridge
42. Arizmendi M, Campa FJ, Fernández J, López de Lacalle LN, Gill A, Bilbao E, Veiga F, Lamikiz A (2009) Model for surface topography prediction in peripheral milling considering tool vibration. CIRP Ann Manuf Technol 58:93–96
43. Olvera D, López de Lacalle LN, Campa FJ, Lamikiz A (2010) In: Proceedings of effect of slenderness on machined surface. application to topography part prediction. Eighth international conference high speed machining
44. Quan L, Yongzhang W, Hongya F, Zhenyu H (2008) Cutting path planning for ruled surface impellers. Chin J Aeronaut 21:462–471
45. Bohez ELJ, Ranjith Senadhera SD, Polc K, Duflou JR, Tar T (1997) A geometric modeling and 5-axis machining algorithm for centrifugal impellers. J Manuf Syst 16:422–436

# Chapter 2
# 5-Axis Flank Milling of Sculptured Surfaces

**Johanna Senatore, Frédéric Moniès and Walter Rubio**

This chapter covers the various free-form surface flank milling strategies available, focusing in particular on those for ruled surfaces as widely used in defining turbomachine parts. All these positionings seek to reduce interference between the cutting tool and the surface to be milled so as to respect the tolerances dictated by the Design Office. The range of strategies presented goes from the simplest, using analytical positioning on a particular rule, through to complex procedures defined using global numerical methods that calculate the toolpath in its entirety. Approaches adapted to conical and half-barrel cutter geometries are also addressed. Machining of free-form surfaces is considered from two differing perspectives: either considering a free-form surface to be a set of ruled surfaces onto which the previously mentioned methods are applied, or studying the differential geometry of the cutters and surfaces.

## 2.1 Introduction

Flank milling involves machining a workpiece with the side part of the cutter. The cutter generatrices are generally straight lines (with a cylindrical or conical cutter) or arcs of circles (barrel cutter) (Fig. 2.1), but in some cases they can take other shapes.

This process is widely used in the industry for all contouring operations. The workpiece parts thus produced take the shape of planes, portions of cylinders or cones and more generally any type of developable surface. The advantage with

J. Senatore · F. Moniès · W. Rubio (✉)
Institut Clément Ader, Université Paul Sabatier,
118 route de Narbonne, Cedex 09, 31062, Toulouse, France
e-mail: rubio@cict.fr

J. P. Davim (ed.), *Machining of Complex Sculptured Surfaces*,
DOI: 10.1007/978-1-4471-2356-9_2, 

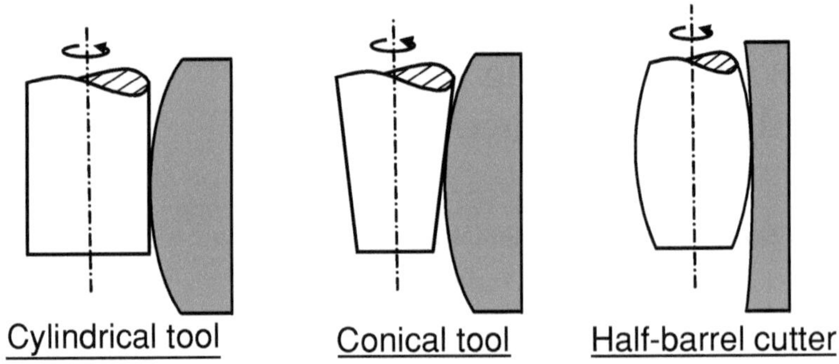


**Fig. 2.1** Cutter geometries

**Fig. 2.2** Impeller-type part

flank milling lies in removing a considerable amount of material and also permitting access to zones that would be inaccessible to end milling.

This milling method applies to other types of surfaces and is especially suited to convex free-form surfaces or surfaces with negative Gaussian curvatures that have one of their principal curvatures small compared with the cutter radius, being subject to only very slight change over the entire surface. Ruled surfaces answer to the previously stated criteria and are generally machined in this way. Such surfaces are mainly encountered on workpieces going into the manufacturing of turbomachines. Given their critical role and the stresses to which they are subjected, such high value added workpieces require special care. It is hardly surprising that such a substantial corpus of scientific works has been devoted to how they are machined with the aim of boosting productivity while respecting the quality levels required by the relevant specifications. Among such workpieces, the following are to be found:

- Impellers (Fig. 2.2): rotating parts used to suck in a liquid, or to diffuse or compress a gas. Impellers are provided with arrays of principal and secondary blades. The concave side and the convex side of these blades are generally modelled by non-developable ruled surfaces with reduced twist on the leading

Fig. 2.3 Inducer type part

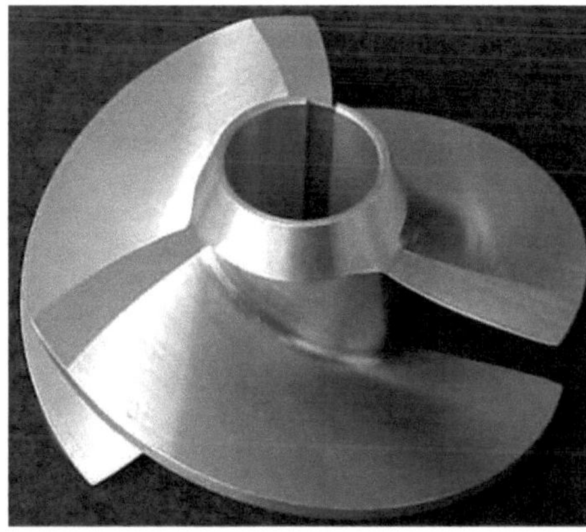

and trailing edges and pronounced twist located towards the middle of the blade. Blade heights remain fairly low.

- Inducers (Fig. 2.3): rotating parts placed at the entry of an impeller with the aim of enhancing performance. They work to provide an overpressure so that the fluid does not cavitate into the main pump. The concave and convex sides are modelled by non-developable ruled surfaces whose twist is fairly high and regular.
- Fans (Fig. 2.4): rotating parts used to accelerate a fluid. They have a large number of short, high blades. The concave and convex sides are modelled by non-ruled free-form surfaces.

Flank milling's main advantage lies in increasing the width of the area machined as the cutter passes through. This leads to shorter machining times, lower production costs and greater productivity. Another positive outcome in increasing the cutting width lies in reducing polishing operations. Where surfaces can be milled in a single pass, no scallop of material is left behind the cutter and polishing operations are therefore much reduced. Before introducing machining strategies, a definition will be provided for ruled surfaces.

## 2.2 Ruled Surfaces

### *2.2.1 Definition*

A surface containing a family of straight lines is called a ruled surface. It is generated by the movement of a straight line, called the rule or generatrix, moving over two curves $\mathbf{C}_0(u)$ and $C_1(u)$ that provide its directrices.

Ruled surfaces are often used to draw functional surfaces in mechanical engineering. They resolve the following problem: given two curves in space $\mathbf{C}_0(u)$ and $\mathbf{C}_1(u)$, both defined on the same parametric interval $u \in [0, 1]$, find a surface $\mathbf{S}$ such that $\mathbf{S}(u, 0) = \mathbf{C}_0(u)$ and $\mathbf{S}(u, 1) = \mathbf{C}_1(u)$. The problem posed has an infinite number of solutions. Here, the following model will be retained:

Fig. 2.4 Fan type part

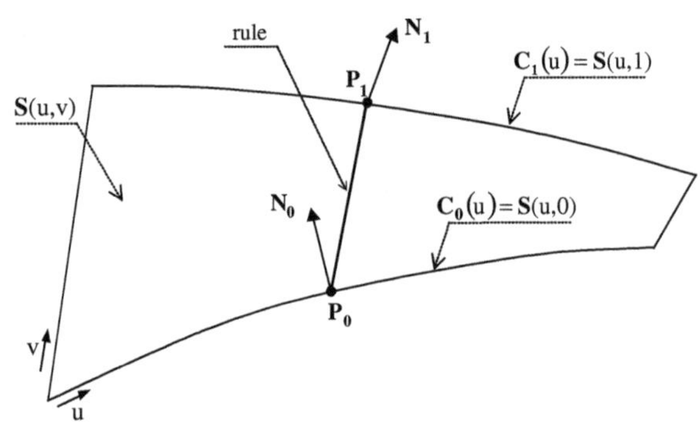


Fig. 2.5 Ruled surface

$$\mathbf{S}(u, v) = (1 - v)\ \mathbf{C_0}(u) + v\ \mathbf{C_1}(u) \quad (u, v) \in [0, 1]^2 \tag{2.1}$$

Ruled surfaces have the following property: each isoparametric line (parameter u constant), is called a rule, and is a straight line segment (Fig. 2.5). One major feature of such surfaces is the authorised generality for curves $\mathbf{S}(u, 0)$ and $\mathbf{S}(u, 1)$ that have practically no restriction other than being defined on the same parameter interval. For example, one of the entry curves can be a cubic polynomial curve, while the other can be a Bezier or a B-spline curve.

This surface definition is widely used when modelling workpieces for the aeronautical, naval and car industry sectors.

Fig. 2.6 Evolution of tangent planes on the rule $[\mathbf{P_0P_1}]$

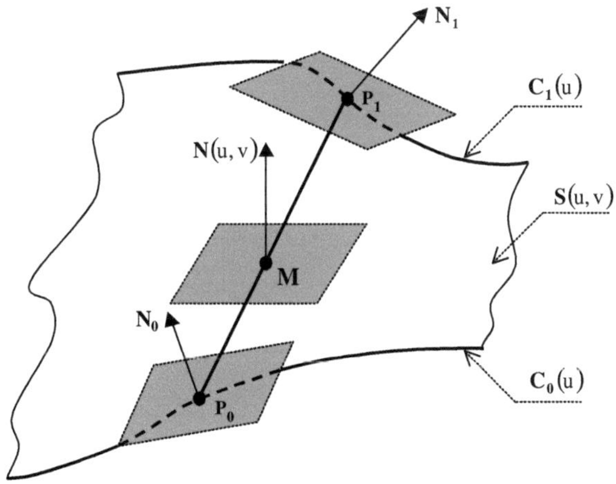


The following notations will be used in the rest of this chapter:

- $\mathbf{S}(u, v)$ the surface to be machined,
- $\mathbf{N}(u, v)$ the normal to the surface $\mathbf{S}(u, v)$,
- $[\mathbf{P_0P_1}]$: the rule considered,
- $\mathbf{C_0}(u), \mathbf{C_1}(u)$: the directrices of the surface $\mathbf{S}(u, v)$,
- $\mathbf{N_0}$,$\mathbf{N_1}$: normal to $\mathbf{S}(u, v)$ on $\mathbf{P_0}$ and $\mathbf{P_1}$ respectively,
- $\alpha$: the twist corresponding to the angle between normals $\mathbf{N_0}$ and $\mathbf{N_1}$:
- $\mathbf{h_p}$: the length of the rule $(\mathbf{P_0P_1})$.

Two types of ruled surfaces can be distinguished:

- Developable ruled surfaces: a surface is developable if its Gaussian curvature K is null and its mean curvature H non-null at any point. Calling $\mathbf{S_{uv}}(u, v)$ the second derivative of $\mathbf{S}(u, v)$ in relation to parameters u and v, this can be expressed by the statement: "$\mathbf{S}(u, v)$ is a developable ruled surface if and only if $\mathbf{N}(u, v) \cdot \mathbf{S_{uv}}(u, v) = 0$ at any point". It can be seen that the twist of a developable surface is null whatever the rule $[\mathbf{P_0P_1}]$ on the surface: $\alpha = 0$.
- Non-developable ruled surfaces: these can be characterised by one of the two following properties:

"Non-developable ruled surfaces have a non-null twist: $\alpha \neq 0$",

"Not all points located on a given rule have the same tangent plane" (Fig. 2.6).

To provide the background to the reasoning adopted in the present section, it will be useful first to conduct a preliminary analysis of the twist of non-developable ruled surfaces.

Evolution of the normals to $\mathbf{S}(u, v)$ along the rule will depend on curves $\mathbf{C_0}(u)$ and $\mathbf{C_1}(u)$. The normal to a surface is calculated by the vectorial product of the first derivatives. For the ruled surface the following will obtain:

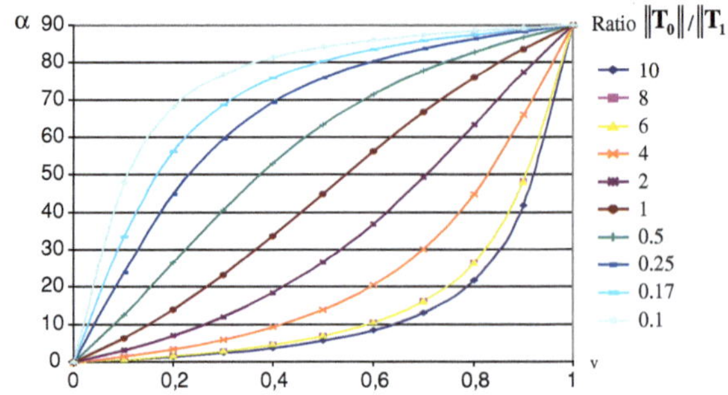


**Fig. 2.7** Evolution of twist as a function of the ratio $\|\mathbf{T_0}\|/\|\mathbf{T_1}\|$

$$\mathbf{N}(\mathrm{u},\mathrm{v}) = \frac{\partial\mathbf{S}(\mathrm{u},\mathrm{v})}{\partial \mathrm{u}} \wedge \frac{\partial\mathbf{S}(\mathrm{u},\mathrm{v})}{\partial \mathrm{v}} = \left((1-\mathrm{v})\frac{\mathbf{dC_0}(\mathrm{u})}{\mathrm{du}} + v\frac{\mathbf{dC_1}(\mathrm{u})}{\mathrm{du}}\right) \wedge \mathbf{P_0P_1} \quad (2.2)$$

Evolution of $\mathbf{N}(\mathrm{u},\mathrm{v})$ along the rule ($\mathrm{v} \in [\mathbf{0},\mathbf{1}]$) will depend on $\frac{\mathbf{dC_0}(\mathrm{u})}{\mathrm{du}}$ and $\frac{\mathbf{dC_1}(\mathrm{u})}{\mathrm{du}}$. To simplify matters, it is assumed that $\frac{\mathbf{dC_0}(\mathrm{u})}{\mathrm{du}}$ and $\frac{\mathbf{dC_1}(\mathrm{u})}{\mathrm{du}}$, noted respectively $\mathbf{T_0}$ and $\mathbf{T_1}$, are perpendicular to $\mathbf{P_0P_1}$.

Let $\mathbf{T}_i = ((1-\mathrm{v})\mathbf{T_0} + \mathrm{v}\mathbf{T_1})$, the evolution of $\mathbf{N}(\mathrm{u},\mathrm{v})$ will follow the evolution of $\mathbf{T_i}$. Let $\alpha_i$ be the angle between $\mathbf{T_0}$ and $\mathbf{T_i}$ as calculated by:

$$\cos(\alpha_i) = \frac{\mathbf{T_0}\cdot\mathbf{T}_i}{\|\mathbf{T_0}\|\|\mathbf{T}_i\|} = \frac{(1-\mathrm{v})\|\mathbf{T_0}\| + \mathrm{v}\|\mathbf{T_1}\|\cos(\alpha)}{\|(1-\mathrm{v})\mathbf{T_0} + \mathrm{v}\mathbf{T_1}\|} \quad (2.3)$$

The following is obtained:

$$\cos(\alpha_i) = \frac{(1-\mathrm{v})\frac{\|\mathbf{T_0}\|}{\|\mathbf{T_1}\|} + \mathrm{v}\cos(\alpha)}{\sqrt{(1-\mathrm{v})^2\left(\frac{\|\mathbf{T_0}\|}{\|\mathbf{T_1}\|}\right)^2 + 2\mathrm{v}(1-\mathrm{v})\left(\frac{\|\mathbf{T_0}\|}{\|\mathbf{T_1}\|}\right)\cos(\alpha) + \mathrm{v}^2}} \quad (2.4)$$

This result is interesting as it shows that the evolution of the normal depends on the ratio between the norms of vectors $\mathbf{T_0}$ and $\mathbf{T_1}$. The evolution of $\alpha_i$ was plotted as a function of v (Fig. 2.7) for several values of the ratio $\|\mathbf{T_0}\|/\|\mathbf{T_1}\|$. With no loss in generality, in this example angle $\alpha$ of the twist equals 90°.

It can be seen that the evolution of the normal along the rule is not linear. However, linearity when the ratio between $\|\mathbf{T_0}\|$ and $\|\mathbf{T_1}\|$ is close to 1 can be considered. Such curves will be analysed later when focusing on strip milling.

### *2.2.2 Machining Non-Developable Ruled Surfaces*

During flank milling of a non-developable ruled surface, the existence of twist $\alpha$ implies that it is impossible to machine the workpiece perfectly using a non-null diameter cutter, with the cutter positioning on the surface leading to inevitable

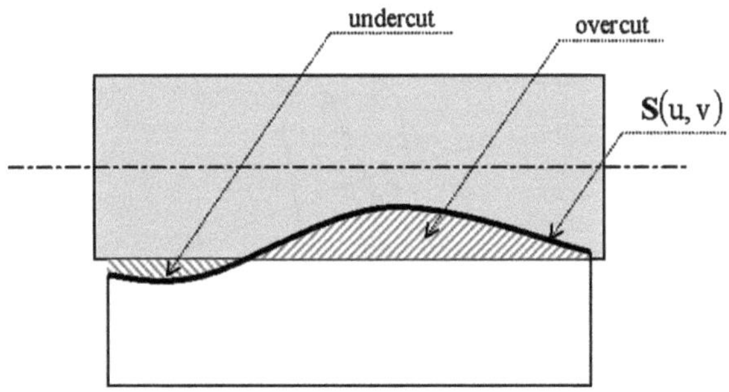


**Fig. 2.8** Overcut and undercut errors

interference. It then becomes interesting to seek the positioning leading to minimal interference. This means that cutters with larger cross-sections can be used; these have good rigidity and ensure maximum material removal while respecting tolerance of the surface shape. This problem of cutter positioning optimisation involves the uses of 5-axis NC machine tools so as to be able to orient the cutter freely.

The interference error caused by the cutter during milling can be of two types: geometrical error corresponding to the maximum interference generated by the cutter for a given positioning, and machining error related to the cutting process environment and expressing the real behaviour of the cutter/workpiece/machine assembly (cutter bending, vibrations, etc.). The remainder of this section will focus on studying geometrical error.

The interference generated by the cutter is of two types (Fig. 2.8): either the cutter removes too much material (overcut), or the cutter leaves too much material behind it (undercut).

To calculate interference error, the distance between a point of the theoretical surface along its normal and the cutter is considered. This corresponds to the definition of the profile tolerance of any surface. To conclude on respect for tolerance of the machined surface, maximum undercut and overcut errors need to be compared with the imposed shape tolerance.

### 2.2.3 *Local Positioning*

When discussing tool positioning, this obviously does not concern developable ruled surfaces. Indeed, for such surfaces, the cutter causes no error due to positioning and problems of visibility (access to the surface) alone are to be considered. Few methods have been developed for cutter positioning during flank milling of non-developable ruled surfaces. Two main approaches can be distinguished:

- Analytical methods: this involves positioning methods that are generally simple to implement but that generate relatively significant errors. Chronologically speaking, these were the first methods to be published.

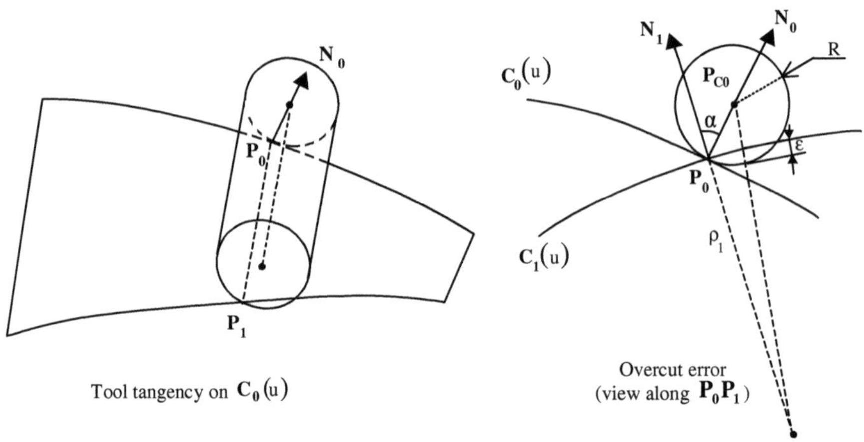


**Fig. 2.9** Software positioning

- Numerical methods: these considerably reduce geometrical machining error but are more complex to implement. Their development was reliant on the enhancement of computing capacity.

#### 2.2.3.1 Analytical Positioning

One of the first forms of analytical positioning to be implemented was of the kind to be found in many CAD/CAM software packages (referred to hereafter as "software positioning"). It was developed for cylindrical cutting tools and involves positioning the cutter axis parallel to the rule considered and the cutter tangent to a directrix (Fig. 2.9). Any interference between the cutter and the workpiece will then be concentrated on the other directrix. The cutter is tangent to directrix $\mathbf{C_0}(\mathrm{u})$, the cutter axis is collinear to vector $\mathbf{P_0P_1}$, and the maximum overcut error will be on $\mathbf{C_1}(u)$. Knowing the orientation of the cutter axis, determining positioning involves calculating a point of the axis $\mathbf{P_{C0}} = \mathbf{P_0} + \mathrm{R}\,\mathbf{N_0}$ (R cutter radius).

Maximum interference error $\varepsilon$ is given by:

$$\varepsilon = \rho_1 + \mathrm{R} - \sqrt{\mathrm{R}^2 + \rho_1^2 + 2\rho_1 \mathrm{R}\cos\alpha} \tag{2.5}$$

with $\rho_1$ the radius of curvature on $\mathbf{P}_1$ of the directrix $\mathbf{C}_1(\mathrm{u})$. $\rho_1$ is assumed to be constant in the zone studied.

If the curvature $\rho_1$ is great compared with R then the error becomes:

$$\varepsilon = \mathrm{R}(1 - \cos(\alpha)) \tag{2.6}$$

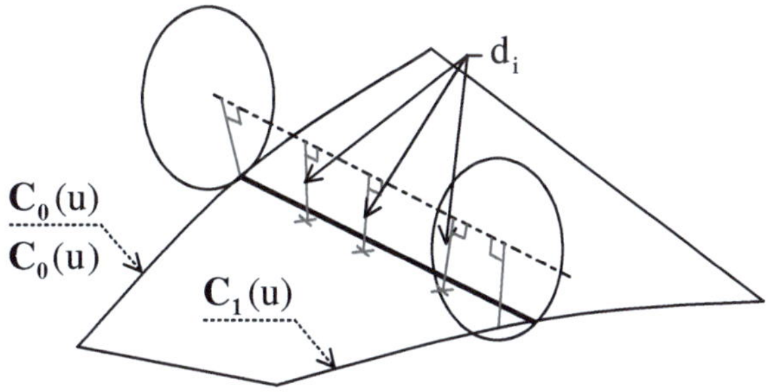


**Fig. 2.10** CATS method, computing distances

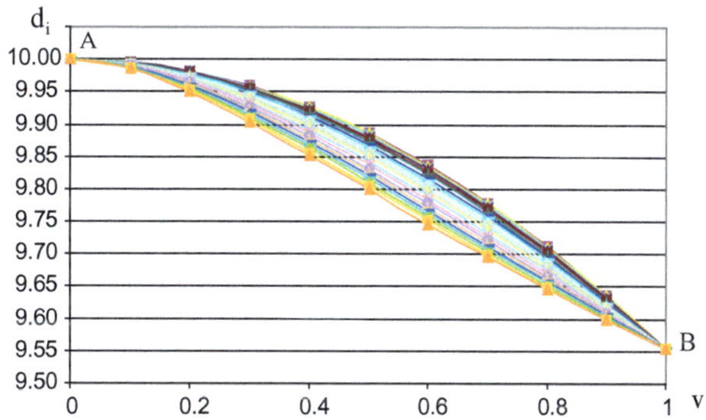


**Fig. 2.11** Profile of the set of distances

The method developed in [1, 2] called Computation of Adapted Tool Shape (CATS) involves using software positioning and adapting it to the geometry of the cutter at the detected interferences. Thus, from evolution in the twist, the minimum distance $d_i$ from the axis to the surface of the workpiece (Fig. 2.10) can be computed at each cross-section of the cutter, taken perpendicular to the axis. The distance from the cutter axis to $\mathbf{C_0}(u)$ is chosen equal to R (10 mm in the following example).

For all rules, distances $d_i$ can be carried over onto a graph (Fig. 2.11). In the example below, all rules are considered to have the same length (30 mm). The cutter geometry is to be chosen from amongst all the curves and will depend on a number of criteria (minimising maximum error, preventing overcut and minimising the volume between the machined surface and the nominal surface). These criteria represent varying difficulties in implementation.

The selected cutter will have barrel geometry (Fig. 2.1) in order to simplify manufacturing. The cutter profile will be defined by a circle with radius Rt going through A and B (Fig. 2.11) and verifying a criterion. This method's advantage lies in the use of elementary positioning while minimising error by its compensation on the cutter diameter; it is especially appropriate to surfaces with a twist that varies little over the set of rules. This applies pretty well to inducers.

Rubio [3] also developed analytical positioning (referred to as standard positioning). The concept developed here involves the distribution of errors on each of the directrices of the surface. The axis of the cylindrical cutter is collinear to the

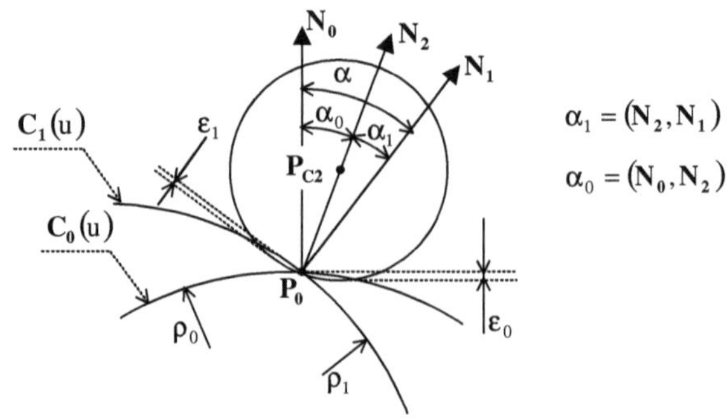


**Fig. 2.12** Standard positioning (view along $\mathbf{P_0P_1}$)

rule considered and its position is defined by a point $\mathbf{P}_{C2}$ calculated such that interferences $\varepsilon_0$ and $\varepsilon_1$ on the directrices are equal.

The radii of curvatures $\rho_0$ and $\rho_1$ of the directrices $\mathbf{C}_0(\mathrm{u})$ and $\mathbf{C}_1(\mathrm{u})$ on $P_0$ and $P_1$ respectively are assumed to be constant locally. The maximum interference error $\varepsilon_i$ ($i = 0, 1$) located on the directrix $\mathbf{C}_i(\mathrm{u})$ can be expressed:

$$\varepsilon_i = \rho_i + \mathrm{R} - \sqrt{\mathrm{R}^2 + \rho_i^2 + 2\rho_i \mathrm{R} \cos \alpha_i} \tag{2.7}$$

The following system results from distribution of the error:

$$\begin{cases} \rho_0 + \mathrm{R} - \sqrt{\mathrm{R}^2 + \rho_0^2 + 2\rho_0 \mathrm{R} \cos \alpha_0} = \rho_1 + \mathrm{R} - \sqrt{\mathrm{R}^2 + \rho_1^2 + 2\rho_1 \mathrm{R} \cos \alpha_1} \\ \alpha = \alpha_0 + \alpha_1 \end{cases} \tag{2.8}$$

Angles $\alpha_0$ and $\alpha_1$ verifying $\varepsilon_0 = \varepsilon_1$, and vector $\mathbf{N}_2$ defined from the normals $\mathbf{N}_0$ and $\mathbf{N}_1$ (Fig. 2.12) defining the point $\mathbf{P}_{C2} = \mathbf{P}_0 + \mathrm{R}\mathbf{N}_2(\mathrm{u})$ can then be determined.

For surfaces showing large radii of curvature as compared with the cutter radius, $\alpha_0 = \alpha_1 = \alpha/2$ is obtained.

The resulting error is then equal to $\varepsilon = \mathrm{R}(1 - \cos(\alpha/2))$. Through a series development of the cosine function, the error is estimated at $\varepsilon \approx \mathrm{R}\alpha^2/8$ representing an error four times smaller than that for software positioning.

In the previous positioning methods, the cutter axis is collinear to the rule considered. Liu [4] proposes a new positioning adopting a different philosophy: here, the cutter swivels in relation to the rule. This positioning is known as "Double Point Offsets" (DPO).

Two corresponding points for the cutter axes, noted $\mathbf{C_A}$ and $\mathbf{C_B}$ (Fig. 2.13) are calculated from two points located at a quarter and three-quarters of the length of the rule, noted respectively **A** and **B**. These points $\mathbf{C_A}$ and $\mathbf{C_B}$ are determined by the following relations:

$$\mathbf{C_A} = \mathbf{A} + \mathrm{R}\mathbf{N_A} \tag{2.9}$$

$$\mathbf{C_B} = \mathbf{B} + \mathrm{R}\mathbf{N_B} \tag{2.10}$$

with: $\mathbf{N_A}$ and $\mathbf{N_B}$ the unit normals to the surface at points **A** and **B** respectively.

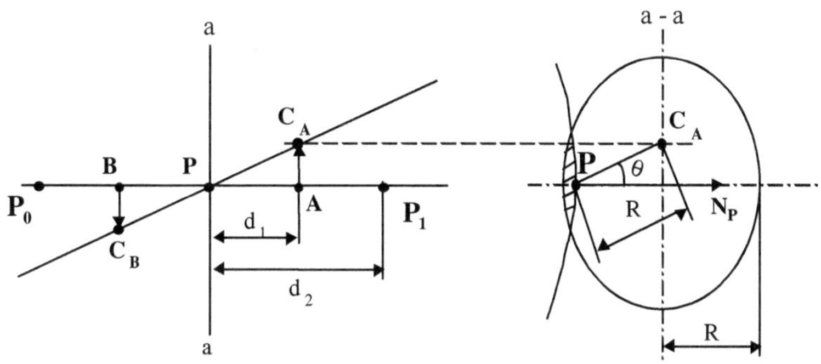


**Fig. 2.13** Positioning Liu

The middle point of the rule is called **P** and the unit normal to the surface at that point is $\mathbf{N_P}$. Maximum interference induced by this positioning is of the undercut type ($\varepsilon_{PO,\,P1}$) at ends $\mathbf{P_0}$ and $\mathbf{P_1}$ of the rule and of the overcut type ($\varepsilon_P$) on **P**. These errors are calculated from the relations given below:

$$\varepsilon_P = R(1 - \cos\theta) \tag{2.11}$$

$$\varepsilon_{P0,P1} = R\left(\sqrt{\cos^2\theta + \frac{d_2^2}{d_1^2}\sin^2\theta} - \sqrt{1 + \frac{R^2 d_2^2 \sin^2\theta \tan^2\theta}{d_1^2\left(d_2^2\tan^2\theta + d_1^2 + R^2\sin^2\theta\right)}}\right) \tag{2.12}$$

The total error generated by this positioning is equal to the sum of these errors:

$$\varepsilon = \varepsilon_P + \varepsilon_{P0,P1} \tag{2.13}$$

The author then proposes to move the cutter along the normal $\mathbf{N_P}$ by a distance $\delta$ as determined algebraically, in order to distribute the error equally on either side of the theoretical surface. This distance can be determined algebraically by:

$$\delta = -\frac{\varepsilon_P + \varepsilon_{P0,P1}}{1 + \sqrt{\frac{d_1^2}{d_1^2 + d_2^2\tan(\theta)}}} \tag{2.14}$$

A first analytical positioning was described initially in [5]. This showed strong similarities with "DPO" positioning. The difference resides in the fact that the points of the cutter axis $\mathbf{P_{C0}}$ and $\mathbf{P_{C1}}$ are calculated from the end points of the rule:

$$\mathbf{P_{C0}} = \mathbf{P_0} + R\mathbf{N_0} \tag{2.15}$$

$$\mathbf{P_{C1}} = \mathbf{P_1} + R\mathbf{N_1} \tag{2.16}$$

The error is of the overcut type between the two ends of the rule. The maximum error is in the middle of the rule and the lowest overcut errors are located on $\mathbf{P_0}$ and $\mathbf{P_1}$. The cutter is then translated to distribute the interference error.

Figure 2.14 shows these different positioning methods.

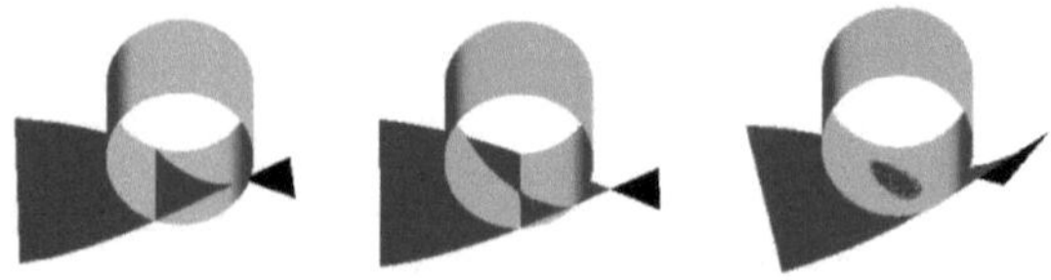

**Fig. 2.14** Software, standard and Liu positioning

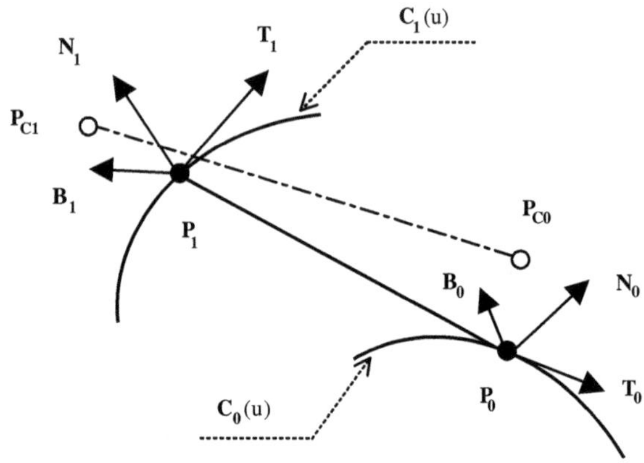


**Fig. 2.15** Positioning Bedi

#### 2.2.3.2 Numerical Positioning

Two types of numerical resolutions to position the cutter can be distinguished:

- methods relying on resolution of a system of $n$ equations with $n$ unknowns,
- iterative methods.

The presentation will now consider positioning obtained by system resolution.

In works by Bedi et al. [6], the cutter is positioned tangent to the two directrices at the ends of the rule (Fig. 2.15). The Frenet frames of curves $\mathbf{C_0}(u)$ and $\mathbf{C_1}(u)$ are placed at points $\mathbf{P_0}$ and $\mathbf{P_1}$. Vectors ($\mathbf{T_i}, \mathbf{N_i}$ and $\mathbf{B_i}$) represent vectors, tangents, normals and binormals. Positioning of the cutter axis is conducted from points $\mathbf{P_{C0}}$ and $\mathbf{P_{C1}}$. that verify the following conditions:

- $\mathbf{P_{C0}}$ et $\mathbf{P_{C1}}$ belong respectively to the planes $(\mathbf{P_0}, \mathbf{N_0}, \mathbf{B_0})$ and $(\mathbf{P_1}, \mathbf{N_1}, \mathbf{B_1})$
- $\|\mathbf{P_{C0}P_0}\| = R$ and $\|\mathbf{P_{C1}P_1}\| = R$
- $\mathbf{P_{C1}P_1} \cdot \mathbf{P_{C0}P_{C1}} = 0$ and $\mathbf{P_{C0}P_0} \cdot \mathbf{P_{C0}P_{C1}} = 0$

This positioning method requires the numerical resolution of a system of four equations with four unknowns and guarantees tangency of the cutter with the directrix curves. At this stage, it is close to positioning Stute [5].

The positioning method known as improved positioning presented in [7–9] involves first setting the cutter in the standard positioning configuration and making a rotation about an axis calculated from the normals to the surface at the

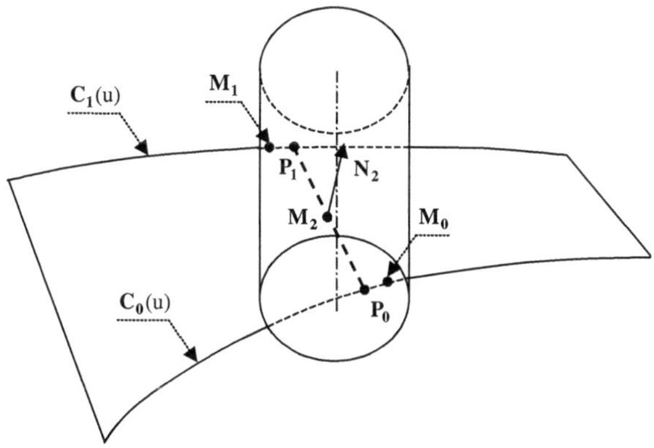


**Fig. 2.16** Improved positioning

ends of the rule. This method is based on searching for two points of tangency between the cutter and the two directrices and a point of contact on the rule considered. This was first developed for a cylindrical cutting tool.

The conditions for bringing into position are as follows (Fig. 2.16):

- contact between the rule $[\mathbf{P_0P_1}]$ and the cutter at point $\mathbf{M_2}$,
- tangency with the two directrix curves $\mathbf{C_0}(\mathrm{u})$ and $\mathbf{C_1}(\mathrm{u})$ on either side of the end points of the rule: points $\mathbf{M_0}$ and $\mathbf{M_1}$.

Working from these imposed parameters and conditions, positioning involves resolution of a non-linear system of seven equations with seven unknowns. The cutter position is then defined by the angle $\gamma$ between the axis of the cutter (with rotation going about the axis $\mathbf{N_2}$ defined on standard positioning) and the rule, and by the position of point $\mathbf{M_2}$ on the rule.

It transpires from this study that improved positioning generates a geometric error considerably less than that produced by other positioning strategies. This can be explained by the fact that the points characteristic of improved positioning $\mathbf{M_0}$, $\mathbf{M_1}$ and $\mathbf{M_2}$ are generator points. It was shown [9] that the points of tangency on the directrices are generator points. Thus, although ascertained purely geometrically, improved positioning has interesting kinematic properties.

Improved positioning as previously presented is based on the axis $\mathbf{y_2} = \mathbf{N_2}$ defined by standard positioning. The axis $\mathbf{y_2}$ was imposed without knowing the influence it could have on machining error. The following evolutions are thus proposed for positioning:

- 1st positioning: the axis of rotation $\mathbf{y_2}$ is made to coincide with the bissectrix of the normals $\mathbf{N_0}$ and $\mathbf{N_1}$, this is known as "improved positioning",
- 2nd positioning: the point of rotation $\mathbf{M_2}$ is made to coincide with the middle of the rule; this positioning is known as centred improved positioning and the axis of rotation $\mathbf{y_2}$ is unknown,
- 3rd positioning: the axis of rotation $\mathbf{y_2}$ is chosen such that undercut and overcut errors are equal.

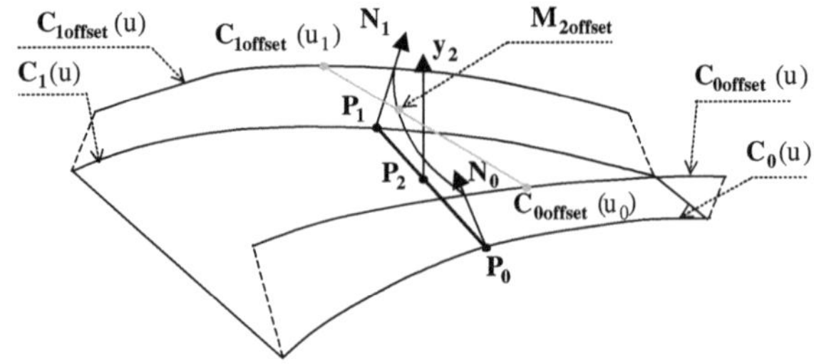


**Fig. 2.17** Improved positioning applied to the offset surface

Among the different versions of improved positioning developed, centred improved positioning is the most efficient from the perspective of the error caused.

In [10, 11], an enhancement of improved positioning was proposed based on the offset surface of $\mathbf{S}(u, v)$. Improved positioning and its different versions were thus applied between the axis of the cutter and the offset surface $\mathbf{S_{offset}}(u, v) = \mathbf{S}(u, v) + R \cdot \mathbf{N}(u, v)$ (Fig. 2.17). The latter positioning method follows the same principle as the previous ones: a rotation of the cutter axis about the axis $\mathbf{y_2}$ is performed and the centre of rotation is located on $\mathbf{M_{2offset}} = \mathbf{M_2} + R \cdot \mathbf{N_2}$. A position of the axis going through $\mathbf{M_{2offset}}$ is sought that cuts through the two offset directrices $\mathbf{C_{0offset}}(u) = \mathbf{C_0}(u) + R \cdot \mathbf{N}(u, 0)$ and $\mathbf{C_{1offset}}(u) = \mathbf{C_1}(u) + R \cdot N(u, 1)$. Geometric positioning of the cutter axis can be expressed as follows:

- vector $\mathbf{M_{2offset}}\mathbf{C_{0offset}}(u_0)$ is perpendicular to the axis of rotation $\mathbf{y_2}$,
- vector $\mathbf{M_{2offset}}\mathbf{C_{1offset}}(u_1)$ is perpendicular to the axis of rotation $\mathbf{y_2}$,
- points $\mathbf{C_{0offset}}(u_0), \mathbf{C_{1offset}}(u_1)$ and $\mathbf{M_{2offset}}$ are aligned as they all belong to the cutter axis

The positioning applied to the offset surface is of interest as it is easy to implement the system to be resolved consisting of three equations and three unknowns. When compared with improved positioning it shows undercut and overcut errors that are extremely close leading to giving preference to positioning on the offset surface if only error values are focused on. However, tangency on the directrices cannot be guaranteed as these conditions are no longer imposed and if tangency is sought, improved positioning offers the best option.

All the above reasoning was developed for cylindrical cutters, but improved positioning can also be used for conical cutters. Conical cutters are frequently used for turbine and fan blade-type workpieces as they involve problems of accessibility and cannot be machined with a cylindrical cutter of a reasonable diameter.

Improved positioning of the conical cutter proposed in [8], [12–14] (Fig. 2.18) is characterised by two points of tangency and one point of contact between the cutter and the surface, as for the cylindrical cutter. This positioning is obtained by a rotation of the cutter by an angle $\gamma$ about an axis $\mathbf{N_2}$ at point $\mathbf{M_2}$ from the rule considered.

As the conical cutter has an evolving radius that has to be taken into account in the search for positioning, a new geometrical parameter $g_0$ called the guard is

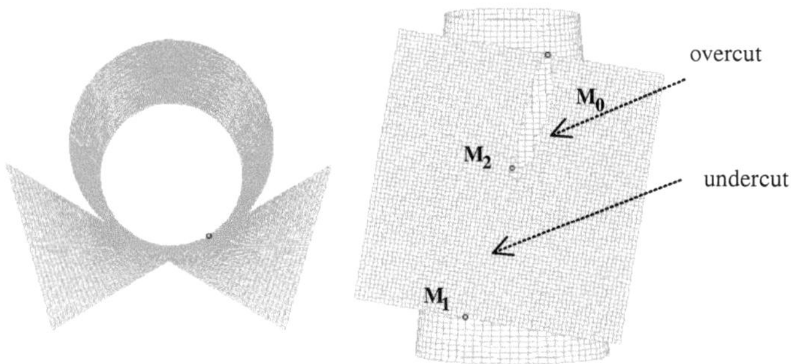


**Fig. 2.18** Positioning the conical cutter

**Fig. 2.19** Conical cutter parameters

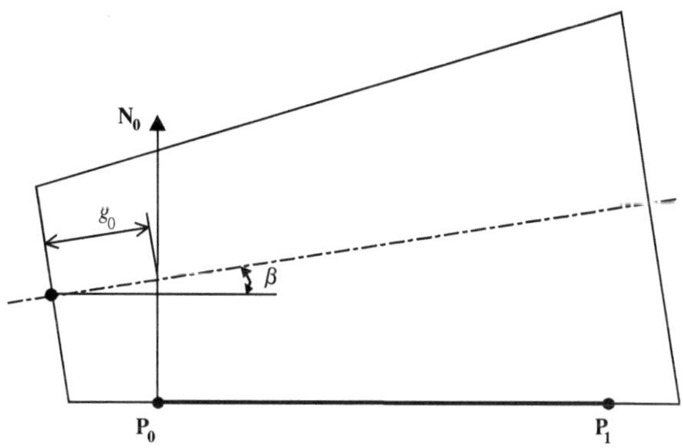


defined (Fig. 2.19). The guard is the distance, along the cutter axis, between the end point of the cutter and the osculating plane on $\mathbf{P_0}$. The guard is initially set by the user. Cutter positioning is thus fully defined by the three following parameters:

- the angle $\gamma$ between the rule and the cutter axis projected in the plane having the vector $\mathbf{N_2}$ as normal,
- the position of point $\mathbf{M_2}$ on the rule,
- the guard $g_0$.

From these parameters and imposed conditions, positioning involves resolution of a non-linear system of seven equations with seven unknowns. This system is resolved numerically by the Newton–Raphson method. Quantitative analysis of error is also proposed by Monies [13], this being addressed in the paragraph "Assisting cutting tool choice".

The numerical positioning methods introduced impose geometric conditions (for distance and sometimes tangency) in order to establish a cutting tool's position. Many other positioning strategies were established using iterative methods

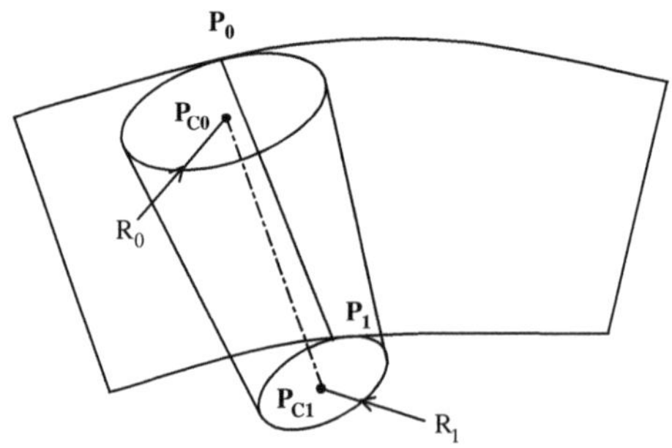


**Fig. 2.20** Positioning Chiou

that involved modifying an initial positioning after error estimation and compensation. These iterative methods will now be presented in what follows.

The application of iterative methods requires knowledge of error so as to compensate for it. The geometrical error between the surface generated by a moving cutter and the nominal surface is determined using the following relation:

$$\mathbf{S}(u, v) + \varepsilon \mathbf{N}(u, v) = \mathbf{S_{env}}(s, t) \tag{2.17}$$

$\mathbf{S_{env}}(s, t)$ represents the envelope surface (or machined surface) corresponding to the trace left by the cutter. $\varepsilon$ represents the distance between the machined surface and the nominal surface along the normal to the nominal surface.

In order to determine $\mathbf{S_{env}}(s, t)$, the successive positions taken by the cutter need to be known so as to allow grazing curves dependent on the cutter speed vector to be established. A certain number of works cover the issue of envelope surfaces, noteworthy amongst which are [15, 16]. The grazing curve cannot be known without knowing the next position of the cutter, meaning that error computations based on a static position of the cutter can only be approximate.

Approximation methods of error calculation used for the development of new strategies were analysed and compared in [17]. The goal with these error calculation methods is to reduce computation times. Indeed, iterative methods that recalculate the envelope surface after each change in tool path are those that give greatest accuracy when calculating error but are detrimental to computation time. A number of error estimation methods have been devised to evaluate a strategy's accuracy rapidly. The various authors of these iterative methods seek a trade-off between positioning effectiveness and speed in obtaining the result. No attempt is made to rank the methods introduced below as they would all have to be implemented on the same computer to compare calculation times, with the error being calculated using Eq. 2.17.

The method developed in [18] is established for roughing and finishing of ruled surfaces. It can be applied to both cylindrical and conical cutters. The study is conducted iteratively and covers use of the cutter's envelope surface. Initially, a set of cutter positions is defined using relations 15 and 16 as already introduced (Fig. 2.20):

The envelope surface is calculated from the set of positions. For each cutter position, the author calculates the distance from the envelope curve to the ruled

Fig. 2.21 Error profile

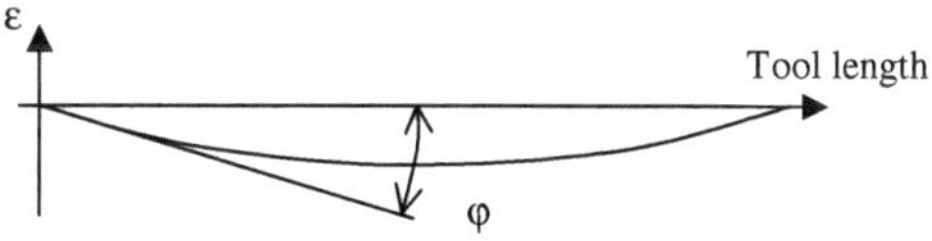


surface of the work-piece. The chosen positioning gives error profiles for overcut and maximum error, noted $\varepsilon_{max}$, is used to reposition the cutter:

$$\mathbf{P_{C0}} = \mathbf{P_0} + (R_0 + \varepsilon_{max})\,\mathbf{N_0} \tag{2.18}$$

$$\mathbf{P_{C1}} = \mathbf{P_1} + (R_1 + \varepsilon_{max})\,\mathbf{N_1} \tag{2.19}$$

The surface envelope and error are recomputed. The cutter will be repositioned until the overcut error becomes null. This procedure is used to rough cut the surface.

The finishing strategy is much in the same spirit. The cutter is first positioned using Eqs. 2.15 and 2.16. Having computed the envelope surface and the error profile between an envelope curve and the ruled surface, the cutter is offset for the error to be null on the directrices. The error profile obtained is shown in Fig. 2.21:

The cutter is then repositioned by a rotation of angle $\varphi$ in the plane containing the cutter axis and the normal $\mathbf{N_0}$ at point $\mathbf{P_{C0}}$. As the envelope surface is modified, this repositioning procedure can be reiterated until there is no further overcut error. Finally, as this strategy leads to a significant undercut error, it is suggested that the surface be milled in several passes breaking down the ruled surface into several ruled surfaces. This study proposes an exact evaluation of error but has an adverse effect on computation time.

Lee and Suh [19] also propose an iterative positioning method where the cutter axis is initially taken to coincide with the considered rule and then translated along the direction $\mathbf{N}(u, 0.5)$ by a distance d before a rotation of angle θ about the same axis is finally performed (Fig. 2.22).

For a set of couples $(\theta, d)$, cutter positioning is evaluated using a performance index $I_u = \sum_{i=1}^{n} \left|d_{u,i} - R\right|$ with $d_{u,i}$ the set of distances between n points of the cutter axis and the surface $\mathbf{S}(u, v)$. The positioning retained will be that which gives the smallest value for the performance index $I_u$.

For each value θ tested, the distance d minimising error is calculated. This calculation is based on an approximation of the surface $\mathbf{S}(u, v)$using triangular patches.

In [20], a new form of iterative positioning is proposed (Fig. 2.23). The method works as follows:

- Initial positioning: the cutter is positioned using the method Bedi et al. [6] as introduced previously. The following hypothesis is then adopted: maximum error will be found at v = 0.5.
- First modification of initial positioning: The positioning established at this stage is identical to the previous one but uses points inside the rule rather than at the

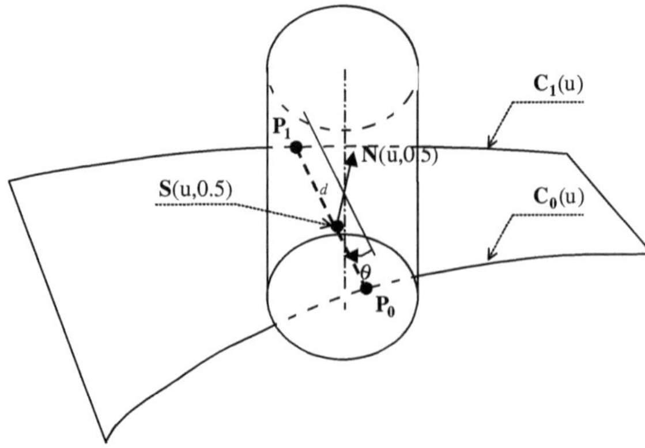


Fig. 2.22 Positioning Lee

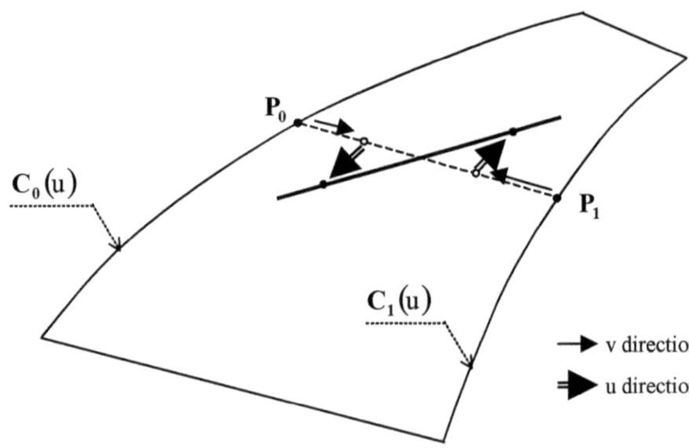


Fig. 2.23 Positioning Menzel

ends. To do so, the two positioning points of the cutter axis are translated in the direction of the rule (from the boundaries $v = 0$ and $v = 1$ up to the parametric value $v = 0.5$) by successive increments. On each iteration, the cutter position leads to a geometric error of the undercut type on the sides of the rule and overcut type in the middle. The error is close to that seen in Fig. 2.14 with Liu's positioning [4]. The position $[(u, v_1), (u, v_2)]$ retained is the one that minimises total error.

- Second modification of positioning*:* from solution $[(u, v_1), (u, v_2)]$, the positioning points of the cutter are translated along the isoparametrics $v = v_1$ and $v = v_2$. The positioning points of the cutter are characterised by: $[(u_1, v_1), (u_2, v_2)]$ : parameters $u_1$ and $u_2$ are such that $(u - u_1)$ and $(u - u_2)$ have opposite signs. The positioning points of the cutter are translated until the minimum distance between the axis and the rule considered is equal to the cutter radius. A third point of tangency is then created on the rule. The rotation induced by this second optimisation is of the order of 5° at most.

[21] adapted a version of their positioning defined in [6] to conical cutters. The method proposed involves first bringing the cutter tangent to the two guide curves. These curves can be other than directrices of the ruled surface. The point where

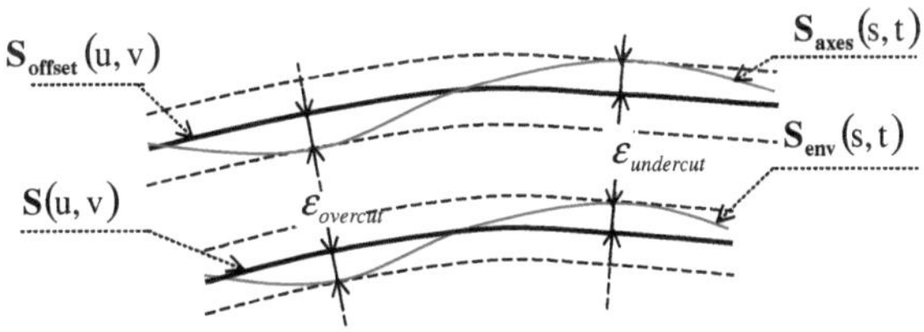


**Fig. 2.24** Properties on offset surfaces

error reaches a maximum value is sought. The second stage allows maximum error to be reduced by moving the guide curves towards the point with the biggest deviation along the rule. This allows positioning error to be reduced by about a half. The third stage involves moving the cutter position along the feed direction to further reduce error between the given surface and the machined surface.

Finally, a last iterative positioning is presented in [22] under the method named "Three Points Offset" (T.P.O.). This method applies only to cylindrical cutters and involves positioning the cutter axis on the offset surface defined by:

$$\mathbf{S_{offset}}(u, v) = \mathbf{S}(u, v) + \mathbf{Rn}(u, v) \tag{2.20}$$

The authors seek to minimise deviation between the offset surface and the axis as they show that maximum error between the offset surface $\mathbf{S_{offset}}(u, v)$ and the surface swept by the axis $\mathbf{S_{axis}}(s, t)$ is equal to the maximum error between the nominal surface $\mathbf{S}(u, v)$ and the envelope surface $\mathbf{S_{env}}(s, t)$ (Fig. 2.24).

"T.P.O." positioning involves positioning the cutter axis by three points belonging to $\mathbf{S_{offset}}(u, v)$. To do so, two points $\mathbf{P_{0offset}} = \mathbf{S_{offset}}(u, 0)$ and $\mathbf{P_{1offset}} = \mathbf{S_{offset}}(u, 1)$ belonging to the two directrices $\mathbf{C_{0offset}}(u)$ and $\mathbf{C_{1offset}}(u)$ of the offset surface (Fig. 2.25) will be considered. Point $\mathbf{P_{0offset}}$ is fixed throughout the argument and is located on the rule considered. Point $\mathbf{P_{1offset}}$ moves along the directrix $\mathbf{C_{1offset}}(u)$. For each position occupied by $\mathbf{P_{1offset}}$, maximum error between the offset surface and segment $[\mathbf{P_{0offset}}\,\mathbf{P_{1offset}}]$ is calculated. The following assumption is made: maximum error is located on $v = 0.5$. The position of $\mathbf{P_{1offset}}$ minimising the value of that error will be kept to position the cutter.

The cutter axis is then defined by the two points $\mathbf{P_{0offset}} = \mathbf{S_{offset}}(u, 0)$ and $\mathbf{P_{1offset}} = \mathbf{S_{offset}}(u', 1)$. The error located at $v = 0.5$ is considered to be small enough to be able to state that $[\mathbf{P_{0offset}}\,\mathbf{P_{1offset}}]$ cuts the isoparametric curve $v = 0.5$.

The authors conclude by asserting that the cutter passes through the three isoparametric curves of the offset surface for values $v = 0$, $v = 0.5$ and $v = 1$. The cutter is then considered to be tangent at three points of the surface $\mathbf{S}(u, v)$.

A second stage is then implemented on all the axis positions calculated. By interpolation on the previously established axis positions, a B-spline surface is calculated. Then, from a cloud of points derived from the offset surface (Eq. 2.20), the B-spline surface is recalculated so as to minimise the quadratic sum of deviations between the cloud of offset points and the B-spline surface. This stage allows the ruled surface of the axis to be better distributed around the cloud of offset points.

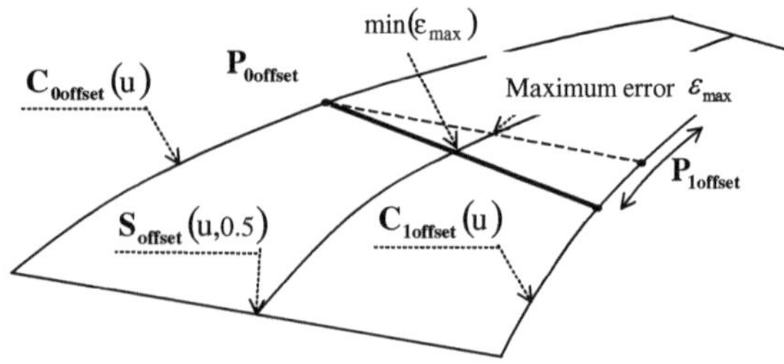


**Fig. 2.25** "T.P.O." positioning for a cylindrical cutter

## 2.3 Free-Form Surfaces

The introductory paragraph to the present section described the gains that flank milling could contribute as compared with end milling. The idea developed in this paragraph is in devising methods for approximation of free-form surfaces into ruled surfaces with the aim of being able to flank mill those surfaces.

### *2.3.1 Discretising into Ruled Surfaces*

Studying the geometry of ruled surfaces shows that the most influential parameter on interference is the twist $\alpha$ of the surface. In order to reduce interference, some authors recommend milling by strips [1, 18]. They break down the surface into several ruled surfaces. On each ruled surface, as the twist is smaller, the interference generated is thus reduced. However, none of the authors describe this breakdown and it is often taken in linear fashion on parameter v. Consider first for example in Fig. 2.26, a surface whose twist is constant and equal to 90° and where the ratio $\|\mathbf{T_0}\|/\|\mathbf{T_1}\|$ is constant and equal to 2 whatever the rule. Taking a linear breakdown of the ruled surface can prove to be relatively ineffective according to the ratio $\lambda = \|\mathbf{T_0}\|/\|\mathbf{T_1}\|$. It was decided to break down the surface into 2 ruled surfaces for v varying between $[0, 0.5]$, $[0.5, 1]$. In Fig. 2.26, it can be seen that the twist of the first surface is equal to 27° while on the other it is 63° (90°−27°). This shows that a breakdown must be adapted to the surface in order to distribute the twist uniformly over the two surfaces. To do so, the breakdown must be made at v = 0.67. The twist for both surfaces will then be 45°. This breakdown can be constant if the ratio λ is constant over the entire surface. However, except for special cases, this remains untrue. When the ratio λ between the tangents is variable over the entire surface, for example between 1 and 2, and if a constant twist is desired on each surface, it will be necessary for any rule to adapt the value of v that cuts through the surfaces. This global positioning leads to having an axial engagement of the cutter that varies over the entire surface, which condition may be technologically undesirable.

Consider now (Fig. 2.26) having a surface with $\|\mathbf{T_0}\|/\|\mathbf{T_1}\|$ that varies from 1 to 2 and twist that is constant over the entire surface. Another solution involves

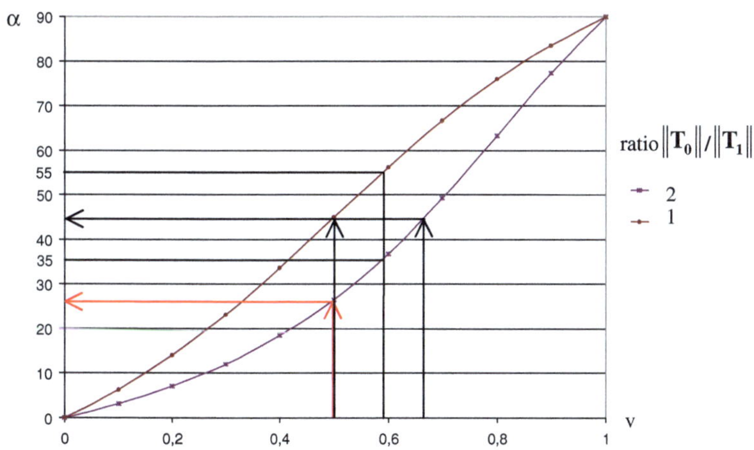


**Fig. 2.26** Evolution of twist in relation to the ratio $\|\mathbf{T_0}\|/\|\mathbf{T_1}\|$

finding the value v that cuts the surface into two such that the maximum twist for each of the two surfaces is equal and minimal. In the example below, this value corresponds to v = 0.58. For the first surface, with v varying over the interval $[0; 0.58]$, the twist will evolve from 35° to 55° while for the second surface for v varying over the interval $[0.58; 1]$, the twist will go from 55° (90−35) to 35° (90−55). The distribution of twists will then be (35°; 55°) for both surfaces.

Breakdown into zones can provide an interesting alternative to reduce error. However, this breakdown has to be precise in order to reduce error and thus allow larger sized cutters to be chosen. The positioning methods employed must absolutely respect the directrices when strip milling, otherwise the entire surface machined is likely to show steps on the transition zones between strips.

With a view to extending ruled surface flank milling methods to free-form surfaces, Elber et al. [23, 24] propose an algorithm for discretising free-form surfaces into ruled surfaces. The authors apply their algorithm on a B-spline surface:

$$\mathbf{S}(u, v) = \sum_{i=0}^{n} \sum_{j=0}^{t} \mathbf{P_{ij}} \, N_{i,m}(u) N_{j,r}(v) \tag{2.21}$$

where the points $\mathbf{P_{ij}}$ constitute the characteristic network of the patch.

The surface with any shape $\mathbf{S}(u, v)$ will be discretised into a number of ruled surfaces $\mathbf{R_k}(u, v)$. Let $\mathbf{R_k^*}(u, v)$ be the representation of $\mathbf{R_k}(u, v)$ in the same B-spline base as that for $\mathbf{S}(u, v)$. After this surface manipulation, the following is obtained:

$$\mathbf{R}^*(u, v) = \sum_{i=0}^{n} \sum_{j=0}^{t} \mathbf{Q_{ij}^*} \, N_{i,m}(u) \, N_{j,R}(v) \tag{2.22}$$

The following inequality provides the limit distance corresponding to the approximation granted between $\mathbf{S}(u, v)$ and the ruled surface $\mathbf{R_k^*}(u, v)$.

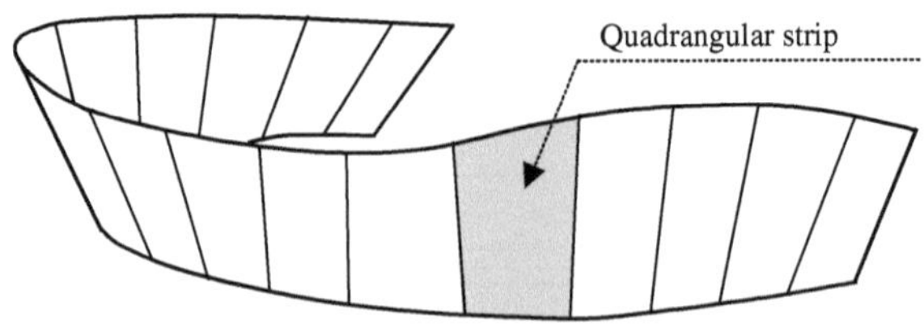


**Fig. 2.27** Approximation of the surface by quadrangular strips

$$\left\| \mathbf{S}(u,v) - \mathbf{R}_k^*(u,v) \right\| \le \mathrm{Max} \left\| \mathbf{P}_{ij} - \mathbf{Q}_{ij}^* \right\| \tag{2.23}$$

If this distance exceeds the accuracy of discretising, the surface $\mathbf{S}(u,v)$ will be discretised in two surfaces to which the previous procedure will be applied. Adopting this methodology, a free-form surface can thus be discretised into ruled surfaces, with a given accuracy. Obviously, it is important after this approximation phase to conduct a study of the visibilities permitting detection of the surfaces that can actually be flank milled.

Other methods exist based on the idea of approximating the surface by developable ruled surfaces rather than seeking to position the cutter locally on the surface to minimise error. Software positioning can be used to machine the developable ruled surface free of interference. The machining error induced is thus managed during the approximation phase for the surface rather than in terms of cutter positioning: the machining error is then the approximation error.

In [25], the authors propose to approximate the surface using developable ruled surfaces of the Bezier patch type connected in tangency. Tsai et al. [26] propose an approximation of the surface $\mathbf{S}(u,v)$ using developable ruled surfaces of the quadrangular type (Fig. 2.27). The authors propose an algorithm for automatic generation of quadrangular strips minimising approximation error.

The algorithm developed is based on the existing Dijkstra algorithm (algorithm minimising the areas of quadrangles) and is adapted to the problematic flank milling to take approximation to its conclusion with the aim of minimising geometrical error.

The construction of quadrangular strips answers to the following constraints (Fig. 2.28):

- the length of the cutter must be greater than the set of distances $[\mathbf{P}_{0,j}, \mathbf{P}_{1,j}]$ (note $\mathbf{P}_{i,j}$ the **j**th quadrangle crest located on the directrix $\mathbf{C}_i(u)$),
- the relative speed between two consecutive crests $\left\| v_{i,j} - v_{i,j+1} \right\|$ must not vary excessively, the aim being to limit cutter acceleration and deceleration phenomena:
- $v_{i,j} = \dfrac{\mathbf{P}_{i,j} - \mathbf{P}_{i,j-1}}{t}$ (t being the time the cutter takes to go between points $\mathbf{P}_{i,j-1}$ and $\mathbf{P}_{i,j}$),
- changes in the direction of speed between two consecutive crests $\dfrac{v_{i,j} \cdot v_{i,j+1}}{\left\| v_{i,j} \right\| \left\| v_{i,j+1} \right\|}$ are limited,
- the twist on each quadrilateral is limited: $\theta_j = a\cos\left( \dfrac{(\mathbf{P}_{1,j} - \mathbf{P}_{0,j}) \cdot (\mathbf{P}_{1,j+1} - \mathbf{P}_{0,j+1})}{\left\| \mathbf{P}_{1,j} - \mathbf{P}_{0,j} \right\| \left\| \mathbf{P}_{1,j+1} - \mathbf{P}_{0,j+1} \right\|} \right)$

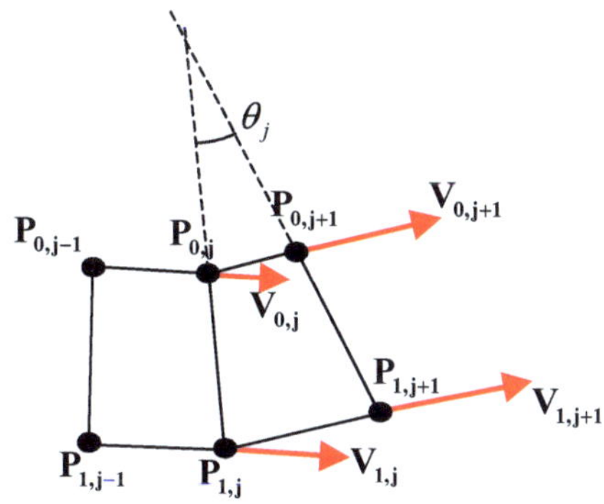


**Fig. 2.28** Constraints applied to quadrangular surfaces

### 2.3.2 Local Positioning

Some authors have taken an interest in flank milling of free-form surfaces without going through discretising into ruled surfaces, whether developable or not. As these works are based on the differential geometry of the surfaces, they have to be of continuity $C^2$ to be able to approximate to the second order, locally at the points of contact, the surfaces of the cutter and the workpiece.

Angular positions for the cutter excluding interference can be determined working from the local topology of surfaces defined by the sign of the main curvatures of the workpiece. These angular positions are those of the principal directions [27, 28] of the cutter in relation to those of the surface.

According to the machining tolerance, the width milled can be calculated through working out the distances between the surfaces of the cutter and the workpiece approximated by their curvatures.

Along similar lines, works [29] address calculation of the Dupin indicatrices for the cutter surface, the cutter envelope surface and the surface of the workpiece in order to determine possible interference. The method applies both to end milling and flank milling. The same authors [30] develop the results of previous works to propose a local optimisation of positioning to maximise the width machined while minimising the normal relative curvature between the envelope surface of the cutter and the surface of the workpiece.

These methods will not be further developed in this chapter as their applications remain limited to relatively tight surfaces (important curvature radius) and the risks of local and global collision remain high. Their application requires a collision management tool.

### 2.3.3 Global Cutter Positioning Methods

All the methods presented allow a tool path comprising two sets of points characterising the positions taken successively by the cutter axis for each rule considered to be defined. Some authors devote attention to the totality of distances between the envelope surface and the nominal surface with a view to optimisation established from a variety of criteria. These works follow from studies conducted into local positioning. The results obtained after resolution of major systems of equations need to be considered circumspectly from a numerical point of view.

Two main objectives justify the formulation of an optimisation problem:

- minimisation of geometric error when the positioning used does not allow the values set by shape tolerance on the surface to be achieved,
- fluidity of the tool path so as to anticipate machining defects due to excessively abrupt cutter movements, especially when using cutter positioning in high speed machining.

In [31–33], the argument developed is based on a compensation of theoretical error. The idea is to position the cutter simply and in a non-optimised fashion while remaining aware that such positioning will be the source of error. Optimisation of positioning is a trade-off between the error induced by cutter placement and the fluidity of the path obtained.

The method proposed can be broken down as follows:

- calculation of the set of cutter positions enabling the surface to be milled. The positioning method is not defined, the only constraint being its simple implementation. The two sets of points characterising positioning of the cutter axis are interpolated into 2 B-spline curves.
- applying a grid on the surface to be machined $\mathbf{S}(u, v)$. At each point of this grid, the distance corresponding to the error between the surface to be machined and the envelope surface generated by the cutter is calculated thus obtaining a grid of positioning errors over the entire surface.
- displacement of the poles of the envelope surface of the cutter in order to minimise the quadratic sum of errors using the small displacement method. By modifying the envelope surface, all cutter positionings are modified.
- a last stage is implemented integrating into the optimisation problem a fluidity criterion based on the deformation energy of each of the guide curves of the axis-cutter surface. This stage was added to find a trade-off between machining precision and fluidity of the path.

In [34], the authors rely on local positioning using the method developed by Liu [4]. They then interpolate this discrete set of positions to generate a NURBS surface. This surface is generated as the envelope of the previously defined cylinders. They then seek to minimise the error between the ruled surface and the envelope surface. To do so, the check points of the NURBS are adjusted. The Downhill Simplex method in Multidimensions is used to optimise its translation and rotation components for each check point so as to minimise error in relation to the ruled surface. No additional information is provided as to how optimisation is obtained.

**Fig. 2.29** Construction of lines of contact

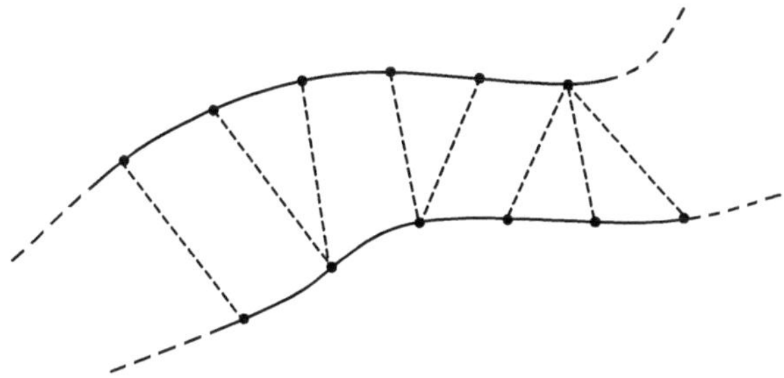

The works presented in [35] focus on global optimisation of positioning from a cloud of points. Initially, a certain number of position settings are calculated on different rules using the method developed by Chiou [18]. This approach has the advantage of being easy to implement. Two sets of points corresponding to the cutter axis are obtained. Each set of points is then interpolated by a cubic B-spline curve in order to obtain an equation of the ruled surface swept by the cutter axis. This equation is then used to calculate the cutter's envelope surface.

Using a cloud of points on the surface of the workpiece, the distance from these points to the envelope surface of the cutter is computed. The authors then pose an optimisation problem that seeks to minimise the maximum distance between the envelope surface of the cutter and the surface of the workpiece taking the cloud of points as a base. Using an approximation to the first order of the distance function, the problem is then linearised and its resolution leads to obtaining new centre–cutter points. To obtain a first positioning that is more effective than that proposed by Chiou [18], they suggest performing a minimisation of the quadratic distance of the points of the workpiece to the envelope surface. The optimisation method leads to a reduction in maximum error but implementation often remains a delicate matter and is costly in computation time without being assured of control over the final result in terms of path fluidity.

In [36], another global positioning approach is proposed. This method is based on approximating the directrices of the surface. The logic can be broken down into stages as follows (Fig. 2.29):

- discretising of the directrices $\mathbf{C_0}(u)$ and $\mathbf{C_1}(u)$ of the surface is performed.
- on one of the two directrices, for each point discretised, contact lines are created between the point concerned and the points of the opposite curve. Two criteria are to be respected in this stage: there must be no intersection between adjacent contact lines and there must be no "point jumping" on the opposite curve.
- each contact line is used as a cutter positioning support. The points for passes of the cutter axis $\mathbf{P_{C0}}$ and $\mathbf{P_{C1}}$ are calculated from the end points of the rule (Eq. 2.15 and 2.16).
- a program is used to define the optimum route (series of contact lines) to minimise error. The physical magnitude to be minimised is the machining error (this is estimated using the Z-buffer method), while the optimisation parameters are four in number:

- the choice of the number of points to be discretised on each directrix may be different between $\mathbf{C_0}(u)$ and $\mathbf{C_1}(u)$,
- the maximum number of contact lines that can be established for each point,
- the number of points the optimisation program chooses to jump between two adjacent lines of contact.
- the last parameter provides an opportunity to discretise between the two directrix curves of the surface. It gives the number of isoparametric curves on v that will be used as a support for approximation. This is equivalent to strip milling.

This original method is difficult to compare with the other global methods whether from the perspective of results in terms of error or tool path fluidity.

## 2.4 Assisting Cutting Tool Choice

All the studies presented using analytical or numerical methods, whether iterative or not, aim to reduce error between the nominal surface and the envelope surface corresponding to the machined surface. The final objective is not to cancel out the error but to be able to use the largest cutting tool while also respecting machining tolerance. Indeed, theoretically it is possible to cancel error if the cutter radius is null. In this case, the cutter will be a line applied to coincide with the rule. If accessibility were possible, wire electrical discharge machining would be the process best suited to obtain the surface sought exactly. From a technological point of view, to limit cutter vibration and bending [37, 38], it is necessary to have the biggest cutter (assuming there is no risk of collision with other surfaces). Now, in all the studies presented, the cutter shape and dimensions are chosen right from the start. If finally, positioning were to lead to an error greater than the tolerance, the entire strategy would have to be thought out again and recalculated using a cutter with other dimensions. With the aim of avoiding this repetition, it seemed advantageous to predict the error with the goal of defining a cutter with dimensions to match the machining tolerance. The influence of the radius R was shown in [13]: it modifies errors in amplitude but does not change the general shape of the error curves. The problem posed thus relates to being able to choose a cutter radius that will lead to an error lower than the tolerance interval imposed by the surface.

The idea of helping to choose the cutter radius does not involve adopting an iterative logic where the algorithm is repeated until the value of the radius giving maximum permitted error is found. On the contrary, the goal is to run the algorithm just once, choosing a cutter radius arbitrarily and then being able, without further calculation, to give the maximum radius to be chosen for the cutter. To achieve this objective, the relation between the cutter radius and the error caused has to be highlighted. Furthermore, it was noted that the zones on the surface where maximum error occurred were those where twist was maximal.

To choose the cutter maximum radius, the position of the cutter on the rule where twist is maximal will need to be calculated.

Some of the authors behind the positioning methods previously introduced propose assistance in choosing the cutter dimensions. Each method presented relates to the positioning developed.

The first analytical positioning mentioned is that implemented in many CAD/CAM software packages. An algebraic calculation of the maximum cutter diameter allowing for the tolerance interval to be respected remains possible:

$$R = \frac{\varepsilon_{max}(\varepsilon_{max} - 2\rho_1)}{2[\rho_1(\cos\alpha - 1) + \varepsilon_{max}]} \tag{2.24}$$

Rubio [3] proposes a numerical calculation of the maximum cutter diameter. The radius is obtained by resolution of the system with three equations and three unknowns $(R, \alpha_0, \alpha_1)$ as shown below:

$$\begin{cases} \rho_0 + R - \sqrt{R^2 + \rho_0^2 + 2\,\rho_0\,R\,\cos\alpha_0} = \varepsilon_{max} \\ \rho_1 + R - \sqrt{R^2 + \rho_1^2 + 2\,\rho_1\,R\,\cos\alpha_1} = \varepsilon_{max} \\ \alpha = \alpha_0 + \alpha_1 \end{cases} \tag{2.25}$$

Liu [4] also proposes a method for numerical resolution to calculate the cutter diameter, but this is an iterative method based on Eqs. 2.11, 2.12 and 2.13.

The work by [10] highlighted the quasi-linear relation existing between the cutter radius and the error caused: $\frac{\varepsilon}{R} \approx$ constant. A study was conducted to show that the linear model was the most effective and that, contrary to what one might have assumed, developing a higher degree model would lead to less accurate results. The effectiveness of the linearity hypothesis can be explained by studying the radius of curvature of the function $\varepsilon(R)$; as the latter is extremely significant; the hypothesis of linearity between the geometric error and the cutter radius is justified. Working from a first cutter radius $R_0$ and the maximum error value $\varepsilon_0$ obtained after positioning, the optimum cutter radius can be calculated by linearising the function $\varepsilon(R)$. The optimum radius is obtained by:

$$R_{max} = \frac{\varepsilon_{max}}{\varepsilon_0} R_0 \tag{2.26}$$

$\varepsilon_{max}$ is the maximum error permissible by the profile tolerance of the surface.

Looking at the error curves published for the various local positioning strategies in the previous works, different curve shapes were identified. Some curves are of the overcut/overcut type, while others are of the undercut/overcut or even overcut/undercut types. The idea developed in this paragraph is to seek to better understand and forecast the type of curve positioning will lead to. The forecasting method described below was developed in relation to improved positioning [10, 39]. It can also be applied to other methods.

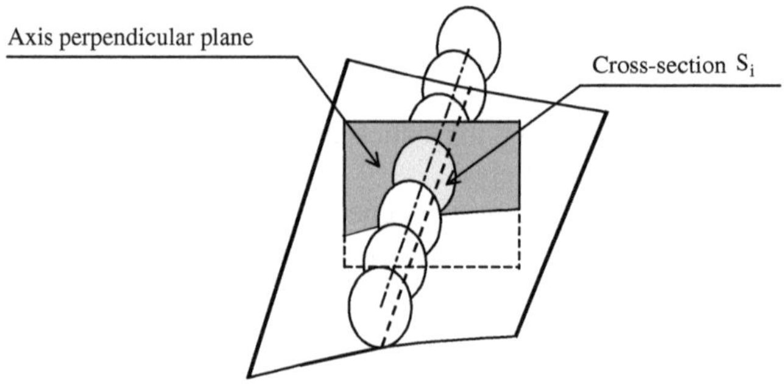


**Fig. 2.30** Cutter discretising

The reasoning here can be summarised as follows:

- the cutter is discretised by parallel planes perpendicular to the axis. Cross-sections $S_i$ (Fig. 2.30) are obtained,
- for each positioning, the real displacement $\Delta_{ri}$ imposed on the cross-section $S_i$ by the adopted strategy is calculated,
- for each cross-section $S_i$, the displacement $\Delta_{ti}$ that would be necessary to ensure tangency with the surface is calculated,
- a conclusion is reached as to the nature of the error induced in each cross-section by comparing $\Delta_{ri}$ with $\Delta_{ti}$.

In order to understand how error forecast functions are implemented, it will first be useful to remind of improved positioning in detail:

- Stage 1: the axis of the cutter is parallel to the rule considered
- Stage 2: the cutter is translated along a direction $\mathbf{y_2}$ by the value of its radius R. Two overcut errors can then be seen on either side of point $\mathbf{M_2}$ where the normal to the surface coincides with $\mathbf{y_2}$.
- Stage 3: the cutter is turned about the axis $(\mathbf{M_2}, \mathbf{y_2})$ by a value $\gamma$. The amplitude $\gamma$ of this rotation is calculated to ensure tangency with $\mathbf{C_0}(u)$ and $\mathbf{C_1}(u)$. When rotation takes place, the circular cross-section becomes an ellipse, and with angle $\gamma$ remaining low, it can be assumed that the cross-section remains circular.

Considering the cross-sections independently of each other, the problem can be modelled simply (Fig. 2.31). The intersection between the surface and the plane containing the cutter cross-section is a curve whose radius of curvature can be considered to be locally constant.

The displacement $\Delta_{ti}$ needed to ensure tangency between the cutter and the surface on the cross-section considered is calculated geometrically:

$$\Delta_{ti} = \sqrt{(R + \rho_i)^2 - (R + \rho_i \cos(\alpha_i))^2} - \rho_i \sin(\alpha_i) \tag{2.27}$$

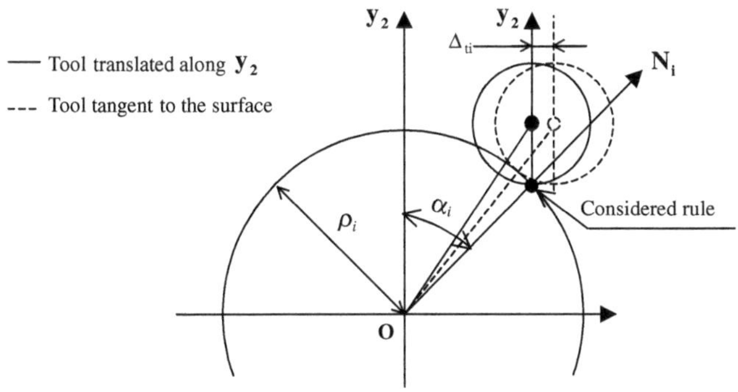


**Fig. 2.31** Modelling the displacement of each cross-section

As the cutter is considered to be non-deformable, the real displacement $\Delta_{ri}$ for each cross-section is calculated by a linear interpolation between the two end cross-sections of the cutter $\Delta_0$ and $\Delta_1$ that are tangent to the surface:

$$\Delta_{ri} = (1 - v)\Delta_0 + v\Delta_1 \tag{2.28}$$

with

$$\Delta_0 = \sqrt{(R + \rho_0)^2 - (R + \rho_0 \cos(\alpha_0))^2} - \rho_0 \sin(\alpha_0) \tag{2.29}$$

$$\Delta_1 = -\left[\sqrt{(R + \rho_1)^2 - (R + \rho_1 \cos(\alpha_1))^2} - \rho_1 \sin(\alpha_1)\right] \tag{2.30}$$

with $\alpha_0 > 0$ and $\alpha_1 > 0$.

To conclude on the type of error obtained on the cross-section, it is sufficient to compare the displacement $\Delta_{ti}$ allowing tangency to be ensured with the real displacement of cross-section $\Delta_{ri}$ (Fig. 2.32).

- if $|\Delta_{ri}| < |\Delta_{ti}|$ and of the same sign, displacement of the cross-section will not be sufficient to ensure tangency and there is overcut,
- if $|\Delta_{ri}| > |\Delta_{ti}|$ and of the same sign, displacement of the cross-section is greater than that needed to ensure tangency and there is undercut,
- if $|\Delta_{ri}| = |\Delta_{ti}|$ there is tangency between the cutter and the surface on that cross-section,
- if $\Delta_{ri} \cdot \Delta_{ti} < 0$ the type of error is difficult to analyse as it depends on the orientation of angle $\alpha_i$.

Geometric error can be estimated (Fig. 2.33) by the relation [39]:

$$\varepsilon_i = (\Delta_{ti} - \Delta_{ri}) \sin(|\alpha_i|) \tag{2.31}$$

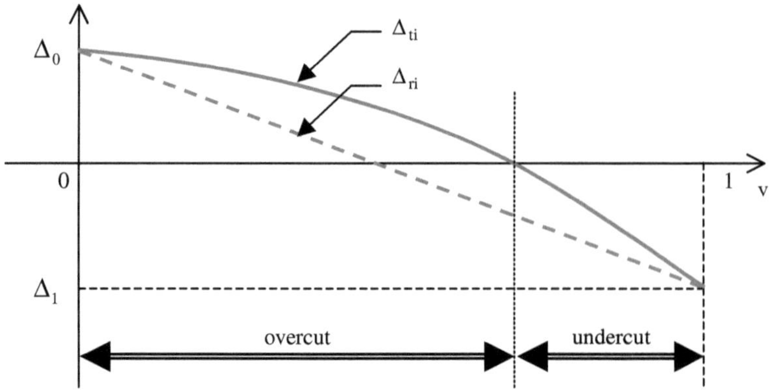


**Fig. 2.32** Analysis of the type of error

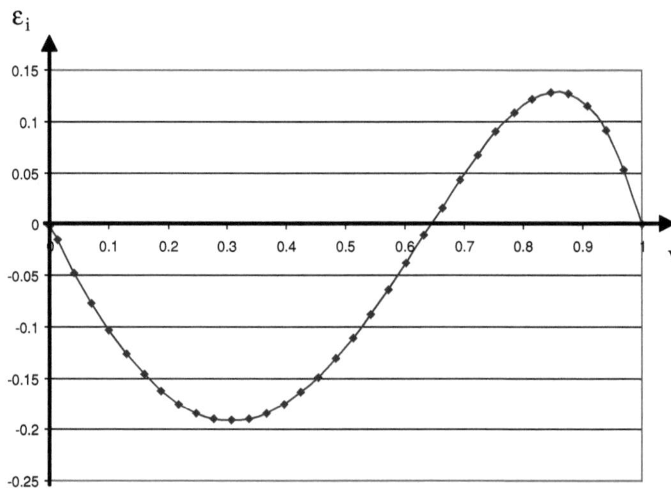


**Fig. 2.33** Error estimation

The error corresponds to the difference in displacement $(\Delta_{ti} - \Delta_{ri})$ projected onto the normal to the surface at the point considered. This method of forecasting is especially advantageous as it allows errors to be evaluated instantaneously and another cutter to be chosen without having to calculate positioning. It can be readily applied to other types of local positionings and to other cutter geometries (conical cutters for example) and allows error to be estimated.

While linearity between the error and the radius of a cylindrical cutter allows the appropriate cutter to be determined, the same does not apply for the conical cutter. The choice of cutter dimensions for the conical cutter is given below.

Two independent parameters define the cross-section of the conical cutter: $R_m$ and $\beta$.

- $R_m$: the minimum radius (at the end of the conical part)
- $\beta$: the half angle at the crest of the conical cutter

Initial entering of a radius $R_{m_0}$ and a half angle at the crest $\beta_0$ taken technologically as large as possible, the algorithm developed by [13] allows optimal couples of cutters ($R_{m_i}$, $\beta_i$) to be obtained with a minimum number of iterations, such that the error obtained $\varepsilon$ verifies:

$$\varepsilon \leq \varepsilon_{max} \tag{2.32}$$

This determination is made for a given position of rule and guard characteristic of maximum errors (significant twist). If the cutter sought keeps a half angle at the crest constant and equal to $\beta_0$ the relation of linearity (Eq. 2.26) can be retained, otherwise a procedure by dichotomy has to be implemented to determine the couples ($R_m$, ,$\beta$) that respect the tolerance. In this manner a series of solution couples is obtained enabling the operator to select the appropriate cutter from the store.

## References


1. Chaves-Jacob J (2009) Développement d'une méthodologie de réduction des défauts géométriques : application à l'usinage five-axis de composants de turbomachine. Thèse de doctorat. Arts et Métiers ParisTech, Cluny
2. Chavez-Jacob J, Poulachon G, Duc E (2009) New approach to 5-axis flank milling of free-form surfaces: computation of adaptated tool shape. Comput Aided Des 41:918–929
3. Rubio W (1993) Génération de trajectoires du centre de l'outil pour l'usinage de surfaces complexes sur machines à trois et cinq axes. Thèse de doctorat, Université Paul Sabatier Toulouse, France
4. Liu XW (1995) Five axis NC cylindrical milling of sculptured surfaces. Comput Aided Des 27:887–894
5. Stute G, Storr A, Sielaff W (1979) NC programming of ruled surfaces for five axis machining. Annals of the CIRP 28:267–271
6. Bedi S, Mann S, Menzel C (2003) Flank milling with flat end milling cutter. Comput Aided Des 35:293–300
7. Redonnet JM, Rubio W, Dessein G (1998) Side milling of ruled surfaces-Optimum positioning of the milling cutter and calculation of interference. Int J Adv Manuf Technol 14:459–465
8. Monies F, Rubio W, Redonnet JM, Lagarrigue P (2001) Comparative study of interference caused by standard and improved positioning of a conical milling cutter working on a ruled surface. J Eng Manuf (Part B) 215:1305–1317
9. Senatore J, Monies F, Redonnet JM, Rubio W (2005) Analysis of improved positioning in five-axis ruled surface milling using envelope surface. Comput Aided Des 37:989–998
10. Senatore J (2007) Analyse qualitative des parametres influents pour la planification des trajectoires sur surfaces gauches. Thèse de doctorat, Université Paul Sabatier Toulouse, France
11. Senatore J, Monies F, Landon Y, Rubio W (2008) Optimising positioning of the axis of a milling cutter on an offset surface by geometric error minimisation. Int J Adv Manuf Technol 37:861–871

12. Monies F, Redonnet JM, Rubio W, Lagarrigue P (2000) Improved positioning of a conical mill for machining ruled surfaces: application to turbine blade. J Eng Manuf (Part B) 214:625–634
13. Monies F (2001) Positionnement hors interférence pour l'usinage en bout et en roulant des surfaces gauches. Thèse de doctorat, Université Paul Sabatier Toulouse, France
14. Monies F, Felices JN, Rubio W, Redonnet JM, Lagarrigue P (2002) Five-axis NC milling of ruled surfaces: optimal geometry of a conical tool. Int J Prod Res 40:2901–2922
15. Peternell M, Pottmann H, Steiner T, Zhao H (2005) Swept volumes. Comput Aided Des Appl 2:599–608
16. Abdel-malek K, Yeh HJ (1997) Geometric representation of the swept volume using Jacobian rank-deficiency conditions. Comput Aided Des 29:457–468
17. Li C, Bedi S, Mann S (2005) Error measurements for flank milling. Comput Aided Des 37:1459–1468
18. Chiou JCJ (2004) Accurate tool position for five-axis ruled surface machining by swept envelope approach. Comput Aided Des 36:967–974
19. Lee JJ, Suh SH (1998) Interference-free tool-path planning for flank milling of twisted ruled surfaces. Int J Adv Manuf Technol 14:795–805
20. Menzel C, Bedi S, Mann S (2004) Triple tangent flank milling of ruled surfaces. Comput Aided Des 36:289–296
21. Li C, Bedi S, Mann S (2006) Flank milling of a ruled surface with conical tools–an optimization approach. Int J Adv Manuf Technol 29:1115–1124
22. Gong H, Cao LX, Liu J (2005) Improved positioning of cylindrical cutter for flank milling ruled surface. Comput Aided Des 37:1205–1213
23. Elber G (1995) Model fabrication using surface layout projection. Comput Aided Des 27:283–291
24. Elber G, Russ F (1997) 5-axis freeform surface milling using piecewise ruled surface approximation. ASME J Eng Ind 119:383–387
25. Chu CH, Chen JT (2006) Tool path planning for five-axis flank milling with developable surface approximation. Int J Adv Manuf Technol 29:707–713
26. Tsai WL, Wang CCL, Chu CH, Tang K (2008) Optimal quadrangulation of a strip for flank milling. Comput Aided Des Appl 5:307–315
27. Marciniak K (1987) Influence of surface shape on admissible tool positions in five-axis face milling. Comput Aided Des 19:233–236
28. Marciniak K (1991) Geometric Modelling for Numerically Controlled Machining. Oxford Science Publications, New York
29. Gong H, Cao L, Liu J (2008) Second order approximation of tool envelope surface for 5-axis machining with single point contact. Comput Aided Des 40:604–615
30. Gong H, Fang FZ, Hu XT, Cao L, Liu J (2010) Optimization of tool positions locally based on the BCELTP for 5-axis machining of free-form surfaces. Comput Aided Des 42:558–570
31. Lartigue C, Duc E, Affouard A (2003) Tool path deformation in 5-axis flank milling using envelope surface. Comput Aided Des 35:375–382
32. Pechard PY (2009) Génération de trajectoires d'usinage grande vitesse 5 axes par flanc d'outil: intégration d'un critère de fluidité. Thèse de doctorat, Ecole Normale Supérieure de Cachan, France
33. Pechard PY, Tournier C, Lartigue C, Lugarini JP (2009) Geometrical deviations versus smoothness in 5-axis high-speed flank milling. Int J Mach Tools Manuf 49:454–461
34. Xia J, Ge QJ (2000) Kinematic approximation of ruled surfaces using nurbs motions of a cylindrical cutter. Proceeding of DETC'00, ASME 2000 Design Engineering Technical Conferences. Baltimore, Maryland
35. Zhang XM, Zhu LM, Zheng G, Ding H (2010) Tool path optimisation for flank milling ruled surface based on the distance function. Int J Prod Res 48:4233–4251
36. Wu PH, Li YW, Chu CH (2008) Optimized tool path generation based on dynamic programming for five-axis flank milling of ruled surface. Int J Mach Tools Manuf 48:1224–1233

37. Larue A, Anselmetti B (2003) Deviation of a machined surface in flank milling. Int J Mach Tools Manuf 14:795–805
38. Landon Y, Segonds S, Lascoumes P, Lagarrigue P (2004) Tool Positioning Error (TPE) characterisation in milling. Int J Mach Tools Manuf 44:457–464
39. Senatore J, Landon Y, Rubio W (2008) Analytical estimation of error in flank milling of ruled surfaces. Comput Aided Des 40:595–603

# Chapter 3
# High Performance 5-Axis Milling of Complex Sculptured Surfaces

**Yaman Boz, S. Ehsan Layegh Khavidaki, Huseyin Erdim and Ismail Lazoglu**

Five-axis milling processes are used widely in various industries such as aerospace, die-mold and biomedical industries where surface quality and integrity is important and the production tolerances are very tight. Therefore, improving surface quality and integrity without sacrificing productivity is crucial in these industries. Improvements in CAD/CAM, cutting tool and the machine tool technologies allow the production of high precision parts with less cycle times. In order to obtain desired quality and productivity, process parameters such as feedrate, spindle speed, axial and radial depth of cut have to be selected appropriately. Most of the time, selection criterion is based on engineering expertise or trial and error methods. Besides, to prevent the cutter or the machine to be damaged, machining parameters are selected conservatively, and therefore, virtual machining simulation for milling processes is an increasing demand before the production of the free-form surfaces.

In this chapter, virtual machining simulation to predict the cutting forces and scheduling of feedrates to achieve the optimum cycle time in 5-axis ball-end milling of free-form surfaces is presented.

---

Y. Boz (✉) · S. Ehsan Layegh Khavidaki · I. Lazoglu
Manufacturing and Automation Research Center, Koc University,
Sariyer, 34450 Istanbul, Turkey
e-mail: yboz@ku.edu.tr

S. Ehsan Layegh Khavidaki
e-mail: slayegh@ku.edu.tr

I. Lazoglu
e-mail: ilazoglu@ku.edu.edu

H. Erdim
Mitsubishi Electric Research Laboratories, Cambridge,
MA 02139, USA
e-mail: erdim@merl.com

J. P. Davim (ed.), *Machining of Complex Sculptured Surfaces*,
DOI: 10.1007/978-1-4471-2356-9_3, 

Moreover, Cutting forces in machining is determined by extracting the Cutter-Workpiece Engagement (CWE) from the in-process workpiece. A discrete method (Three-Orthogonal Dexelfield) of obtaining CWE maps for 5-axis ball-end milling is developed. The results of the Three-Orthogonal Dexelfield method is compared with the solid-modeler-based CWE calculation method. Cutting force modeling is performed in the fixed coordinate frame (for table type dynamometer) and in the rotating coordinate frame (rotating coordinates dynamometer). Finally, an enhanced Force-model-based Feedrate Scheduling (FFS) technique in 5-axis machining of parts with complex free-form surfaces is presented. Several validation tests are presented in this chapter and the validation tests demonstrate that presented cutter-workpiece engagement model is accurate and force predictions are in good agreement with the measured data.

## 3.1 Introduction

Five-axis machining has been used in aerospace applications for many years. Automotive, die-mold, toolmaking and biomedical industries have shown similar interest. Parts manufactured with 5-axis machine tools constitute main components of the high-level systems which can be manufactured using today's technology.

Aircraft parts typically have walls that are not perpendicular to the floor of the part. Cutting these walls with 3-axis machining requires multiple-pass milling operations using cylindrical end mills or special form tools. 5-axis machining, eliminates the use of multiple setups since tool can tilt and lead yielding one-pass cut using a standard end mill.

Blisks and Integrally Bladed Rotors (IBR) are used in jet-powered aircraft engines and in high pressure compressors for military purposes (Fig. 3.1). Blades can be produced with 3-axis machining, however, this requires multiple setups and the use of longer tools. Blisks and IBRs can also be manufactured on profilers by tracing templates to control the $X$ and $Y$ axis of the machine and following cams to control the $Z$ axis. Furthermore, the fixtures have to be manually rotated prior to cutting of each blade. Hence, each blade differs in shape due to manual alignment of the fixtures. Similarly, impellers (Fig. 3.1) used in compressors and turbines have very complex shapes and demands the use of 5-axis machining for reducing machining times and improving part quality and uniformity.

More recently, biomedical industries have started using 5-axis machining technology for knee joint implants, dental implants, heart pumps and spine implants (Fig. 3.2). Generally, medical implants have free-form surfaces demanding 5-axis machining.

In these industries different workpiece materials are used in order to provide desired mechanical or other properties. Impellers, blades and blisks are commonly produced from Titanium alloys (e.g. Ti6Al4V) and Inconel. For the purpose of jet engine parts Titanium alloys and Inconel are preferred due to their strength at high temperatures and low thermal conductivity. Aluminum alloys (e.g. 7,000 series)

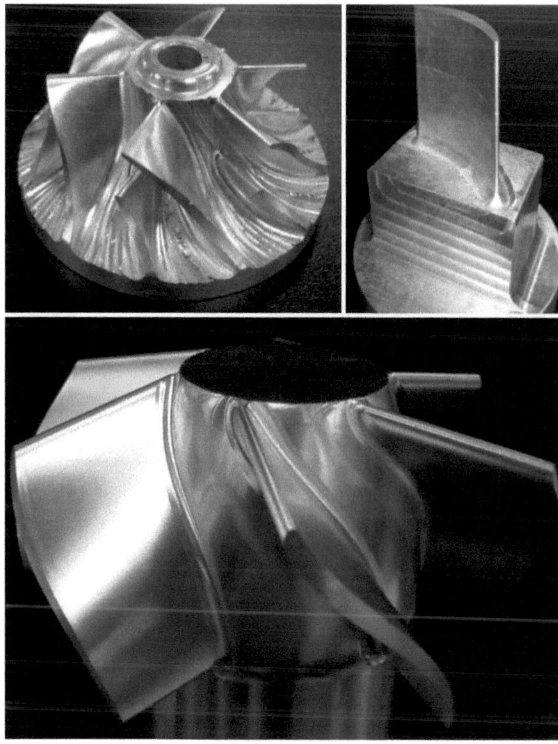

**Fig. 3.1** Compressor impeller, blade, and blisk

are preferred for moderate working environments such as transport applications, including marine, automotive and aviation applications, due to their high strength-to-density ratio. Biomedical applications require several properties such as high wear resistance, toughness, ductility and biocompatibility. For this reason, bio-compatible Titanium alloys (e.g. Ti6Al4V) or Stainless Steels (e.g. AISI 316LVM) are used.

Additionally, in almost all of the 5-axis machining processes the feedrate is conservatively kept constant to avoid damage to machine tool and deteriorate the surface quality of products. Besides, commonly used commercial CAM programs and feedrate scheduling methods are just based on the geometry and volumetric analysis of the process and the mechanics of the operation has not been considered in feedrate scheduling of those methods. Not considering cutting forces in feedrate scheduling may lead to high magnitude of cutting forces, vibrations, decreasing the quality of finished surface and damaging the tool and machine tool. However, using a reliable force model for complex 5-axis ball-end milling of free-form surfaces, it is possible to schedule the feedrate effectively so that the cycle time will reduce safely while surface quality is kept acceptable.

In this chapter, a cutting force model and a force-based feedrate scheduling method for the 5-axis ball-end milling of free-form surfaces is presented. The process methodology which is used in this chapter is given in Fig. 3.3.

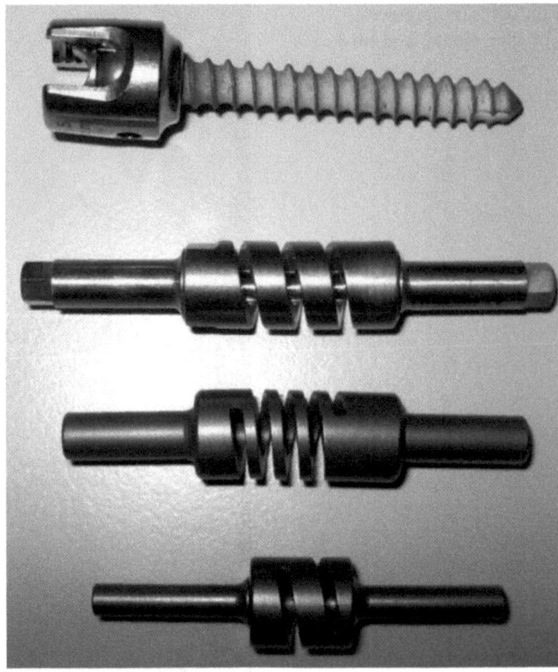

**Fig. 3.2** Medical implants

This chapter is organized as follows:

In Sect. 3.2, the Cutter-Workpiece Engagement (CWE) is determined from the in-process workpiece in the form of start and exit angles as a function of axial height along the tool axis. Furthermore, Sect. 3.2 describes a discrete method (Three-Orthogonal Dexelfield) of obtaining CWE maps for 5-axis ball-end milling. It is compared with the solid-modeler-based CWE calculation method. The results of this CWE model are used in the validation tests.

A cutting force prediction model for 5-axis ball-end milling is developed in Sect. 3.3. Cutting force modeling is performed in the fixed coordinate frame (for table type dynamometer) and in the rotating coordinate frame (rotating coordinates dynamometer). The approach developed based on this model is modular. Therefore, different cutter and workpiece geometries and tool motions can be incorporated into the model without additional analysis.

In Sect. 3.4, an enhanced force-based feedrate scheduling strategy is presented. Using this strategy, the process engineers are able to keep the resultant cutting forces below a preset threshold in order to increase the productivity of the machining process.

Finally, validation tests are presented in Sect. 3.5, and it is shown that there is a good agreement between the theoretical models and experimental investigations.

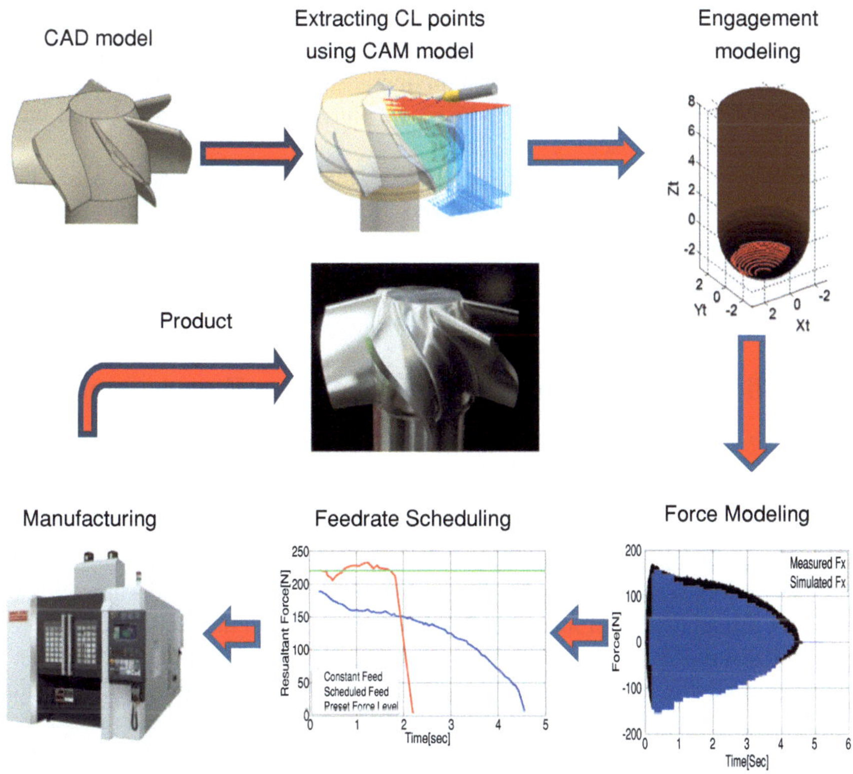


**Fig. 3.3** Process methodology

## 3.2 Cutter-Workpiece Engagement Model

In sculpture surface machining, the cutter/workpiece engagement region does vary along the cutter path and in general, unless some specific and very simple work-piece geometry is machined, it is difficult to find an exact analytical representation for the engagement region. Chip load and force calculations are based on the cutter/workpiece engagements; therefore the output of the engagement model is very critical. In the literature, NC machining simulation can be mainly categorized into three major approaches. The first approach is the exact Boolean, the second approach is the spatial partitioning and the third approach is the discrete vectors. The direct Boolean subtraction approach is an exact and analytical approach. It directly performs the Boolean subtraction operation between a solid model and the volume swept by a cutter between two adjacent tool positions. Although this approach can provide accurate verification and error assessment, the computation cost is known to grow too much for a large number of tool-paths. The second approach uses spatial partitioning representation to define a cutter and the work-piece. In this approach, a solid object is decomposed into a collection of basic geometric elements, for example Z-map (Z-buffer), voxel and ray representation, thus simplifying the processes of regularized Boolean set operations. However, its computation time and memory consumption are increased drastically to get better accuracy. One of the other widely used NC simulation methods is based on the vector-clipping approach.

In this section, both Solid-based (Direct Boolean) method and Spatial partitioning method (Depth Buffer) is utilized. Solid-Modeler-based cutter-workpiece engagement method is presented in Sect. 3.3 and Depth Buffer method is presented in Sect. 3.4.

## 3.3 Solid-Modeler-Based Cutter-Workpiece Engagement

Currently the most popular schemes used in solid modelers are the Boundary representation and Constructive Solid Geometry (CSG). In the B-rep methodology an object is represented by both its boundaries defined by faces, edges, vertices and the connectivity information.

The prototype program is implemented using the commercial Parasolid solid modeler kernel. The tool movements are subtracted from the workpiece model by using Parasolid 'PK_BODY_sweep' and 'PK_BODY_boolean_2' functions in order to find the in-process machined surface. Figure 3.4 shows the resultant machined surfaces for the airfoil and penguin surface examples used in Sect. 3.12.

Once the in-process workpiece is obtained for each CL point, the contact patch surface between the tool and workpiece can be extracted by using Parasolid 'PK_BODY_boolean_2' function. Then, the resulting 3D contact surface, as illustrated in Fig. 3.5, is projected to the plane perpendicular to the cutter axis by using parasolid 'PK_BODY_make_curves_outline' function. This step finds the enclosing boundaries and curves of the contact patch. Since the force model discretizes the cutter into slices perpendicular to the tool axis and to perform force calculation for each slice, the disks at each level are projected to the plane perpendicular to the cutter axis.

Since engagement domain is simply the combination of start and exit angles of each discrete disc located on the cutter, the next step is to assign the start and exit angles to each respective projected disc by intersecting the 2D disks with the boundaries of the contact patch in plane by using Parasolid 'PK_CURVE_intersect_curve' function. A final step is required to convert the intersection points into start and exit angles that are required for the force prediction model. Cutter-workpiece engagement geometry extraction for ball-end mill is shown in Fig. 3.5.

The procedure described above is implemented in Visual Studio.NET using the Parasolid solid modeling Kernel and Parasolid Workshop on a Windows Core2Duo, 1.8 GHz/4 GB Personal Laptop. The output of the program is processed in Matlab and the engagement angles are shown together with the contact patch for CL point #25 in Fig. 3.6 for the airfoil geometry test. The computation

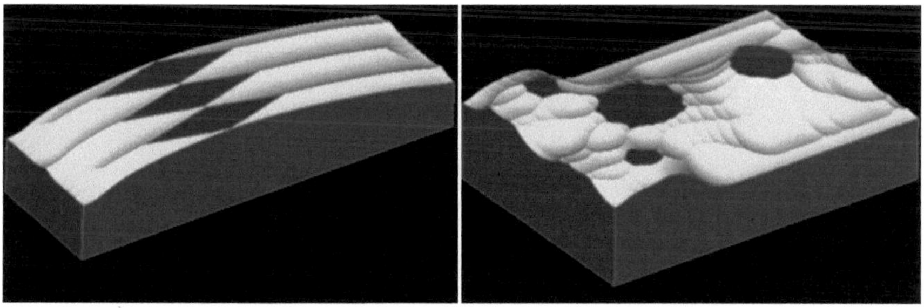

**Fig. 3.4** Simulated machined surfaces for airfoil and penguin surface geometry

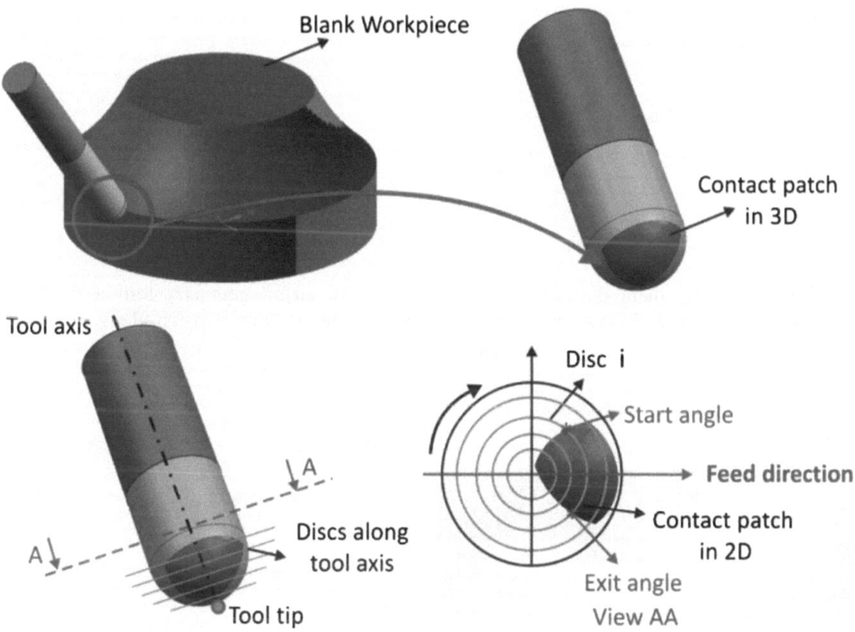


**Fig. 3.5** Cutter-workpiece engagement geometry extraction for ball-end mill

time for the engagement domain for the corresponding examples are 21 and 48 s for airfoil (137 CL points) and penguin (415 CL points) surfaces, respectively.

For 5-axis machining of impeller workpiece, forces are compared for one-pass where the tool moves from the lower rim of the cylinder workpiece to the upper rim of the cylindrical workpiece. This one pass has 119 CL points, the lead and tilt angles for one-pass is shown together with the simulated workpiece in Fig. 3.7. The computation time for the engagement domain for this one-pass is 58 s for 119 CL points.

The engagement domain can have more than one piece because of the complexity of the geometry of the workpiece and the tool motion as shown simply

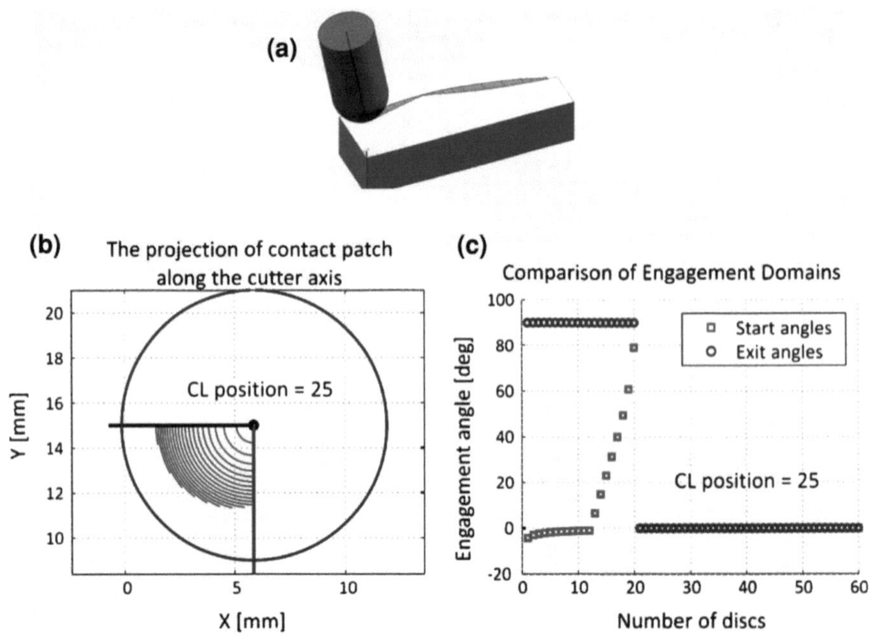


**Fig. 3.6** The engagement domain for CL point #25 for airfoil geometry test: **a** Previously machined surface with the tool instance. **b** Projected view of contact patch along cutter axis. **c** Start and exit angles for the disks along the cutter axis

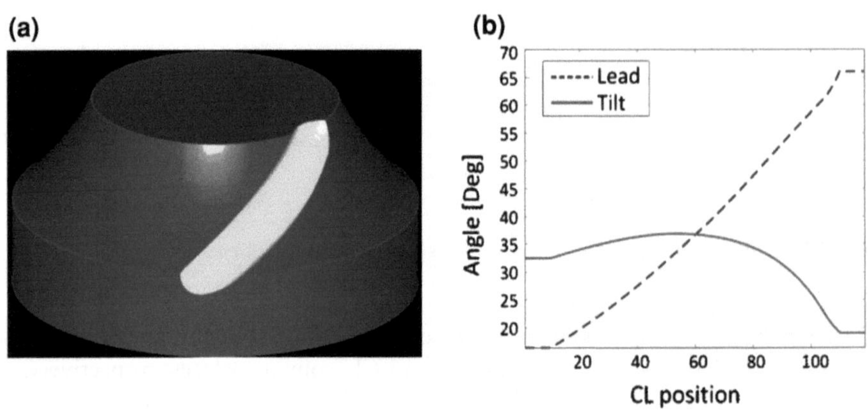


**Fig. 3.7** **a** Simulated workpiece for one-pass from Parasolid. **b** Lead and tilt angles for one-pass of impeller toolpath

in Fig. 3.8. In this case, the engagement domain will have more than two intersections for the disks along the cutter. Engagement domains containing more than one piece for the impeller toolpath are shown in Fig. 3.9.

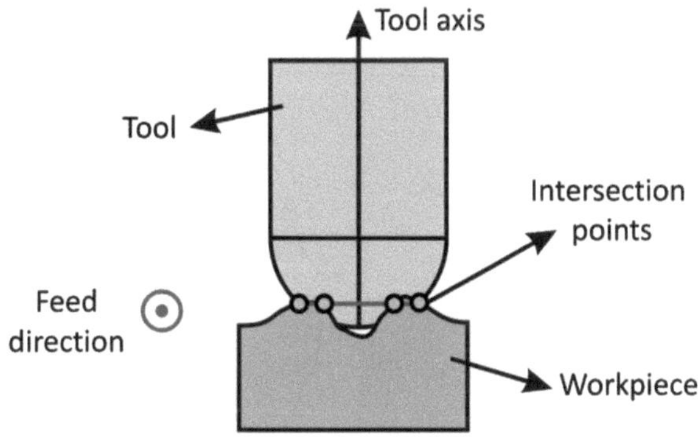


**Fig. 3.8** Illustration of multiple contact regions and intersections

## 3.4 Depth Buffer Based Cutter-Workpiece Engagement

There are several discrete methods used for the representation of the in-process workpiece such as Octree, Voxel, ray representation and Depth buffer (Dexel) approaches.

Main advantage of discrete approaches is that they are computationally simpler than the solid modeling approach. Typically, discrete methods require intersection calculations between simple geometric primitives, allowing simple and robust analytical or algebraic solutions. This simplicity provides robust behavior, and also increases the computational efficiency.

Discrete representation of the geometry may result in the loss of geometric accuracy. However, if the simulation parameters are selected properly, considering both workpiece and toolpath tolerances, the error introduced by the discrete representation may be kept in an acceptable level.

In Octree and Voxel approach, workpiece is modeled as volume cells (Voxels), for instance cubes for the Octree data structure. Octree method is based on the divide-and-conquer principle that recursively subdivides a cube into octants up to specified resolution. Coordinates of each vertex (node) in a voxel is stored and by checking the inner–outer nodes stock workpiece is obtained. During NC simulation tool swept volume between two CL points is subtracted from the stock workpiece and machined workpiece is obtained. This method is simple and fast, however, main drawback is the excessive memory requirements (especially at high resolutions) due to the large amount of data stored.

The most popular and commonly used Depth Buffer scheme in the literature and in the CAM software is Z-Buffer method. Z-buffer method is usually referred as Z-map method. In conventional Z map method [1], workpiece is represented as the intersection points of the Z direction vectors (ZDV) with the workpiece surface on a 2D grid of ZDVs. These intersection points are also upmost part of the workpiece surface where only one intersection of the workpiece with a ZDV is permitted.

In 5-axis machining, conventional representation is not sufficient because almost all of the parts have walls with negative inclination angle and undercut

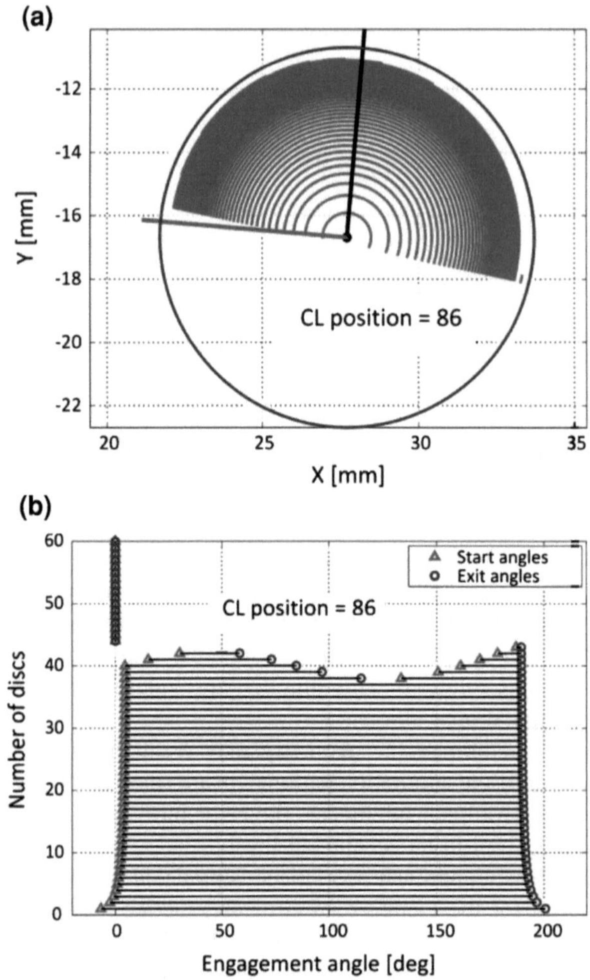


**Fig. 3.9** The engagement domain for CL point #86: **a** Projected view of contact patch along cutter axis. **b** Start and exit angles for the disks along the cutter axis

machining is required. Hence, for 5-axis machining NC simulation extended *Z*-map approach is utilized. In extended *Z*-map approach, for one ZDV, multiple intersections and gap elements between the intersection points can also be stored most of the time using linked list data structure. Conventional and extended *Z*-map approaches are shown schematically in Fig. 3.10.

Material removal simulation in extended *Z*-map approach is performed through calculation of intersections between ZDVs and the geometric representation of the tool swept volume (envelope). Resulting intersection points are compared to previously machined surface and updated for each tool movement along the toolpath.

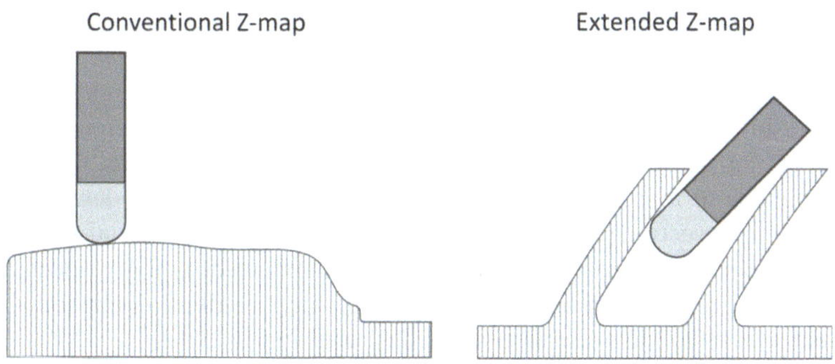


**Fig. 3.10** In-process workpiece representations

**Fig. 3.11** Illustration of three-orthogonal dexelfield

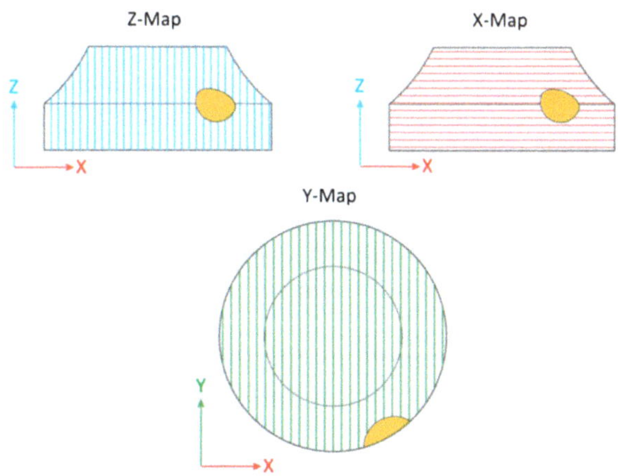


## *3.4.1 Three-Orthogonal Dexelfield (Depth Buffer)*

In this section, a commercial NC verification kernel [2] is used for machining simulation in order to obtain the contact patch between the tool and the workpiece. This verification kernel provides the use of three-orthogonal dexelfield which is similar to extended *Z*-map approach, however, the depth buffer is applied in three orthogonal directions. In other words, three-orthogonal dexelfield approach utilizes *Z*-map, *Y*-map and *X*-map simultaneously.

The use of three-orthogonal dexelfield is quite critical since in extended *Z*-map approach engagement region may not be obtained accurately due to the perpendicular intersection regions of the tool with the ZDVs. Therefore, in these regions contact patch is truncated from the actual contact patch even if the verification resolution is high. Moreover, contact patch may be obtained more accurately at lower resolutions. In Fig. 3.11, example tool position and corresponding three-orthogonal dexelfield obtained from the intersection points are shown; respectively.

In the previous section it is stated that, material removal is calculated by intersecting the tool swept envelope with the workpiece. For a given tool movement, if tool swept volume intersects with the workpiece (material removal occurs) cutting result elements are generated in three-orthogonal directions. In 5-axis machining, for an individual dexel element multiple intersections with the tool may occur due to tool axis rotation. For this reason, it has to be determined that whether the intersection point belongs to a point of the initial material-surface or to a point of the material-surface, after the cut was conducted. In this respect, cutting result entry elements are defined using a height value distribution method. Cutting result entry height value distribution is illustrated in Fig. 3.12.

### *3.4.2 Cutting Result Entry Elements*

As it is illustrated in Fig. 3.12, there are four possible cases for a cutting result entry element which can be given as; FIRST_IS_INITIAL_SECOND_IS_CUT, FIRST_IS_CUT_SECOND_IS_INITIAL, BOTH_ARE_CUT and BOTH_ARE_INITIAL. Height1 and Height2 fields represent the possible intersection points of the tool with the workpiece where Height1 is the upper intersection and Height2 is the lower intersection point. According to this distribution each cutting result entry is processed and "CUT" points are collected into a vector for the candidate contact patch points.

Another important property regarding the cutting result entry is transformation information with respect to workpiece coordinate frame. For the given example on the *Z*-map, the indices of the dexel block on the *X*–*Y* grid is stored internally and used for obtaining the cut point coordinates.

### *3.4.3 Calculation of Engagement Domain from Contact Patch (Points)*

#### 3.4.3.1 Coordinate Frames and Movement Vectors

Once contact region between the tool swept envelope and workpiece is obtained they are transformed (translated) from the workpiece coordinate frame to the tool tip of the cutter as shown in Fig. 3.13.

In Fig. 3.13, $\overrightarrow{P_i}$ represents the coordinates of the *i*th CL point in the toolpath in the workpiece coordinate frame. Contact patch is the surface obtained by combining all of the intersection points and is transformed into tool tip origin as,

$$\overrightarrow{\mathrm{CV}} = \overrightarrow{\mathrm{CP}} - \overrightarrow{P_i} \tag{3.1}$$

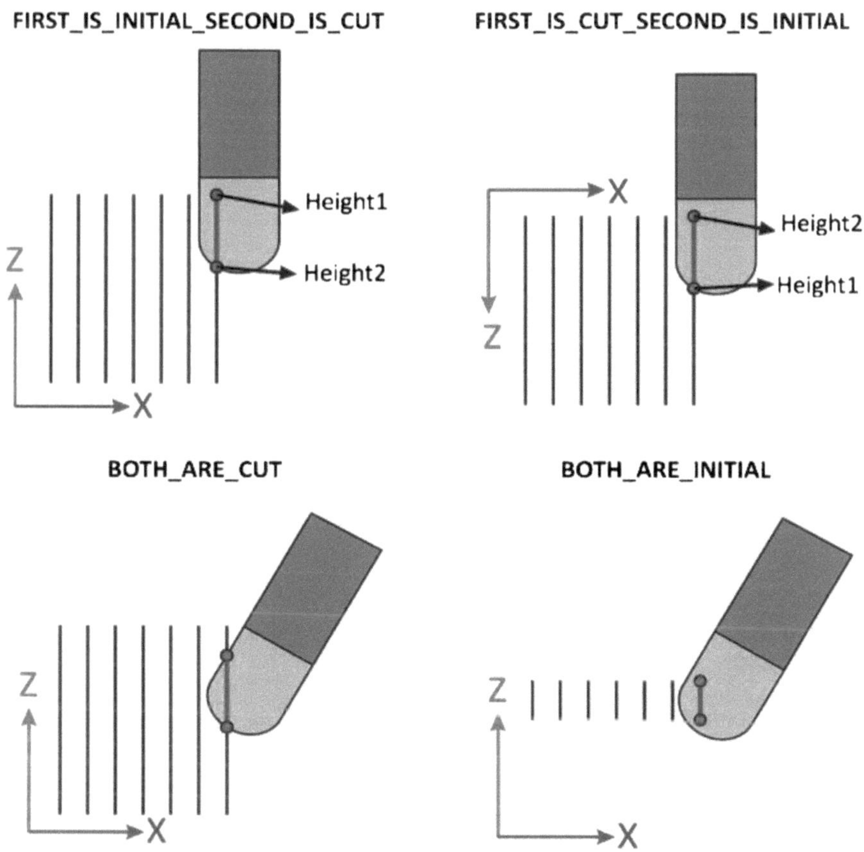


**Fig. 3.12** Cutting result entry height value distribution

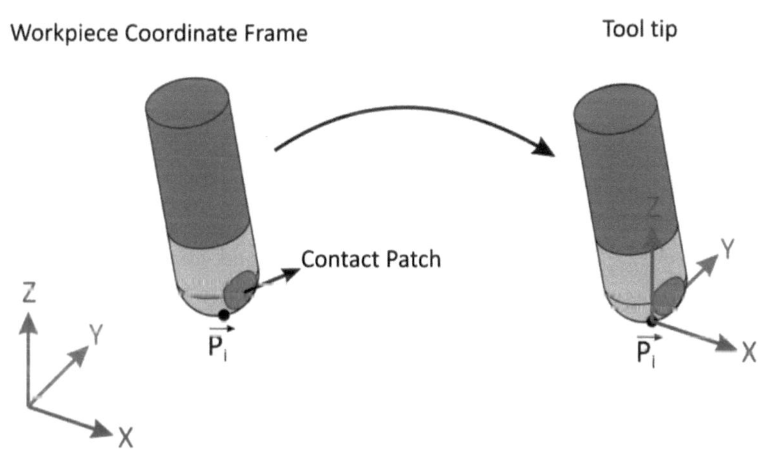


**Fig. 3.13** Transformation from workpiece coordinate frame to tool tip

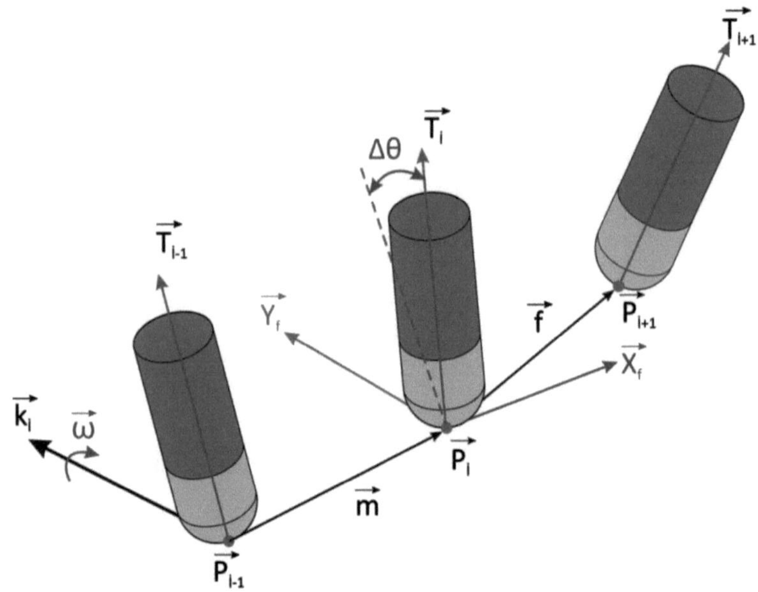


**Fig. 3.14** Basis of the tool movement vector, feed vector and feed coordinate frame

where $\overrightarrow{CV}$ is cut vector in the tooltip origin and $\overrightarrow{CP}$ is the cut point in workpiece coordinate frame.

After the transformation given in Eq. 3.1 is applied, basis of the tool movement vector, feed vector and feed coordinate frame is defined. Consider two consecutive 5-axis tool movements shown in Fig. 3.14, where tool is first moving from $(i-1)$th CL point to the $i$th CL point then to the $(i+1)$th CL point.

For determining the engagement domain for the $i$th CL point, tool swept volume from the $(i-1)$th CL point to the $i$th CL point has to be calculated. Therefore tool movement vector is defined as,

$$\vec{m} = \frac{\left(\vec{P_i} - \vec{P}_{i-1}\right)}{\left\|\vec{P_i} - \vec{P}_{i-1}\right\|} \tag{3.2}$$

In 5-axis machining tool can rotate as well as translate. Therefore, during translation tool axis rotates from $\vec{T}_{i-1}$ to $\vec{T}_i$ around an arbitrary axis $\vec{k}_i$ an amount of $\Delta\varphi$ where rotation axis is both orthogonal to $\vec{T}_{i-1}$ and $\vec{T}_i$. Rotation axis $\vec{k}_i$ and rotation angle $\Delta\varphi$ are calculated as,

$$\vec{k}_i = \frac{\vec{T}_{i-1} \times \vec{T}_i}{\left\|\vec{T}_{i-1} \times \vec{T}_i\right\|} \tag{3.3}$$

$$\Delta\varphi = a\tan2\left(\frac{\left\|\vec{T}_{i-1} \times \vec{T}_i\right\|}{\vec{T}_{i-1} \cdot \vec{T}_i}\right) \tag{3.4}$$

Cutting force model use the engagement domain for the *i*th CL point while cutting tool is moving from the *i*th CL point to $(i + 1)$th CL point assuming that in this toolpath segment cutting forces are the same until tool arrives to $(i + 1)$th CL point.

In this respect, feed coordinate frame is constructed by using these CL points. Feed vector $\vec{f}$ can be expressed as,

$$\vec{f} = \frac{\left(\vec{P}_{i+1} - \vec{P}_i\right)}{\left\|\vec{P}_{i+1} - \vec{P}_i\right\|} \tag{3.5}$$

Another important property regarding feed coordinate frame is that, the feed direction and the cross feed direction denoted as $\vec{X}_f$ and $\vec{Y}_f$ respectively have to lie in a plane where the normal of this plane is $\vec{T}_i$. In other words, an orthogonal basis is defined using tool axis vector $\vec{T}_i$ and feed vector $\vec{f}$,

$$\vec{Y}_f = \vec{T}_i \times \vec{f} \tag{3.6}$$

$$\vec{X}_f = \vec{Y}_f \times \vec{T}_i \tag{3.7}$$

Here it is necessary to remind that Z axis of the feed coordinate frame $\vec{Z}_f$ is coincident with the tool axis orientation $\vec{T}_i$ at cutter location *i*.

#### 3.4.3.2 Tool Swept Volume (Envelope)

NC verification kernel gives all of the cut points while tool is moving from one CL point to the other CL point. This yields that cut points have to be trimmed before the engagement region is determined. Schematic illustration of a ball-end mill sweep is shown in Fig. 3.15.

As shown in Fig. 3.15 tool swept volume of a ball-end cutter comprises three regions which are Egress points, Ingress points and Grazing points region. While obtaining the swept volume most important parameter is cutting direction since it determines the grazing points together with the geometric properties of the cutter. Simply, egress points represent the front side, ingress points represents back side and grazing points represents the swept envelope of the cutter with respect to cutting direction. Hence, possible engagement domain of the cutter lies in the egress points region meaning that front side of the tool swept volume.

According to solid sweep theory [3, 4], three regions shown in Fig. 3.16 can be obtained as follows,

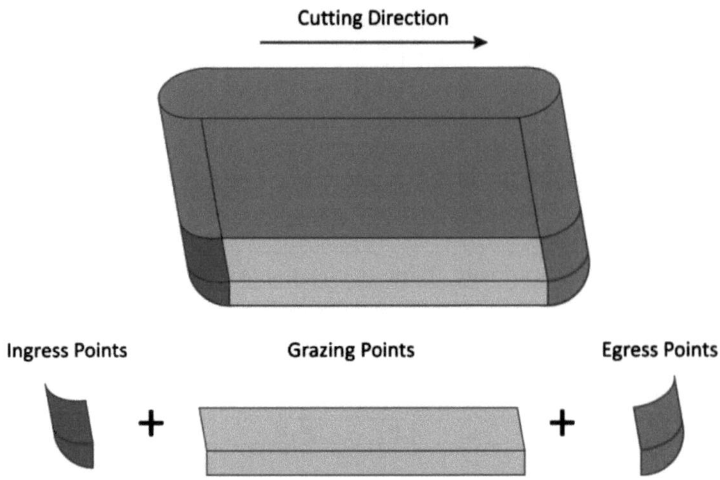


**Fig. 3.15** Tool swept volume of a ball-end mill

**Fig. 3.16** Illustration of swept regions on a ball-end mill

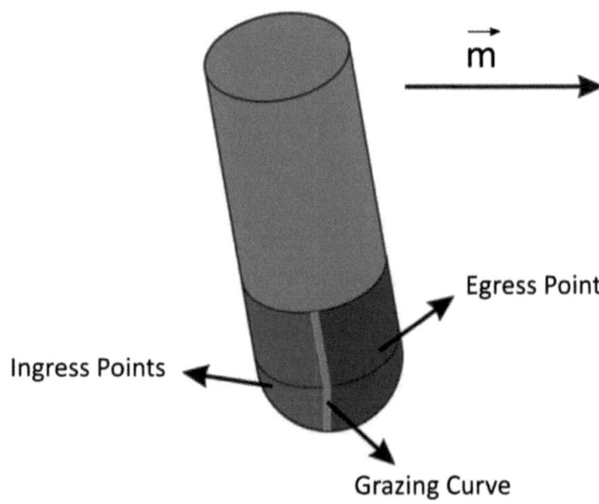


$$\begin{aligned} n \cdot \vec{m} &> 0 \quad \text{Egress points} \\ \vec{n} \cdot \vec{m} &= 0 \quad \text{Grazing Curve (points)} \\ \vec{n} \cdot \vec{m} &< 0 \quad \text{Ingress points} \end{aligned} \tag{3.8}$$

where $\vec{n}$ is the surface normal of an arbitrary point on the cutter surface and $\vec{m}$ is the movement vector given in Eq. 3.8.

Tool motion in 5-axis machining also includes an arbitrary rotation and this effect must be taken into account if swept volume of the cylinder part of the cutter is in cut, on the other hand in free-form surface machining distance and angular

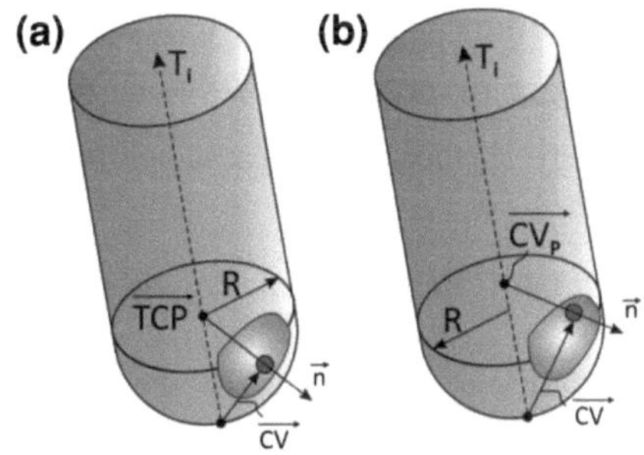


**Fig. 3.17** Surface normal calculation: **a** Sphere part. **b** Cylinder part

rotation between is relatively small, therefore instead of applying exact 5-axis tool motion swept volume is modeled assuming 3 + 2 axis tool motion.

#### 3.4.3.3 Surface Normals for a Ball-End Mill

Surface normal of a general milling cutter can be expressed analytically [4], however, in this study tool swept volume is not directly calculated. Primary aim of the swept volume modeling is to identify the cut points which are in the egress region. Therefore, surface normals are found geometrically.

Figure 3.17 shows the calculation principle of the surface normals for a ball-end mill. Surface normal for the sphere part can be found as,

$$\overrightarrow{\mathrm{TCP}} = R \cdot \vec{T}_i$$

$$\vec{n} = \frac{\overrightarrow{\mathrm{CV}} - \overrightarrow{\mathrm{TCP}}}{\left\| \overrightarrow{\mathrm{CV}} - \overrightarrow{\mathrm{TCP}} \right\|} \tag{3.9}$$

where $\overrightarrow{\mathrm{TCP}}$ is the tool center point, $R$ is the radius to the cutting tool, $\vec{T}_i$ is tool axis vector for the CL point $i$ and $\overrightarrow{\mathrm{CV}}$ is the cut vector given in Eq. 3.1.

Surface normal calculation for the cylinder part is quite different from the sphere part since cylinder part is not symmetric around tool center point. For this reason, first $\overrightarrow{\mathrm{CV}}$ is projected on to the tool axis in order to determine the axial height from the tooltip, then surface normal is found.

$$\overrightarrow{\mathrm{CV}}_p = \left( \overrightarrow{\mathrm{CV}} \cdot \vec{T}_i \right) \cdot \vec{T}_i$$

$$\vec{n} = \frac{\overrightarrow{\mathrm{CV}} - \overrightarrow{\mathrm{CV}}_p}{\left\| \overrightarrow{\mathrm{CV}} - \overrightarrow{\mathrm{CV}}_p \right\|} \tag{3.10}$$

Before calculating the tool surface normal, each point has to be verified that the point is on the sphere part or on the cylinder part. If $\left\|\overrightarrow{\mathrm{CV}}_p\right\| < R$, point belongs to sphere surface otherwise it belongs to the cylinder surface.

#### 3.4.3.4 Calculation of Engagement Angles

In order to determine the engagement angle, cut vector is first projected onto $\overrightarrow{X}_f - \overrightarrow{Y}_f$ plane as follows,

$$\overrightarrow{\mathrm{CV}}_{xy} = \left(\overrightarrow{T}_i \times \overrightarrow{\mathrm{CV}}\right) \times \overrightarrow{T}_i \tag{3.11}$$

Final step is necessary for determining the engagement angles by projecting $\overrightarrow{\mathrm{CV}}_{xy}$ vector onto feed and cross-feed direction, then angles are calculated using arctangent function.

$$\begin{aligned} \mathrm{CV}_x &= \overrightarrow{\mathrm{CV}}_{xy} \cdot \overrightarrow{X}_f \\ \mathrm{CV}_y &= \overrightarrow{\mathrm{CV}}_{xy} \cdot \overrightarrow{Y}_f \\ \theta_e &= a\tan2(\mathrm{CV}_x, \mathrm{CV}_y) \end{aligned} \tag{3.12}$$

where $a$tan2 is a four quadrant arctangent function.

Since engagement angles are defined in clockwise direction, in Eq. 3.12 the order of the $X$ and $Y$ components of the projected cut vector is reversed. This process is presented in Fig. 3.18.

#### 3.4.3.5 Determination of Engagement Quadrants

Start and exit angles of the engagement domain can be identified correctly by checking the quadrant of the engagement angle. The method for the quadrant determination can be given as follows,

$$\begin{aligned} \mathrm{CV}_x > 0, \quad & \mathrm{CV}_y > 0 \Rightarrow \text{Quadrant I} \\ \mathrm{CV}_x > 0, \quad & \mathrm{CV}_y < 0 \Rightarrow \text{Quadrant II} \\ \mathrm{CV}_x < 0, \quad & \mathrm{CV}_y < 0 \Rightarrow \text{Quadrant III} \\ \mathrm{CV}_x < 0, \quad & \mathrm{CV}_y > 0 \Rightarrow \text{Quadrant IV} \end{aligned} \tag{3.13}$$

In 3-axis machining engagement region is determined only by the tool movement direction since tool axis direction is fixed in vertical or horizontal direction. In contrary, tool orientation for 5-axis tool motion changes spatially, therefore possible engagement region is related to move direction and tool axis orientation.

If $\overrightarrow{T}_i \cdot \overrightarrow{m} > 0$, Quadrant I and Quadrant II is valid. If $\overrightarrow{T}_i \cdot \overrightarrow{m} < 0$, Quadrant I, Quadrant II, Quadrant III and Quadrant IV is valid. The method explained above is illustrated in Fig. 3.19.

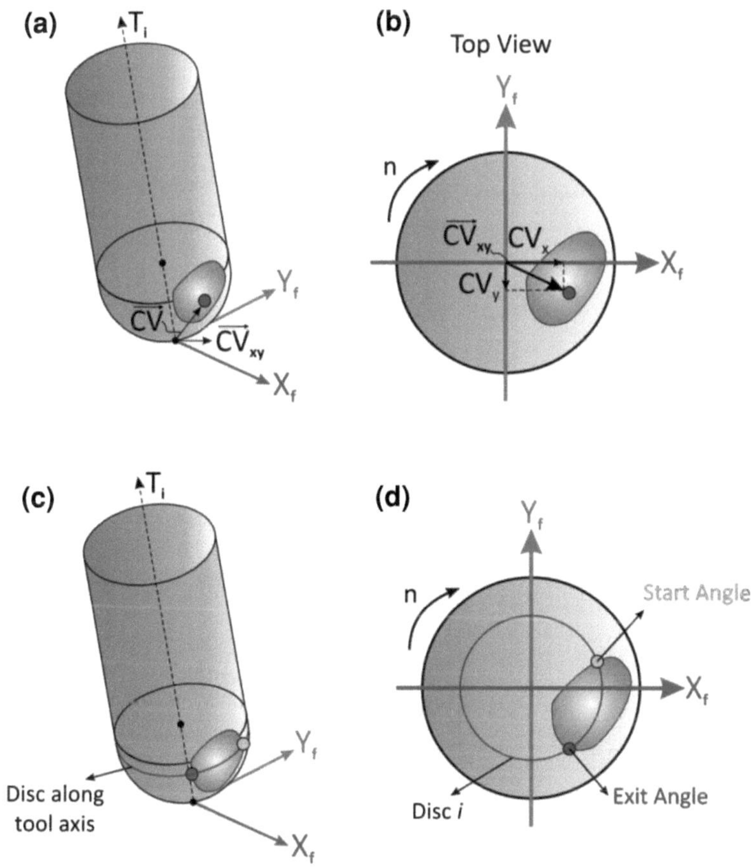


**Fig. 3.18** Engagement angle calculation: **a** Projection on $\overrightarrow{X}_f - \overrightarrow{Y}_f$ plane, **b** Projection on feed and cross feed directions, **c**, **d** Start and exit angles

#### 3.4.3.6 Engagement Domain Results

Three-orthogonal dexelfield approach explained in Sect. 3.4.3 is implemented in Microsoft Visual Studio using C++ programming language. For the verification, impeller roughing toolpath shown in Fig. 3.7 is simulated with the three-orthogonal dexelfield approach. Simulated machined workpiece is shown in Fig. 3.20.

In Fig. 3.21, engagement domains with their projections along the tool axis for CL point 86 are shown. The engagement domain CL point 86 consists of two pieces. This method can only determine the multiple contacts if the gap region occurs between the quadrants. If the gap region lies in one quadrant, all engagement angles have to be checked whether sudden increase or decrease occurs between consecutive engagement angles. However, this method comes with a decrease in computational efficiency since for the example CL point 86 there are 669780 intersection points in other words engagement angles. Therefore, determining the in-quadrant gap regions

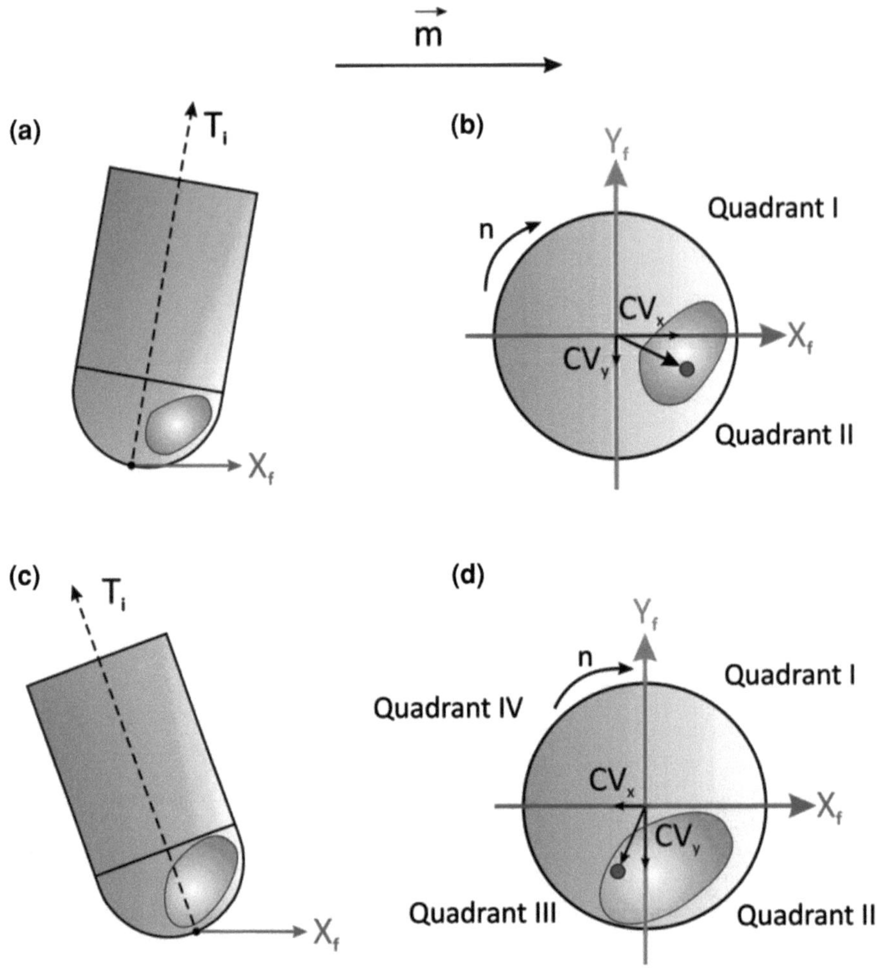


**Fig. 3.19** Engagement quadrant determination: **a** Tool move for $\overrightarrow{T}_i \cdot \overrightarrow{m} > 0$, **b** Valid engagement regions for $\overrightarrow{T}_i \cdot \overrightarrow{m} > 0$, **c** Tool move for $\overrightarrow{T}_i \cdot \overrightarrow{m} < 0$, **d** Valid engagement regions for $\overrightarrow{T}_i \cdot \overrightarrow{m} < 0$

costs considerable amount computation time; as a result, a simplified approach is utilized for obtaining the multiple piece engagement domains along the toolpath.

## 3.5 Sample Results for Three-Orthogonal Dexelfield Engagement Model

Toolpath shown in Fig. 3.22 which consists of 1572 CL points is simulated for the extraction of the Cutter-Workpiece Engagement maps. Impeller hub is machined using a 6 mm ball-end mill therefore in the simulation aforementioned cutting tool

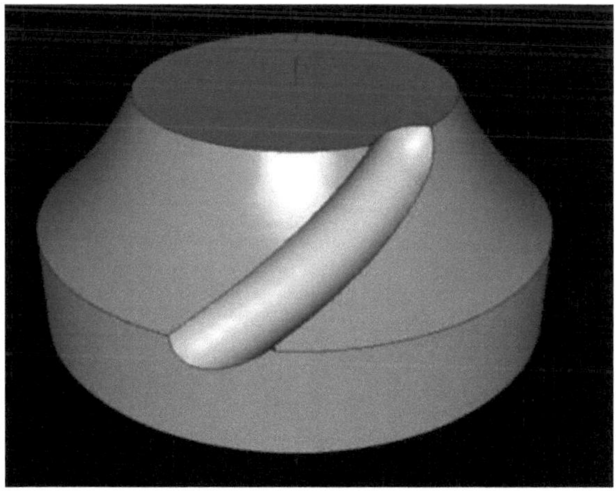

**Fig. 3.20** Machined workpiece using three-orthogonal dexelfield approach

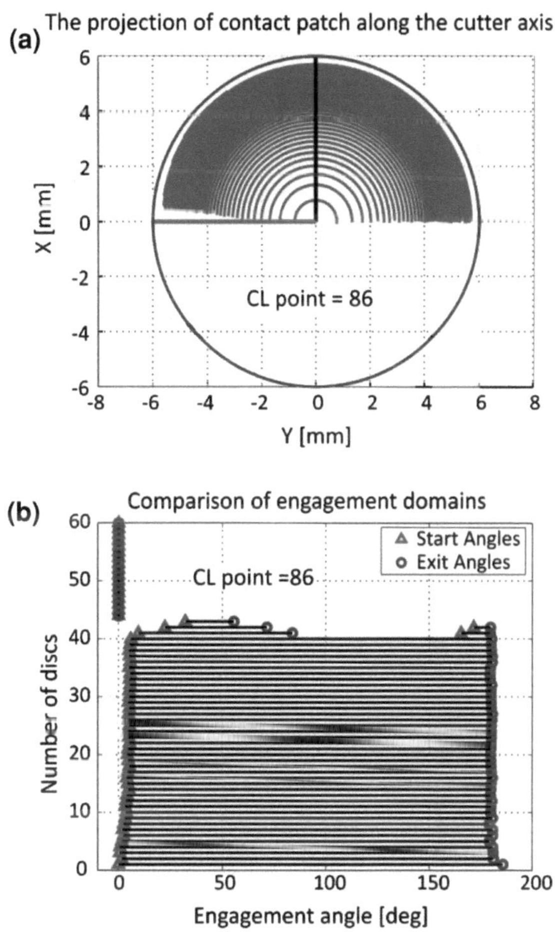


**Fig. 3.21** Three-orthogonal dexelfield engagement domain for CL point #86: **a** Projected view of contact patch along cutter axis. **b** Start and exit angles for the disks along the cutter axis

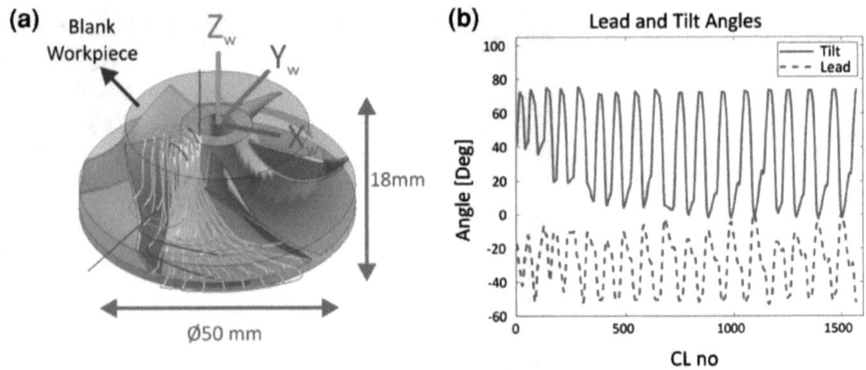


**Fig. 3.22** **a** Simulated impeller roughing toolpath. **b** Lead and tilt angles for the toolpath

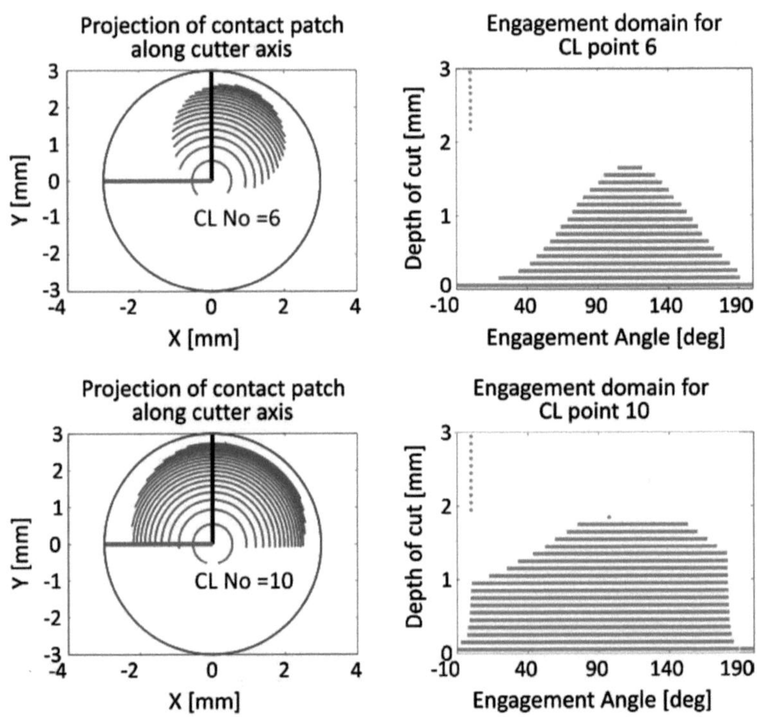


**Fig. 3.23** Engagement results for sixth and tenth CL points

is selected. In order to demonstrate the complex and spatially changing engagement regions in 5-axis machining several 2D engagement maps is shown from Figs. 3.23 and 3.24.

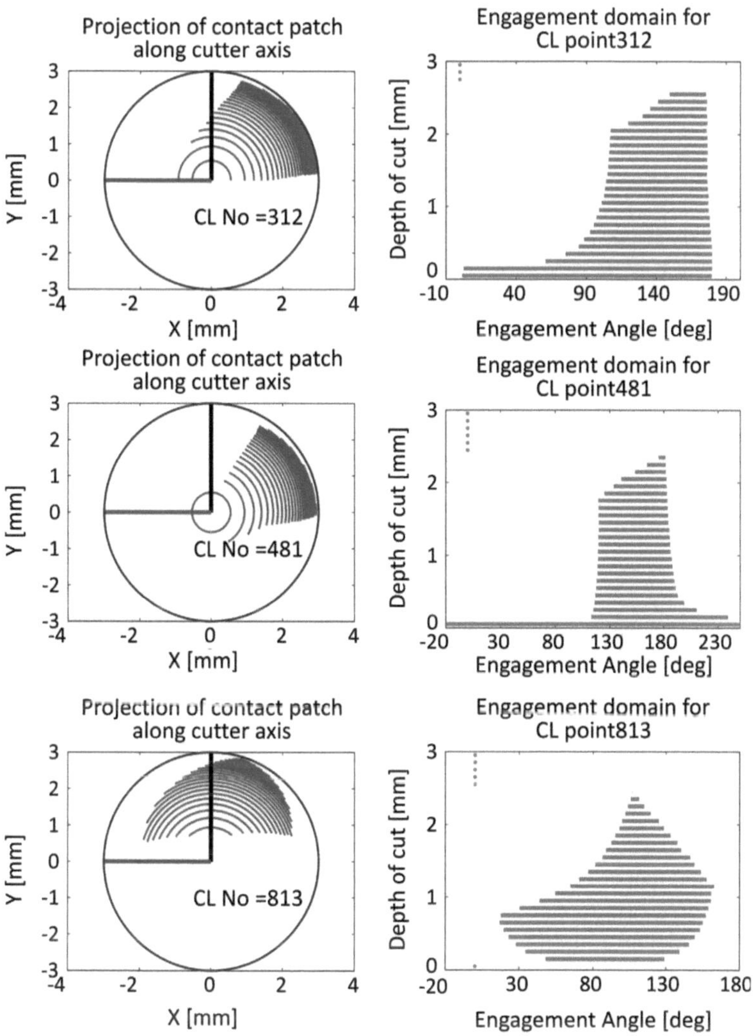


**Fig. 3.24** Engagement results for 312th, 481st and 813th CL points

## 3.6 Comparison of Cutter-Workpiece Engagement Approaches

This section presents two different cutter-workpiece engagement calculation schemes. Namely, solid-modeler-based and discrete vector-representation (dexel, depth buffer)-based cutter-workpiece methods.

The efficiency of these two methods is compared considering the computation time and the simulation accuracy. Regarding the computation time, it is

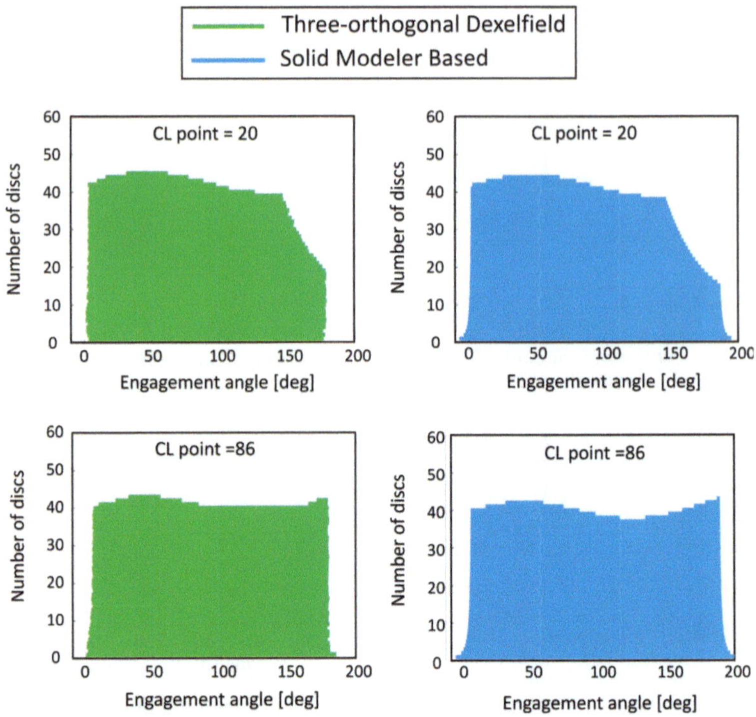


**Fig. 3.25** Comparison of engagement maps from the three-orthogonal dexelfield (*green*) and the solid-modeler-based (*blue*)

demonstrated that solid-modeler-based approach is approximately four times faster than the three-orthogonal dexelfield approach. As it is stated before, for the example impeller roughing toolpath computation time in solid-modeler-based approach is 58 s and for the three-orthogonal dexelfield approach computation time is 4 min 57 s.

This computational inefficiency of the three-orthogonal dexelfield approach arises from the use of three dexelfields simultaneously. Instead of using three dexelfields, if the conventional *Z*-map technique is used computation time can be reduced to approximately one-third of the presented approach meanwhile losing the simulation accuracy.

Second important criteria while judging the efficiency of the cutter-workpiece engagement model is the accuracy of the model since precise determination of the engagement region is crucial for the cutting force model.

In this respect, solid-modeler-based engagement model is superior to the three-orthogonal dexelfield approach. As it is illustrated in Fig. 3.25 there are slight differences between two methods and it is observed that solid-modeler-based approach is more accurate than the three-orthogonal dexelfield approach. Solid-modeler-based approach uses exact Boolean operations between the cutter swept

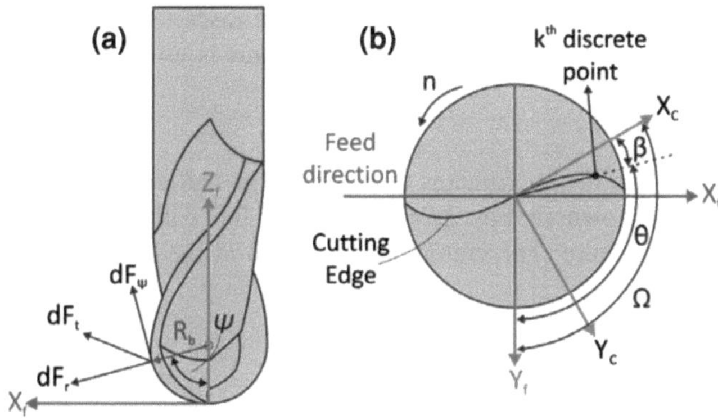


**Fig. 3.26** Illustration of cutting force vectors and angular relationships

envelope and the workpiece, for this reason, surface patch boundaries are exact and smooth. In three-orthogonal dexelfield approach due to numerical instability and the nature of the intersections, fluctuations in the engagement angles and the trimming of the engagement region occur.

In conclusion, for five-axis milling, solid-modeler-based cutter-workpiece engagement approach calculates the engagement angles more accurately and faster compared to three-orthogonal dexelfield approach. Furthermore, multiple engagement regions can be handled effortlessly and correctly whereas in three-orthogonal dexelfield approach in-quadrant contact regions are not considered, hence this may cause the incorrect interpretation of the engagement information.

## 3.7 Cutting Force Model

End mills can be characterized with many aspects, such as the macro geometry (helical-end, ball-end, bull-nose, flat-end, tapered, etc.), the micro geometry (helix, rake, clearance angles), the tool material (WC, PCD, HSS), the coating material (nanocomposite PVD, TiAlN, TiCN, etc.), the area of usage (rough, semi-finish, finish, super-finish milling) and the workpiece material (steel, Aluminum, titanium, etc.).

Helical-end mills have constant radius and helix angle along the depth of cut. Ball-end mills differ from flat end mills by their ball part such that their radius varies along the ball end. This varying radius $r$ affects the cutting forces because cutting speed changes with varying $r$. On the contrary, flat-end mills have constant cutting speed along the tool axis.

The detailed geometry of a ball-end mill is shown in Fig. 3.26. It can be observed that each flute lies on the surface of the hemisphere, and has a changing helix angle. Due to the decreasing radius toward the tip of the cutter, the local helix angle changes with varying cutting velocity for a discrete point along the cutting flute. The equation of the geometry of the ball part is given by,

$$x^2 + y^2 + (R_b - z)^2 = R_b^2 \tag{3.14}$$

where $x$, $y$ and $z$ are the coordinates of cutting edge ball-end mill according to the coordinate axes shown in Fig. 3.26, and $R_b$ is the ball radius measured from the center of the sphere. The cutter radius is zero at the tip and at axial location $z$, in plane $x$–$y$,

$$r^2 = x^2 + y^2 \tag{3.15}$$

The ball-mill cutter used in calibration and validation tests is 12 mm diameter, two fluted ball-end mill. The cutting edge is measured with a Coordinate Measuring Machine (CMM). Obtained data points were used to obtain following third degree polynomial that represents the cutting edge geometry for the cutting force model.

$$\beta = 0.0036 \times r^3 - 0.0205 \times r^2 + 0.0547 \times r - 0.041 \tag{3.16}$$

where $r$ [mm] is the radius of an arbitrary point on the cutting edge perpendicular to the cutter axis, and $\beta$ [deg] is the lag angle between the line which connects this arbitrary point to the tip and the line tangent to the cutting edge at the tip. Details of the cutting edge geometry can be seen in Fig. 3.27. The cutting geometry can be represented in the cutter coordinate system as the following;

$$x_c = r \times \cos(\beta),\ y_c = r \times \sin(\beta),\ z_c = R_b - \sqrt{R_b^2 - r^2} \tag{3.17}$$

In this approach lag angle is calculated spatially on the cutter considering the helix angle of the tool and the height from the tool tip. For practical applications this approach can also be used with acceptable error.

## 3.8 Geometry of 5-Axis Milling

Five-axis milling geometry differs from 3-axis milling geometry. Hence transformation from 3-axis milling to 5-axis milling has to be defined. In this section, important concepts and parameters which define geometry of 5-axis machining is introduced. Then, these formulations are used in mechanistic cutting force modeling of 5-axis machining.

In 3-axis milling tool movement is given as three translational motions along the $X$–$Y$–$Z$ coordinate frame axes. In 5-axis milling two additional rotary axes are present. Consequently, tool motion is defined as a combination of three translational motions and two rotational motions. There are several different kinematic configurations in 5-axis machine tools [5]. In this study all of the formulations are presented according to a table tilting-rotating 5-axis vertical machine tool.

**Fig. 3.27** **a** Cutting edges. **b** Third degree polynomial fitting for $\beta(r)$

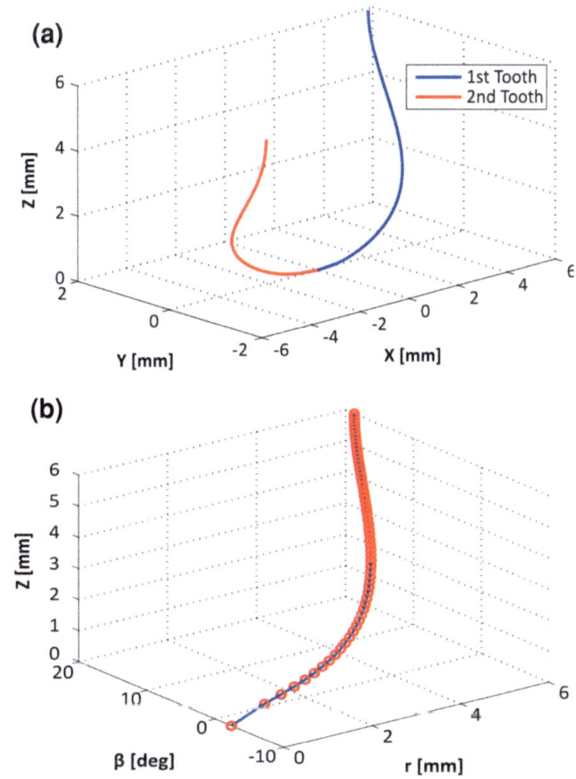


Contrary to 3-axis milling, tool orientation vector in 5-axis milling is not constant. Therefore, tool coordinate frame (TCF) has to be mapped on to the workpiece coordinate frame (WCF).

Two additional rotational motion in 5-axis milling is given as lead and tilt angles. Lead angle is defined as the rotation angle about $Y_w$ which is $Y$ axis of workpiece coordinate frame. Tilt angle is the rotation angle about $X_w$ which is $X$ axis of the workpiece coordinate frame. Definition of the lead and tilt angles is shown in Fig. 3.28.

It is worthwhile to state that there are other conventions [6] which use feed and cross feed vectors as reference frames; however, if these angles are calculated relative to these vectors, reference for angles naturally becomes drive surface and if the surface normal has a large angle with machine tool axis there may be unrealizable lead and tilt angles.

In order to extract lead and tilt angles from toolpath data CL (Cutter Location) output of Siemens NX6 is used. CL file is parsed via a preprocessor, and then CL points and tool orientation vectors in the form of direction cosines are obtained. Example CL block is shown in Fig. 3.29. GOTO/ keyword states the beginning of a CL point block, three numbers after the keyword gives $X$, $Y$ and $Z$ coordinates of the tooltip in the workpiece coordinate frame, respectively. Remaining three

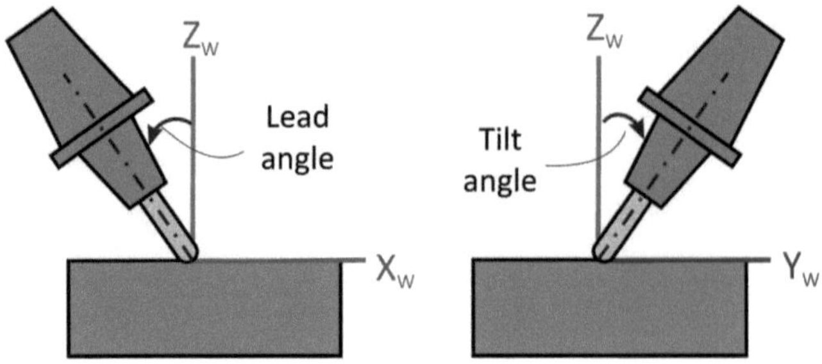


Fig. 3.28 Definition of lead and tilt angles

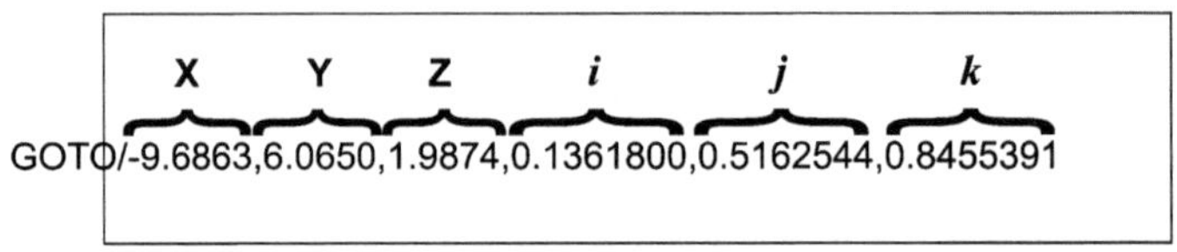


Fig. 3.29 Example cutter location block

numbers gives the tool orientation vectors $i$, $j$, $k$, respectively, relative to workpiece coordinate frame.

Lead and tilt angles can be calculated as [7]:

$$\text{lead} = \text{atan2}\left(i, \sqrt{j^2 + k^2}\right) \tag{3.18}$$

$$\text{tilt} = \text{atan2}(-j, k) \tag{3.19}$$

Since, transformation from the workpiece coordinate frame to the tool coordinate frame is necessary for inverse transforming the calculated cutting forces in cutting force model; the rotation matrix from workpiece coordinate frame to tool coordinate frame has to be calculated.

Illustration of coordinate frames is shown in Fig. 3.30, where $(X_f - Y_f - Z_f)$ is the feed coordinate frame. Transformation matrix from workpiece coordinate frame to tool coordinate frame is given as:

$$T = \begin{bmatrix} \cos(\text{lead}) & 0 & \sin(\text{lead}) \\ \sin(\text{tilt})\sin(\text{lead}) & \cos(\text{tilt}) & -\sin(\text{tilt})\cos(\text{lead}) \\ -\cos(\text{tilt})\sin(\text{lead}) & \sin(\text{tilt}) & \cos(\text{tilt})\cos(\text{lead}) \end{bmatrix} \tag{3.20}$$

Inverse of transformation matrix $T$ gives the necessary transformation from TCF to WCF as:

$$T_I = T^{-1} = T^T \tag{3.21}$$

Cutting forces calculated in TCF can be transformed to WCF as follows:

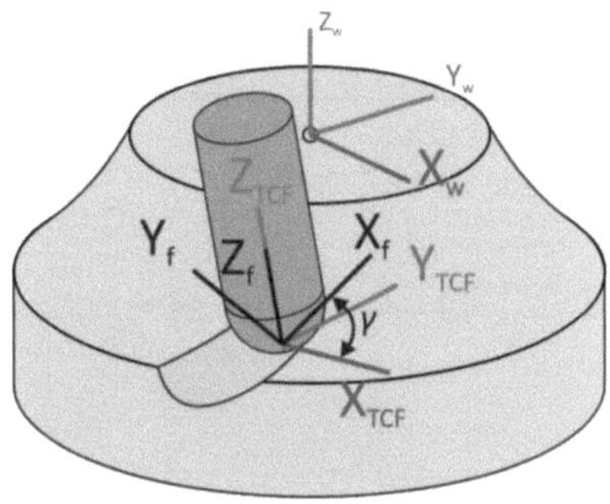


**Fig. 3.30** Illustration of coordinate frames

$$\begin{bmatrix} F_x \\ F_y \\ F_z \end{bmatrix}_{\text{WCF}} = T_I \begin{bmatrix} F_x \\ F_y \\ F_z \end{bmatrix}_{\text{TCF}} \tag{3.22}$$

## 3.9 Cutting Force Model in Fixed Coordinate Frame

In milling, cutting forces depend on the instantaneous chip thickness. Hence, for 5-axis machining cutting force predictions, accurate calculation of the chip thickness is quite critical since tool can rotate as well as translate within a toolpath segment.

In free-form surface machining the distance and the rotation angle between two CL points are relatively small, therefore the effect of rotational velocities of the tool is negligible. On the other hand, the effect of the lead and tilt angles on the cut geometry, and horizontal and vertical feed components has to be considered.

For ball-end mill tool, instantaneous undeformed chip thickness is obtained as follows [8, 9, 10];

$$(t_c)_k = t_x \times \sin(\theta) \times \sin(\psi) \times \cos(\alpha) \pm t_x \times \cos(\psi) \times \sin(\alpha) \tag{3.23}$$

where $(t_c)_k$is the chip thickness $t_x$ is the feed per tooth, $\theta$ is the immersion angle of the cutting point, $\psi$ is the cutting element position angle and $\alpha$ is the feed inclination angle measured with respect to horizontal feed direction. Distribution of horizontal and vertical chip thickness are shown in Figs. 3.31 and 3.32.

The immersion angle of a discrete cutting point on the flute of the cutter is given as:

$$\theta = \Omega + \frac{2\pi(n-1)}{N_f} - \beta_k \tag{3.24}$$

where $\theta$ is the immersion angle for flute $n$, $k$ represents the number of discrete point on a cutting edge, $\Omega$ is the cutting edge rotation angle, $N_f$ is the total number

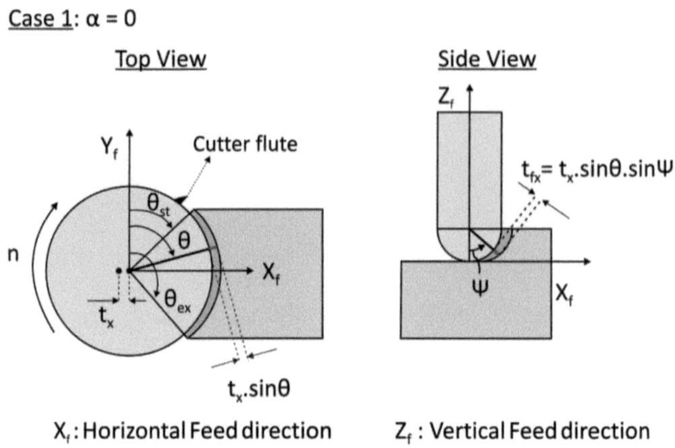


**Fig. 3.31** Chip thickness due to horizontal feed

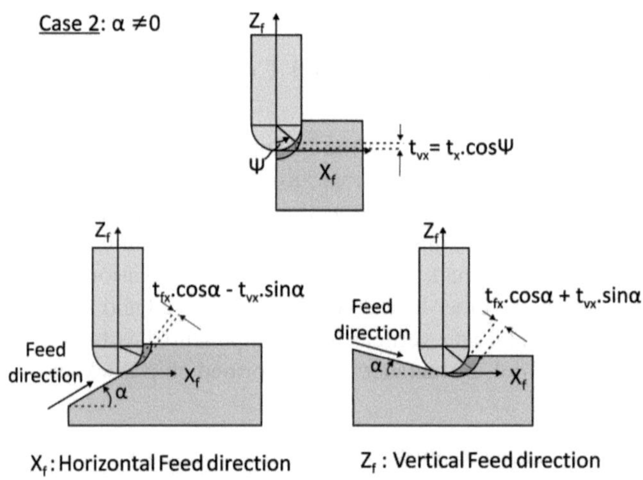


**Fig. 3.32** Chip thickness due to horizontal and vertical feed

of flutes and $\beta_k$ is the lag angle due to helix angle of the cutter in the respective $k$th disk.

The instantaneous infinitesimal chip load is written as follows:

$$dA_c = (t_c)_k \times (dz)_k \tag{3.25}$$

For a differential chip load $dA_c$ in the engagement domain, the differential cutting forces in radial, axial, and tangential directions $(r, \psi, t)$ are written as follows;

$$\begin{aligned} dF_r &= K_{rc} \times dA_c + K_{re} \times dz \\ dF_\psi &= K_{\psi c} \times dA_c + K_{\psi e} \times dz \\ dF_t &= K_{tc} \times dA_c + K_{te} \times dz \end{aligned} \tag{3.26}$$

where $K_{rc}$, $K_{\psi c}$ and $K_{tc}$ are radial, axial and tangential cutting force coefficients and $K_{re}$, $K_{\psi e}$ and $K_{te}$, are cutting edge coefficients, respectively. Cutting force and edge coefficients are determined by mechanistic calibration procedure where these coefficients vary along tool axis direction [8, 11].

Transformation matrix $A$ transforms the cutting forces into feed coordinate frame which is initially coincident with TCF. If the angle between feed direction and $X_{\text{TCF}}$ is not zero, $B$ matrix transforms the cutting forces into tool coordinate frame.

$$A = \begin{bmatrix} -\sin(\psi) \times \sin(\theta) & -\cos(\psi) \times \sin(\theta) & -\cos(\theta) \\ \sin(\psi) \times \cos(\theta) & \cos(\psi)\cos(\theta) & -\sin(\theta) \\ \cos(\psi) & -\sin(\psi) & 0 \end{bmatrix} \tag{3.27}$$

$$B = \begin{bmatrix} \cos\gamma & -\sin\gamma & 0 \\ \sin\gamma & \cos\gamma & 0 \\ 0 & 0 & 1 \end{bmatrix} \tag{3.28}$$

In this formulation, a table type dynamometer (fixed coordinate frame) is used. Therefore, cutting forces in feed coordinate frame are transformed into WCF which is also dynamometer coordinate frame. By using transformation matrix $T_I$ given in (3.20), cutting forces in WCF is written as:

$$\begin{bmatrix} dF_X \\ dF_Y \\ dF_Z \end{bmatrix} = [T_I][B][A] \times \begin{bmatrix} dF_r \\ dF_\psi \\ dF_t \end{bmatrix} \tag{3.29}$$

Finally calculated cutting forces are summed for the all of the axial disks and the cutting flutes in order to obtain total cutting force for an immersion angle of $\theta(z,k)$.

$$\begin{bmatrix} F_X \\ F_Y \\ F_Z \end{bmatrix} = \sum_{n=1}^{N_f} \sum_{k=1}^{K} \begin{bmatrix} dF_X \\ dF_Y \\ dF_Z \end{bmatrix} \tag{3.30}$$

### 3.9.1 Cutting Force Model in Rotating Coordinate Frame

Cutting force measurement in 5-axis machining is a challenging task due to the varying orientation of the tool axis with respect to the workpiece. In 5-axis milling cutting forces can be measured in two ways.

**Fig. 3.33** Rotating cutting force dynamometer

First one is using a table-type dynamometer which is attached to the rotary table of the machine tool. In this method dynamometer coordinate frame is fixed and transformation from tool coordinate system to dynamometer coordinate frame is difficult. Furthermore, due to the rotation of the rotary axis of the machine tool measured data is affected by the weight of the workpiece and forces induced by the cutting torque.

Second method requires the use of a rotary dynamometer. Rotary dynamometer is directly attached to the spindle of the machine tool and cutting tool is attached to the dynamometer. In other words, cutting forces are directly measured with respect to tool coordinate frame and the effects of the machine tool rotary axes are eliminated. Detail of the rotating cutting force dynamometer is shown in Fig. 3.33.

With this rotating cutting force dynamometer (RCD) cutting forces in three orthogonal directions ($X_D - Y_D - Z_D$ directions) and the cutting moment about the $Z_D$ axis can be measured. $X_D - Y_D - Z_D$ represents the rotating dynamometer coordinate frame.

Transformation to Rotating Coordinate Frame

In Sect. 3.9, cutting force modeling in feed coordinate frame and its transformation to fixed coordinate frame is given. In this section, transformation from feed coordinate frame $(X_f - Y_f - Z_f)$ to rotating dynamometer coordinate frame $(X_D - Y_D - Z_D)$ is introduced.

Consider a two fluted ball-end mill where the cutting flute of the cutter is supposed to be aligned with the $X$ axis of the dynamometer and is traveling along an arbitrary direction. In this position, the angle between $Y$ axis of the dynamometer ($Y_D$) and the first cutting flute of the cutter is represented as the reference rotation angle $\Omega_R$ The rotation angle $\Omega$ is the angle between the cross feed direction and the cutting flute. Reference rotation angle is constant unless tool alignment changes with respect to dynamometer coordinate frame. In contrary, rotation angle is updated during the simulation at each time step by the given rotation increment angle. For inserted milling cutters and end mills alignment of the cutting flute with the $X_D$ axis can be performed easily on the other hand for ball mills due to the complex cutting flute geometry perfect alignment may not be achieved. In this case, a misalignment angle $\Omega_a$ is defined in order to compensate

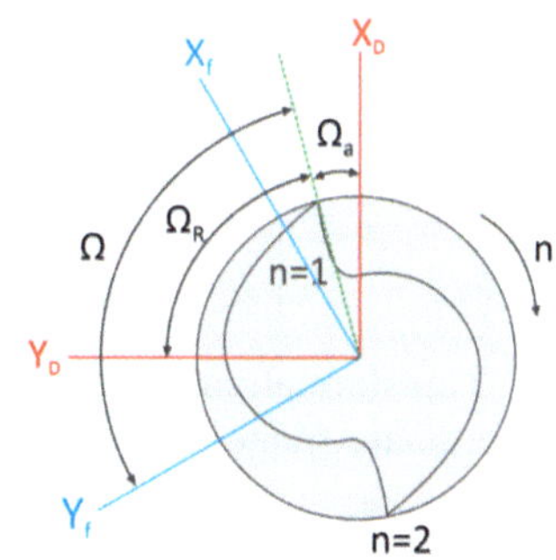


**Fig. 3.34** Rotating coordinate frame transformation angles

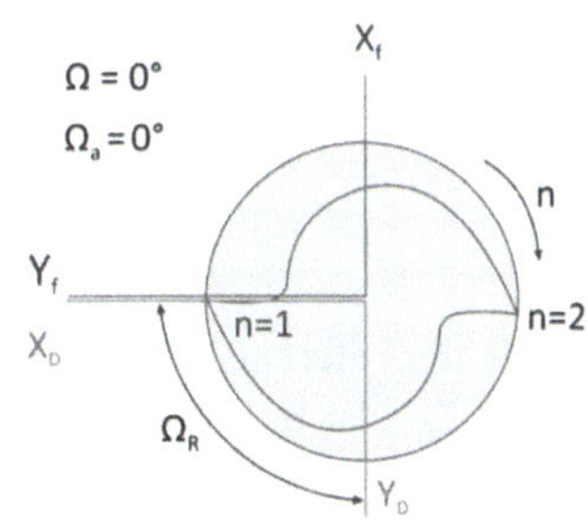


**Fig. 3.35** Special case 1 $\Omega_R = 90°$

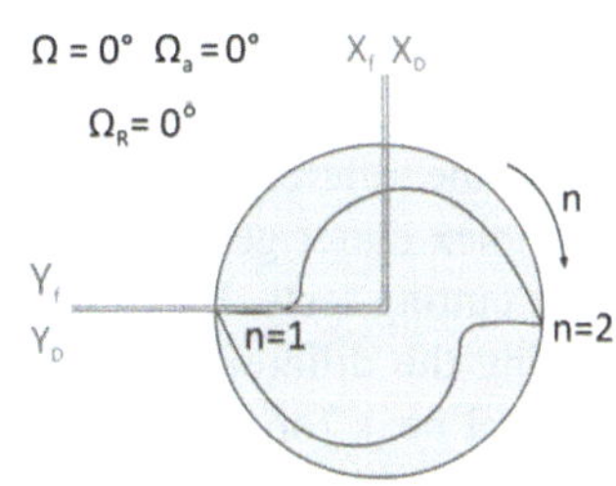


**Fig. 3.36** Special case 2 $\Omega_R = 0°$

the error introduced. Detailed illustration of the transformation angles is shown in Fig. 3.34.

General definition of the rotating coordinate frame is given above. For clarity two special cases are shown in Figs. 3.35 and 3.36.

In the first case, at the instant shown consider that cutting edge of the cutter is perfectly aligned with the $X$ axis of the rotating dynamometer therefore reference rotation angle $\Omega_R$ is 90° and rotation angle $\Omega$ is 0°, meaning that feed coordinate system must be rotated 90° counter clockwise for mapping the feed coordinate frame onto dynamometer coordinate frame.

For the second case, at the instant shown consider that cutting edge of the cutter is perfectly aligned with the $Y$ axis of the rotating dynamometer therefore reference rotation angle $\Omega_R$ is 0° and rotation angle $\Omega$ is 0°, therefore feed coordinate frame and dynamometer coordinate frames are coincident. In both cases, since it is assumed that tool cutting edge is perfectly aligned with the desired rotating coordinate frame axis misalignment angle $\Omega_a$ is equal to 0°.

In order to obtain the transformation from the $(r - \psi - t)$ coordinate frame to feed coordinate frame transformation matrix given in Eq. 3.27 is modified as,

$$A = \begin{bmatrix} -\sin(\psi) \times \sin(\theta) & -\cos(\psi) \times \sin(\theta) & -\cos(\theta) \\ -\sin(\psi) \times \cos(\theta) & -\cos(\psi)\cos(\theta) & \sin(\theta) \\ \cos(\psi) & -\sin(\psi) & 0 \end{bmatrix} \tag{3.31}$$

Transformation from the feed coordinate system to rotating dynamometer coordinate frame can be obtained using these matrices,

$$B = \begin{bmatrix} \cos(\Omega_R + \Omega) & -\sin(\Omega_R + \Omega) & 0 \\ \sin(\Omega_R + \Omega) & \cos(\Omega_R + \Omega) & 0 \\ 0 & 0 & 1 \end{bmatrix} \tag{3.32}$$

$$C = \begin{bmatrix} \cos(\Omega_a) & -\sin(\Omega_a) & 0 \\ \sin(\Omega_a) & \cos(\Omega_a) & 0 \\ 0 & 0 & 1 \end{bmatrix} \tag{3.33}$$

If the reference rotation angle is known or can be measured, forces in $X_D - Y_D - Z_D$ can be calculated as,

$$\begin{bmatrix} dF_X \\ dF_Y \\ dF_Z \end{bmatrix}_{\text{RCD}} = [B][A] \times \begin{bmatrix} dF_r \\ dF_\psi \\ dF_t \end{bmatrix} \tag{3.34}$$

If the reference rotation angle is not known or cannot be measured due to complex cutter geometry, misalignment angle can be extracted by running simple slot cutting tests. By examining the force magnitudes for one tool revolution and taking the difference of the simulated peak force rotation angle with the experimental peak force rotation angle.

If this is the case, it is assumed that cutting edge of the cutter is perfectly aligned with the intended coordinate axis of the dynamometer and the misalignment matrix accounts for the misalignment in the calculation.

$$\begin{bmatrix} dF_X \\ dF_Y \\ dF_Z \end{bmatrix}_{\text{RCD}} = [C][B][A] \times \begin{bmatrix} dF_r \\ dF_\psi \\ dF_t \end{bmatrix} \tag{3.35}$$

## 3.10 Feedrate Scheduling

### *3.10.1 Introduction*

Because of the complex geometry of the 3D free-form surfaces, CAM processors and CNC operators tend to select conservative cutting parameters to avoid tool breakage, low surface quality, tool deflections, machine tool damages, etc. On the other hand, process planner engineers want to increase the productivity by reducing the cycling time.

One of those cutting parameters that affect the productivity in a direct manner is feedrate. Conservative constant feedrate values have been widely used in industry for 5-axis free-form surface machining due to lack of a comprehensive model that is able to take into account the physics of the process. Besides, currently used CAM programs are based on only the geometry of the process and volumetric analysis. Therefore, the next step in developing smart CAM packages is introducing efficient algorithms to update the cutting parameters according to different cutting conditions during free-form surface machining.

In this section a feedrate optimization strategy is introduced to increase the productivity. Typically, there are two methods for scheduling feedrate value namely the volumetric and force-based strategies. According to previous studies, volumetric-based feedrate strategy is insufficient for determining the optimum feedrate values [8, 12, 13, 14, 15, 16]. Therefore, force-based feedrate strategy is presented in this section.

### 3.10.2 Force-Based Feedrate Scheduling Strategy

The offline feedrate scheduling regulates the original constant feedrate values according to the reference cutting force which is determined before. In order to increase the resultant force to the desired reference value, a simple linear relation is found between the feedrate and the reference limiting cutting force.

From the slot cutting tests, it is observed that the both the simulated and measured resultant cutting forces increase linearly with increasing feedrate values as seen in Fig. 3.37, therefore, this simple relation is found for the feedrate scheduling formula and it is also verified for non-slot cuts also known as non-horizontal feed directions.

The feedrate and cutting force relation can be seen for the slot cutting tests for various depths. As it is seen from the Fig. 3.37, the equation of line is changed for each depth of cut value, so that the feedrate scheduling formula should change for each depth of cut.

When the feederate is zero, it seems to be zero that there is no force acting on the cutting tool, since the tool is not moving. However there is a force which occurs due to edge coefficients and rubbing effect between tool and workpiece. This effect causes shearing and plowing mechanisms at the tip of the cutter. The relation can be expressed from the cutting force and chip load module. For ball-end mill cutter the instantaneous undeformed chip thickness is obtained in Sect. 3.3 as follows:

$$(t_c)_k = t_x \times \sin(\theta) \times \sin(\psi) \times \cos(\alpha) \pm t_x \times \cos(\psi) \times \sin(\alpha) \tag{3.36}$$

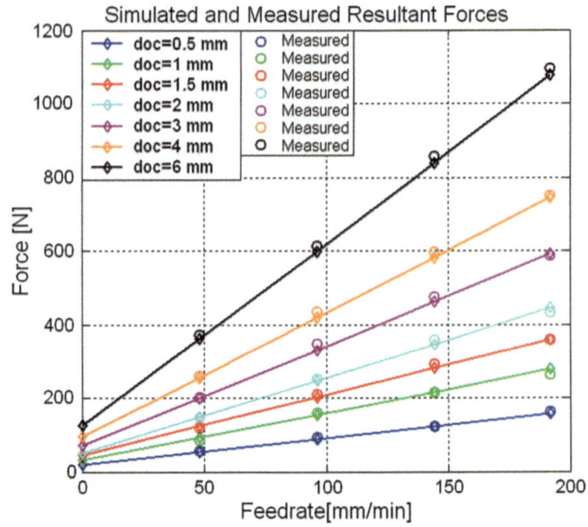


**Fig. 3.37** Slot cutting resultant forces for different depth of cuts and feedrate values

As it is seen from Eq. 3.36, the undeformed chip thickness is found from feed per tooth value which can be found from federate (mm/min). The instantaneous infinitesimal chip load is written as follows;

$$dA_c = (t_c)_k (dz)_k \tag{3.37}$$

For a differential clip load ($dA_c$) in engagement domain, the differential cutting forces in $r, \psi$ and $t$ directions is written as follows:

$$dF_i = dF_{\text{cutting}} + dF_{\text{edge}} = K_{ri} dA_c + K_{ri} dz, \quad i = r, \psi, t \tag{3.38}$$

As it is seen from Eqs. 3.37 and 3.38, a linear relation between these equations and feedrate exists. It is also seen from the differential cutting force equation, the differential cutting force consists of cutting and edge parts like calibration constants. This provides to derive a linear relation depending on feedrate.

As it can be inferred, there is a linear relation between cutting forces and feedrate values. This provides to derive a linear relation depending on feedrate as follows:

$$dF_i = A \cdot f + B \tag{3.39}$$

where $dF_i$ is differential cutting force, $f$ is feedrate value and A and B are constant values. The model is processed to keep the resultant force at the desired constant limit level along the tool path for the CL points; the model uses the contact region defined for each CL point. The limiting feederate formula for the $i$th CL point is given as follows:

$$f_{\text{lim},i} = \left(F_{\text{lim},i} - F_{1,i}\right) \cdot \frac{f_2 - f_1}{F_{2,i} - F_{1,i}} + f_1 \tag{3.40}$$

where $= 1, 2 \ldots c$, is the total number of CL points in the tool path, $f_1$(mm/min) is the original constant feedrate for the tool path; $f_2$(mm/min) is twice of $f_1$ in order to obtain the linear relation for the $i$th CL point's depth of cut value. $F_1$(Newton) is the maximum resultant force value for the $i$th CL point for $f_1$ feedrate value, $F_2$ (Newton) is the maximum resultant force value for $i$th CL point for $f_2$ feedrate value, $F_{\text{lim},i}$ is the limiting constant resultant force threshold value which the cutting forces will be under this threshold value and $F_{\text{lim},i}$ is obtained as scheduled in (mm/min) for the $i$th CL point.

In 3D free-form surface milling because of permeate fluctuation in depth of cut and engagement angles, the resultant cutting force is not fixed. By anticipating the cutting forces using above-mentioned force model, appropriate feedrate values can be selected to keep the cutting forces below a threshold magnitude which is determined by user.

## 3.11 Experimental Validation

Force validation tests are gathered in two groups since cutting force modeling is performed using different types of coordinate frame conventions. In the first part, cutting force modeling for the fixed coordinate frame convention and the in the second part for the rotating coordinate frames cutting force modeling is validated. In fixed coordinate frame modeling cutting forces are measured using a table-type dynamometer which is fixed to the rotary table of the machine. For the rotating coordinate frame cutting force modeling a rotating coordinate dynamometer is used which is directly attached to the spindle of the machine tool (Fig. 3.38).

## 3.12 Validation Experiments Using Fixed Coordinate Frame

Preliminary study for the validations is performed on 5-axis slotting cases. Four different toolpaths with different lead and tilt angles are simulated and compared with the measured forces. These slotting cases are given in Table 3.1. In these tests, a two fluted carbide ball-end mill with a diameter of 12 mm, nominal helix angle of 30° and projection length of 37 mm is used on Al7039 workpiece material. The spindle speed and the feedrate are 600 rpm and 48 mm/min, respectively. Simulated and measured forces are given through Figs. 3.39, 3.40, 3.41, 3.42, 3.43, 3.44, 3.45 and 3.46.

As it is demonstrated in the figures, simulated and measured forces match quite well for slotting cases; however,main discrepancy in the forces can be attributed to the cutter runout since used model is the simplified model of the true chip formation kinematics.

Free-form surface validation tests are performed on three different toolpaths. First one is airfoil geometry and the other one is the penguin free-form surface.

**Fig. 3.38** Five-axis machining center and data acquisition system used in the experimental setup

**Table 3.1** Five-axis slotting experiments

| Case | Lead angle (°) | Tilt angle (°) |
|---|---|---|
| 1 | 10 | 0 |
| 2 | 10 | −10 |
| 3 | 10 | 10 |
| 4 | 0 | 10 |

For airfoil geometry test nominal 10° lead angle with smoothing, for penguin free-form surface constant 15° lead and 5° tilt angle is simulated. Details of first two toolpaths are shown in Fig. 3.47.

A table-type dynamometer is used for measuring forces which is attached to the rotary table of the machine. Although the cutting forces for whole toolpaths are measured and simulated, one passes of both toolpath simulations are compared against experiments for better illustration of the comparison.

The spindle speed and the feedrate for these toolpaths are kept constant at 600 rpm and 48 mm/min, respectively. A two fluted ball-end mill with a diameter of 12 mm, nominal helix angle of 30°, and projection length of 37 mm is used on

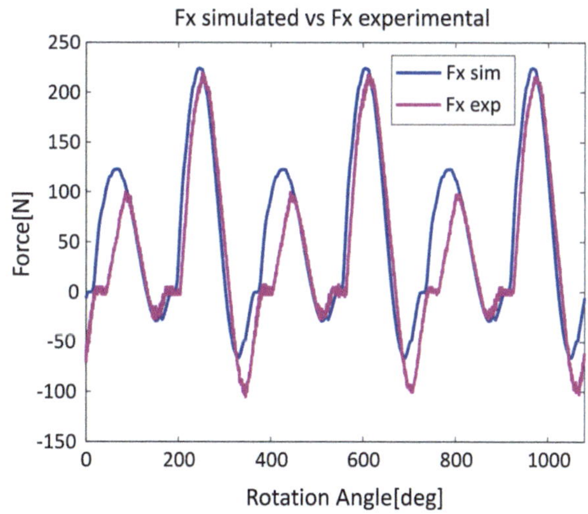


**Fig. 3.39** Case #1, simulated and experimental Fx forces

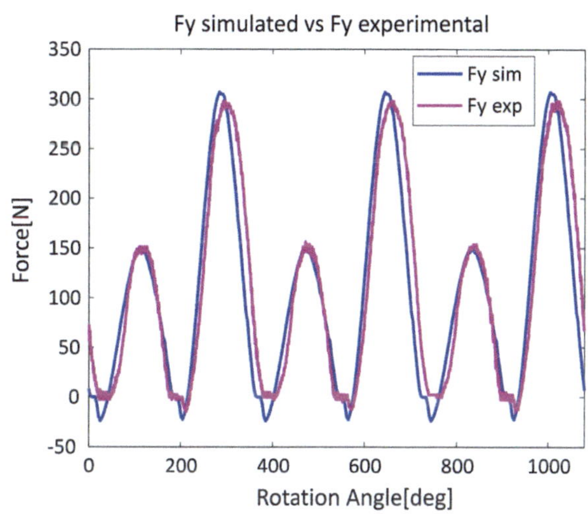


**Fig. 3.40** Case #1, simulated and experimental Fy forces

Al7039 workpiece material. Depths of cut during two toolpaths vary approximately between 0 and 5 mm along tool axis.

Figures 3.48 and 3.49 show the comparison for the simulation and the experimental cutting forces. As it is demonstrated in the figures, simulated and experimental cutting forces match quite well not only in their trends but also in their amplitudes. In most of the regions, the error between simulation and the experimental force amplitudes is below 15% which can be considered as acceptable for 5-axis milling process simulations.

The main differences in cutting force predictions can be attributed to the unequal cutter radius of the flutes which may change the force amplitudes with a

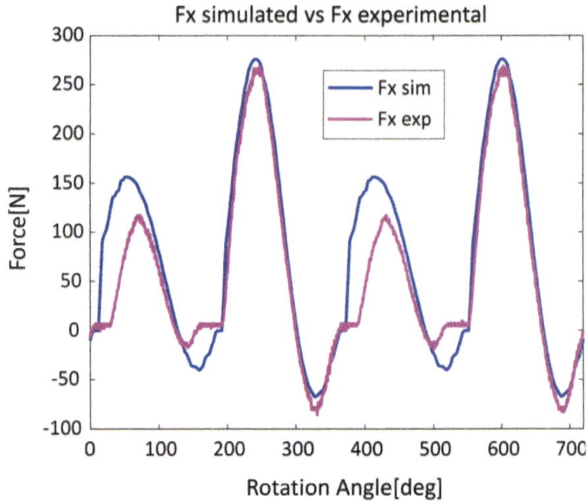


**Fig. 3.41** Case #2, simulated and experimental Fx forces

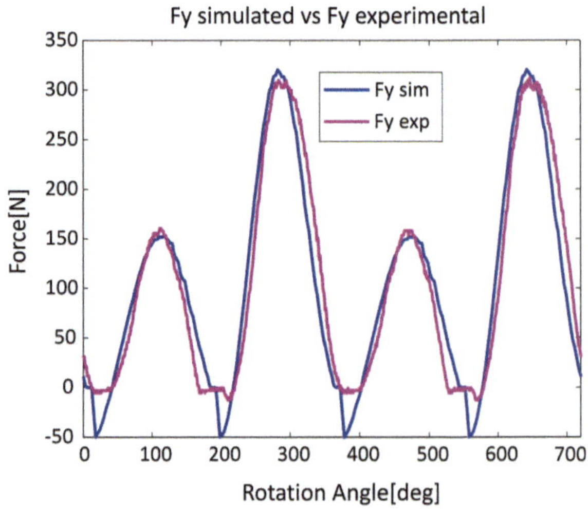


**Fig. 3.42** Case #2, simulated and experimental Fy forces

phase difference in peak forces. This phenomenon is observed in the cutting tool, although a set of the same tool is used. Another reason can be stated as; penguin surface has free-form geometry, in some regions tooltip contact with the workpiece occurs. Therefore, cutting edge of the tool may be rubbing the workpiece material rather than cutting due to zero cutting velocity at the tooltip.

For the third validation test an impeller roughing toolpath is simulated for the cutting force prediction of 5-axis ball-end milling. Lead angles vary between 17 and 66°, and tilt angles vary between 32 and 19°. The simulated toolpath is shown in Fig. 3.50. Dimensions of the blank workpiece and the workpiece coordinate frame are also illustrated in Fig. 3.50.

**Fig. 3.43** Case #3, simulated and experimental Fx forces

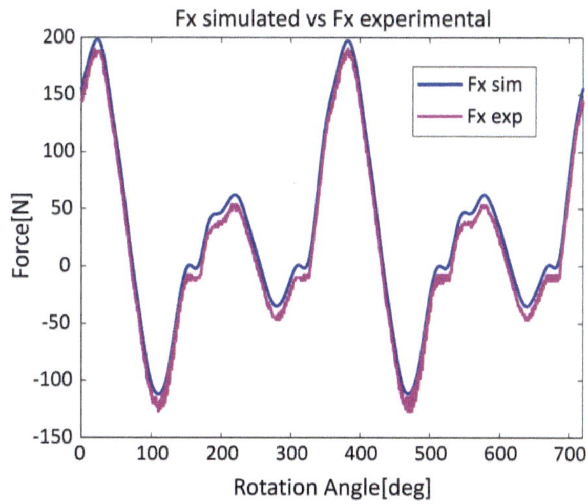


**Fig. 3.44** Case #3, simulated and experimental Fy forces

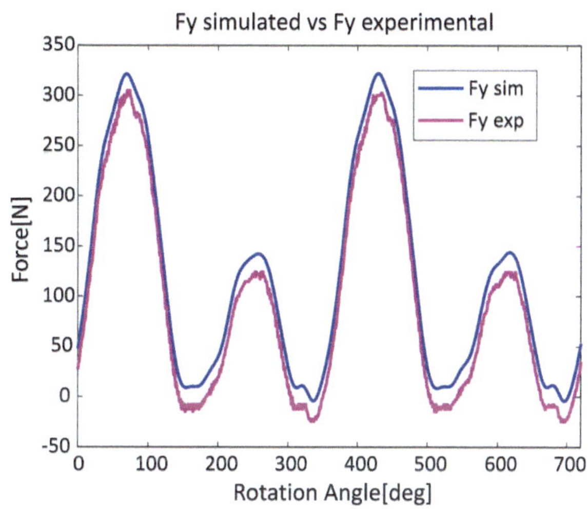


Cutting parameters such as feedrate and spindle speed are selected to be the same as the first two validation tests. Sandvik two fluted ball-end mill with a diameter of 12 mm, nominal helix angle of 30° and projection length of 37 mm is used as the cutting tool and Al7075 as workpiece material. Depths of cuts during the toolpath vary approximately between 0 and 5 mm along tool axis.

Figure 3.51 shows the comparison of the resultant cutting forces for simulated and measured cases. As it is demonstrated in Fig. 3.51 simulated and experimental cutting forces match quite well. In general, the difference between simulation and the experimental force amplitudes is below 20% which can be considered as

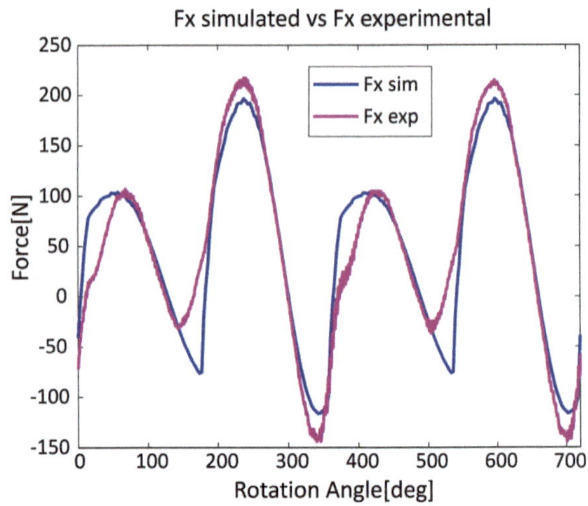


**Fig. 3.45** Case #4, simulated and experimental Fx forces

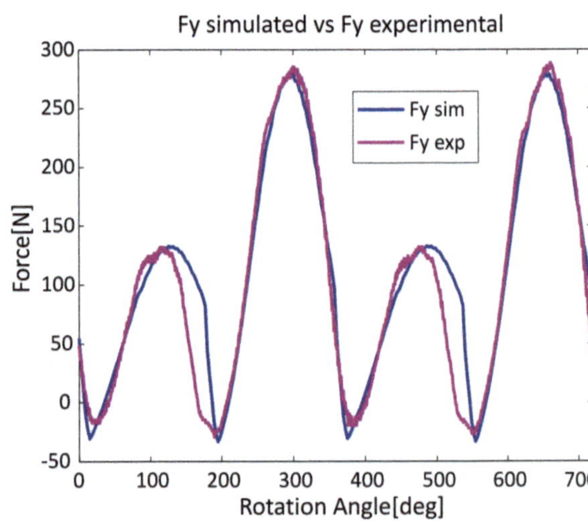


**Fig. 3.46** Case #4, simulated and experimental Fy forces

acceptable for simultaneous 5-axis milling process simulations. Figure 3.52 represents the envelope of the cutting forces for this case.

The discrepancy in cutting force predictions can be attributed to use of a table-top dynamometer for this test. While machining the workpiece on the machine tool the workpiece has a contribution on the cutting forces due to its weight, furthermore this force component is not constant and changes relative to rotation of the rotary axes of the machine tool. Experimental setup of impeller machining test is shown in Fig. 3.53.

Another reason for the discrepancy in the cutting forces is due to induced cutting force component by the cutting torque. Workpiece length is adjusted to be

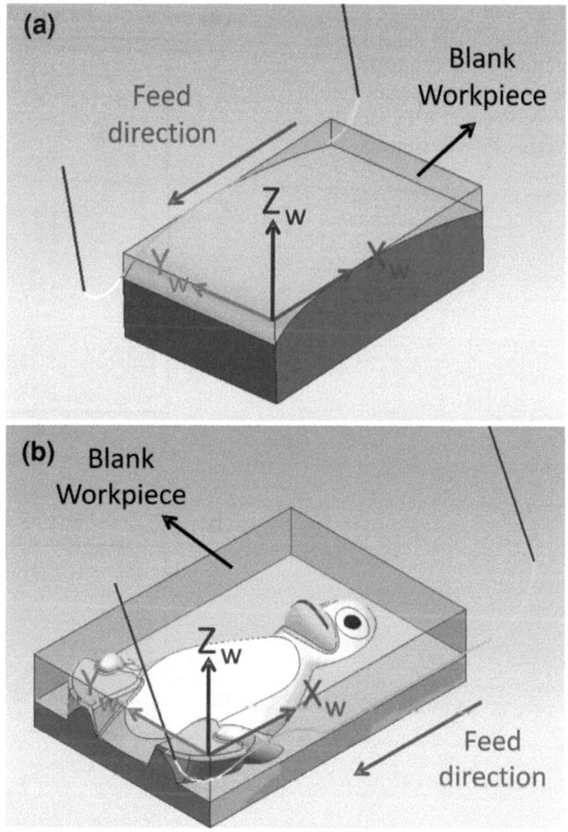


**Fig. 3.47** **a** Airfoil geometry. **b** Penguin free-form surface toolpaths

approximately 150 mm in order to avoid the risk of collisions; therefore this induces the forces due to cutting torque.

## 3.13 Validation Experiments Using Rotating Coordinate Frame

For the validation of the cutting force model in rotating coordinate frame, an impeller roughing toolpath pass similar to path shown in Fig. 3.50 is simulated and tested. Details of the toolpath are illustrated in Fig. 3.54.

In the test Al7075 workpiece material and a 6 mm diameter carbide cutting tool from Sandvik Plura series are used. Simulated toolpath is generated in NX6 CAM software. In the toolpath, impeller hub surface is used as the drive surface and the orientation of the tool axis is set to the normals of the drive surface (normal to drive). Consequently, a simultaneous 5-axis toolpath is obtained.

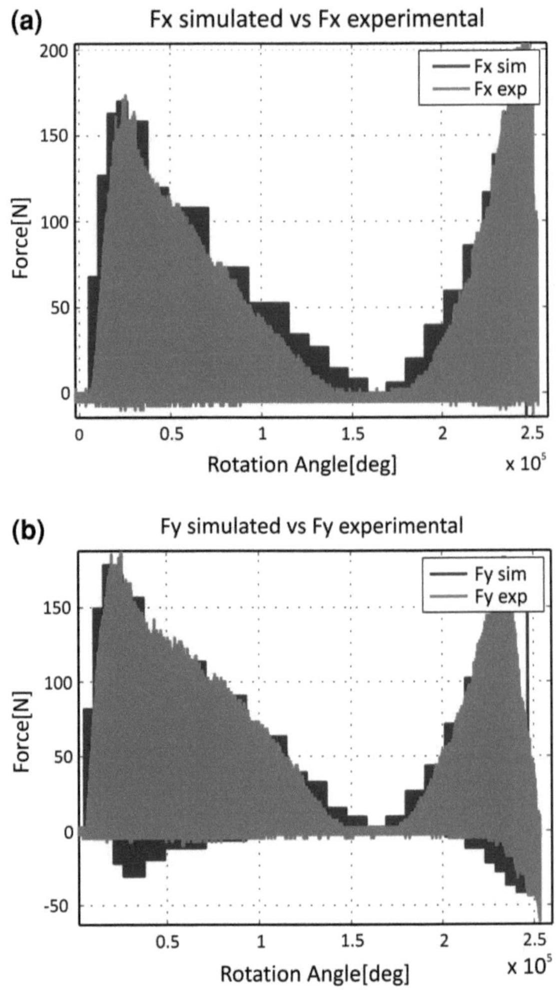


**Fig. 3.48** Airfoil geometry simulation and experimental cutting force comparison. **a** Fx simulated versus Fx experimental, **b** Fy simulated versus Fy experimental

Spindle speed and the feedrate are selected as 5000 rpm and 250 mm/min, respectively. Axial depth of cut varies approximately between 0 and 1.1 mm. Validation test is performed on Mori Seiki NMV5000 DCG machine tool using the Tool Center Point (TCP) control. In simultaneous 5-axis machining, movement of the translational axes and the rotary axes must be synchronized in order to keep the relative feedrate between the tool and the workpiece constant. Thus, TCP control (G43.4) makes the machining feedrate constant at the tool center point. In TCP control machining feedrate is specified as the constant, programed feedrate and controller adjusts the feedrate automatically. For the simulated toolpath, in TCP control mode instantaneous feedrate change between 200 and 1,250 mm/min. Conventional method for the constant feedrate implementation Inverse Time Feed (G93) function could also be used.

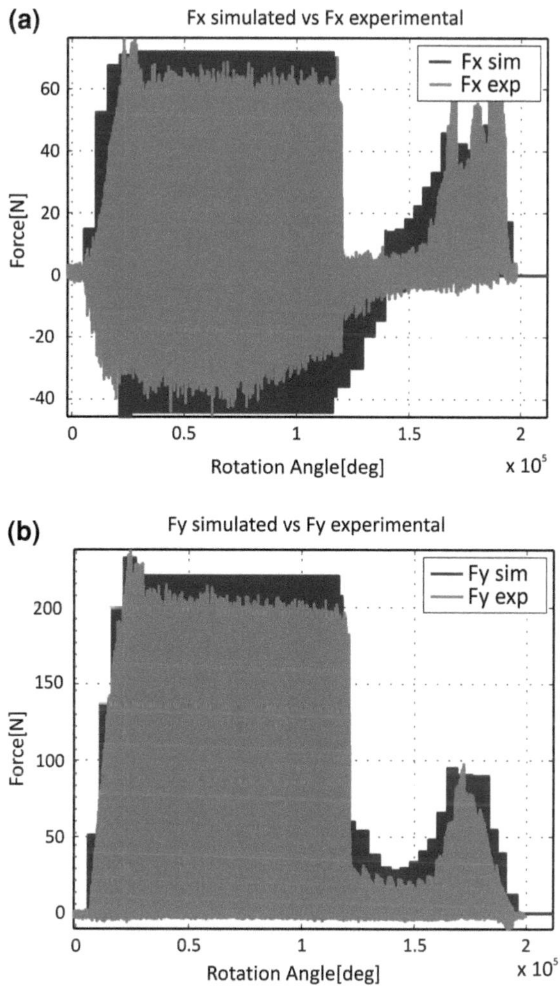


**Fig. 3.49** Penguin surface simulation and experimental cutting force comparison. **a** Fx simulated versus Fx experimental, **b** Fy simulated versus Fy experimental

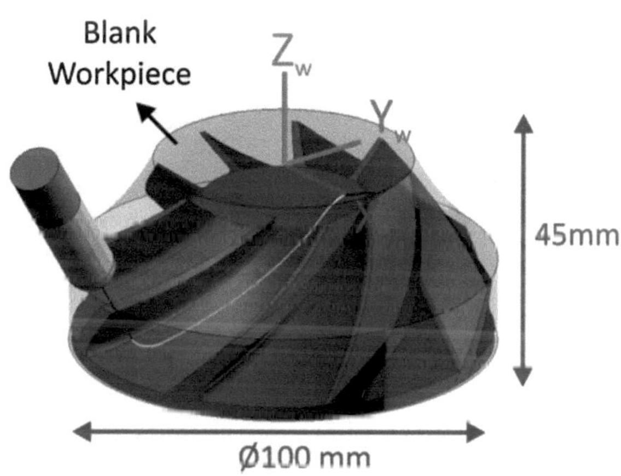


**Fig. 3.50** Simulated impeller roughing toolpath

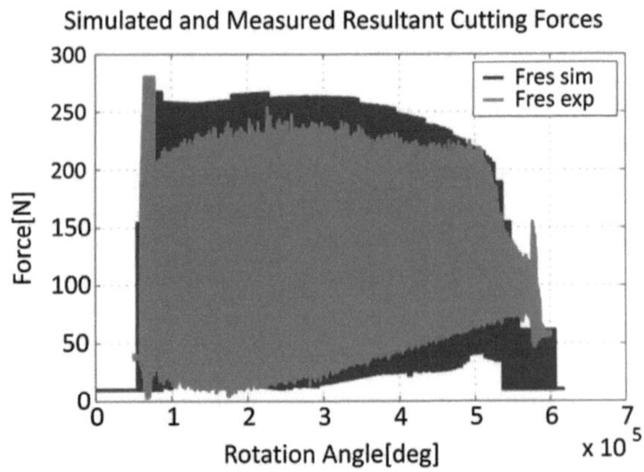


**Fig. 3.51** Impeller toolpath simulation and experimental resultant cutting force comparison

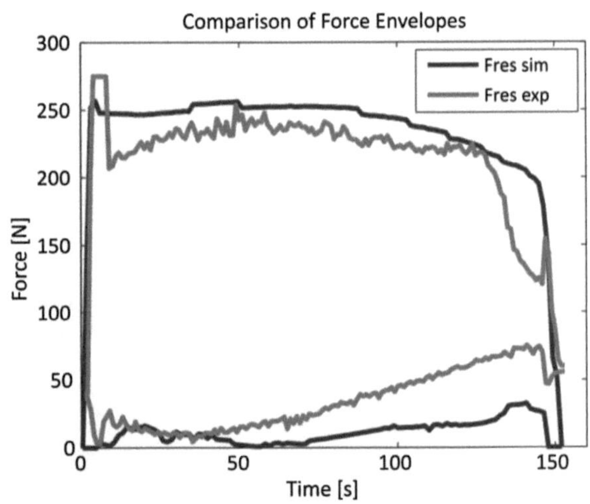


**Fig. 3.52** Comparison of force envelopes for impeller toolpath

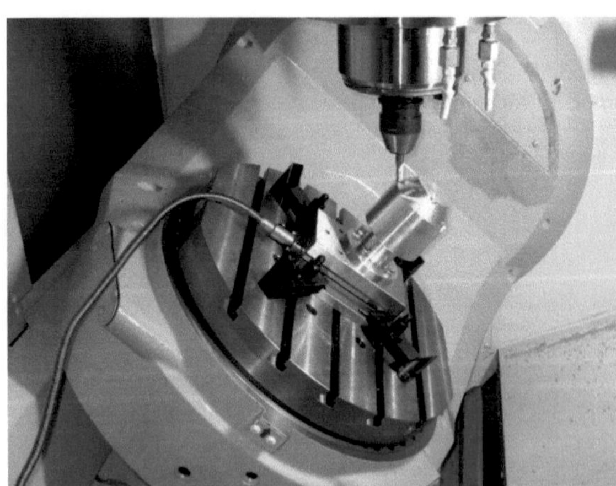

**Fig. 3.53** Experimental impeller machining test

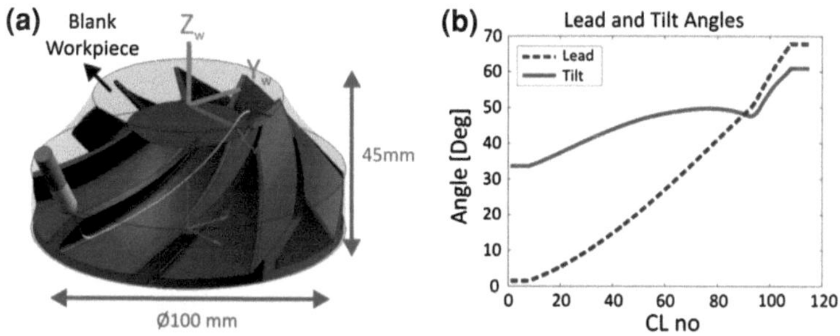


Fig. 3.54 a Simulated impeller roughing toolpath. b Lead and tilt angles for one-pass of impeller toolpath

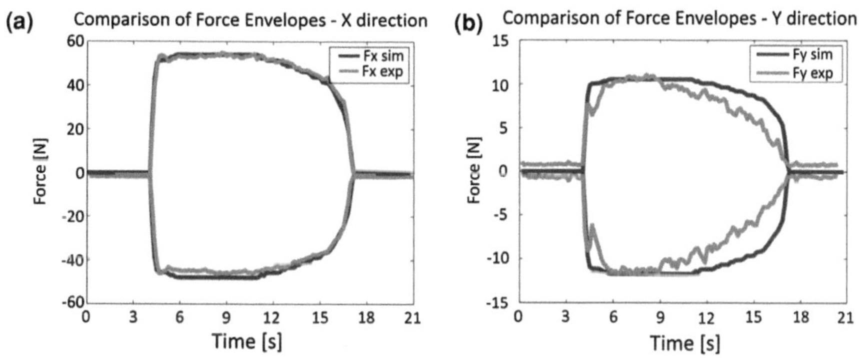


Fig. 3.55 Comparison of cutting force envelopes, a Comparison of force envelopes-*X* direction, b Comparison of force envelopes-*Y* direction

The results of the predicted cutting forces with the experimental data are shown from Figs. 3.55, 3.56 and 3.57. From figures, it can be concluded that, predicted cutting forces in *X* direction match well with the experimental data. In *Y* direction, there are slight differences in the predicted cutting forces with the experimental data; however, the agreement is still reasonably well.

Second validation test is performed for the full roughing toolpath of an impeller hub. Simulated toolpath is generated in NX7.5 CAM software using the Multi Blade Rough method. Generated toolpath consists of 1572 CL points employing zig-zag toolpath with lifts. Axial depth of cut varies approximately between 0 and 3 mm Details of the toolpath are shown in Fig. 3.58.

In the test Al7075 workpiece material and a 6 mm diameter carbide cutting tool are used. Spindle speed and the feedrate are selected as 5000 rpm and 500 mm/min, respectively. Cutting tests are performed in TCP control mode and instantaneous feedrate changes approximately between 500 and 6,600 mm/min. Forces are collected for every 4° during the machining operation. Output of the

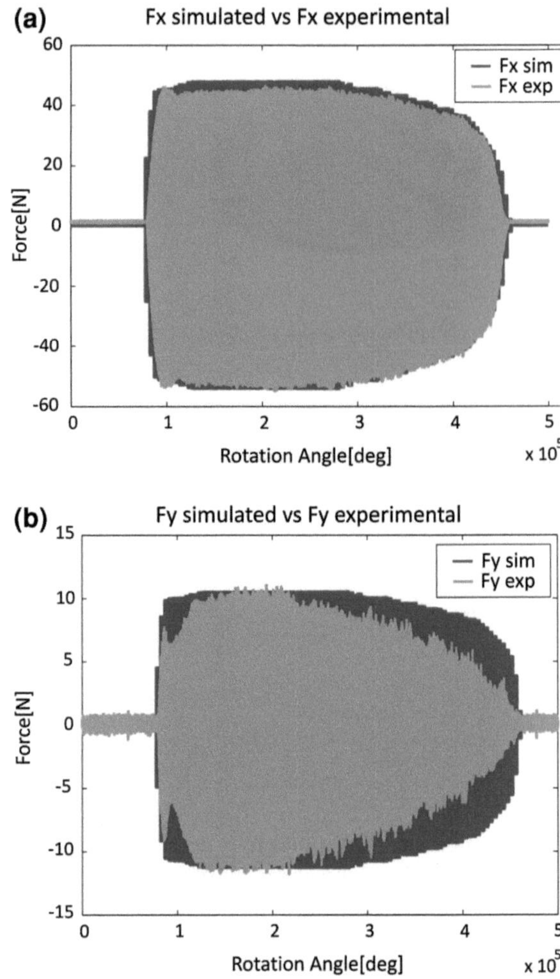


**Fig. 3.56** Predicted versus experimental data for the toolpath. **a** Fx simulated versus Fx experimental, **b** Fy simulated versus Fy experimental

simulated machined workpiece and the experimental machined workpiece are shown in Figs. 3.59 and 3.60, respectively.

Cutter-Workpiece engagements are extracted using the three-orthogonal dexelfield engagement model. In the simulation disk height is set to 0.1 mm and the element spacing between individual depth buffers is set to 0.05 mm. Computation of the engagement result took 56 min and 25 s on a Windows 7 64-bit, Core2Duo 3.16 GHz/8 GB ram desktop PC.

Results of the validation tests are presented through Figs. 3.61, 3.62, 3.63 and 3.64. Figures 3.61 and 3.62 show the comparison of the cutting forces for full toolpath and comparison of the cutting force close-ups are illustrated in Figs. 3.63 and 3.64.

Fig. 3.57 Close-ups of cutting forces for $X$ and $Y$ directions

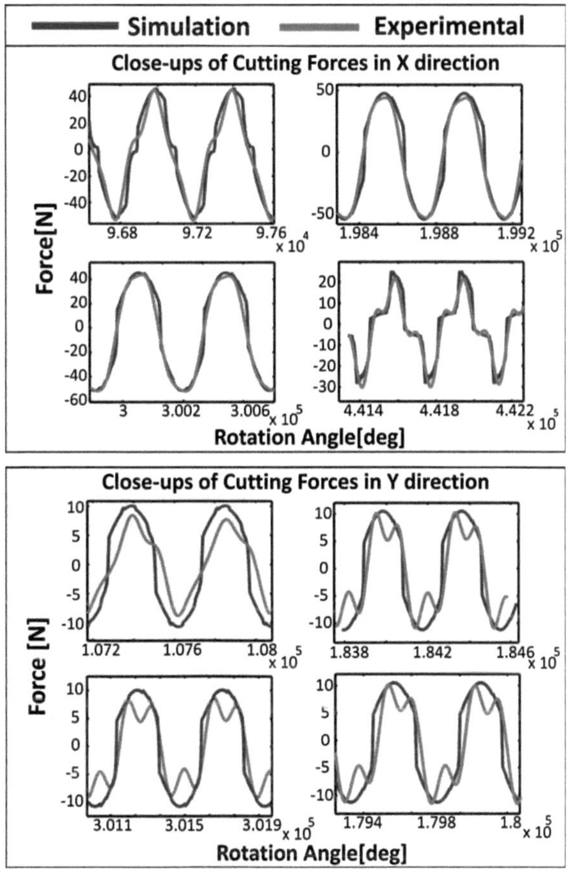


Close-ups are shown for six regions for better illustration of the simulated data with the measured data. According to the results of the validation tests, it can be stated that simulated and measured cutting forces match reasonably well although there are minor discrepancy in the predictions.

Trend of the predicted forces are in good agreement with the measured data. Discrepancy of the cutting forces is observed especially in the low axial immersion regions; however, this is likely due to low resolution at smaller depth of cuts. The change in chip thickness is high due to the geometry of the ball-end cutter. Hence, increasing the simulation resolution from 100 microns to higher resolutions may increase the simulation accuracy which also causes an increase in the computation time. Another reason for the discrepancies may be attributed to high feedrates during the toolpath that exhibits an aggressive machining. This may also degrade the prediction accuracy since in high feedrates dynamic effects may alter the static forces and proposed model is a static force model.

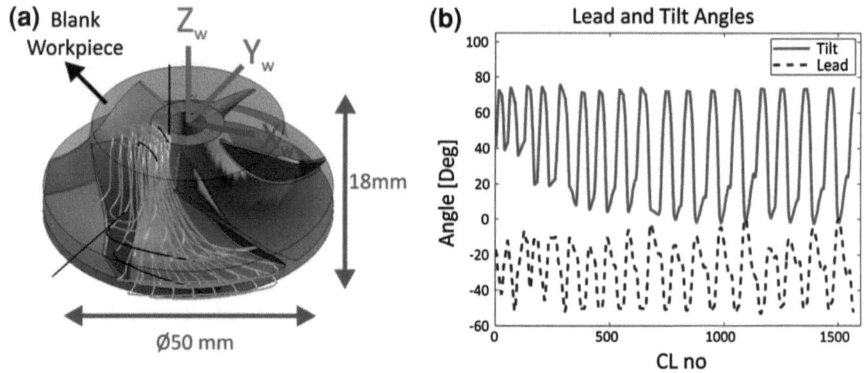


**Fig. 3.58** **a** Simulated impeller roughing (Hub Roughing) toolpath, **b** Variation of the lead and tilt angles along the toolpath

**Fig. 3.59** Simulated machined workpiece using three-orthogonal dexelfield engagement model

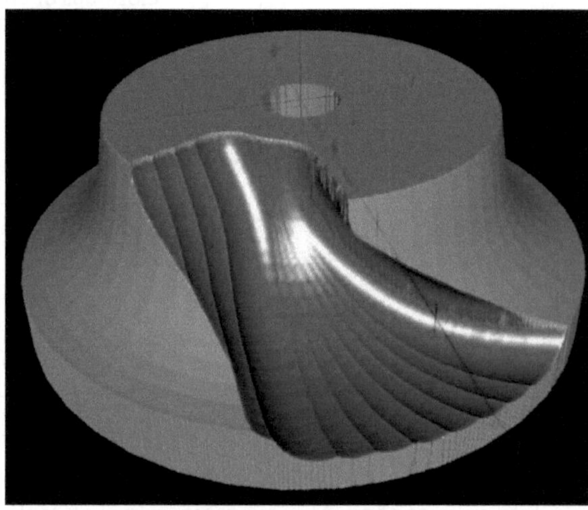

**Fig. 3.60** Machined workpiece

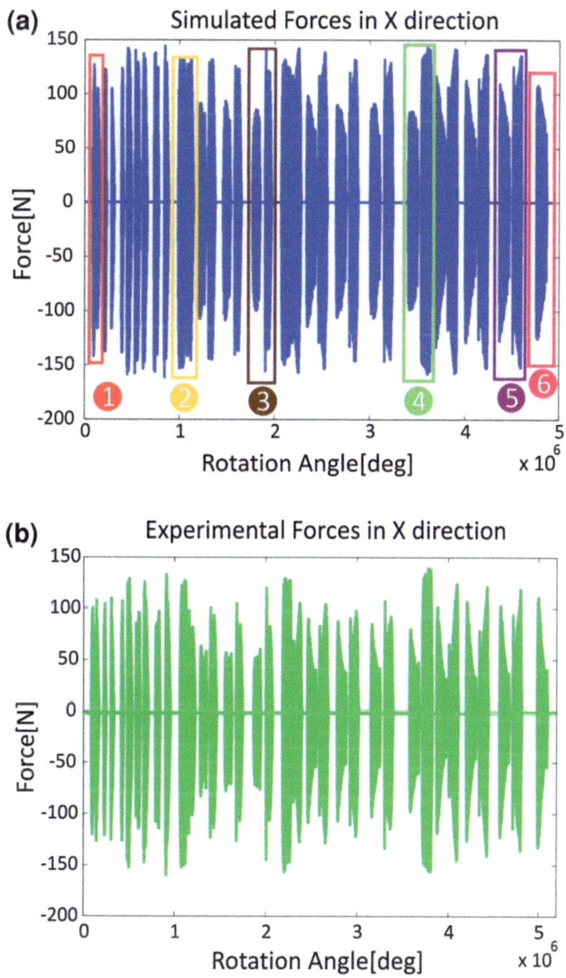


**Fig. 3.61** Comparison of simulated (**a**) and experimental (**b**) cutting forces for the impeller roughing toolpath in *X* direction

## 3.14 Feedrate Scheduling Validation Test

For the validation of offered feedrate scheduling model in Sect. 3.4, an impeller roughing toolpath pass which is shown in Fig. 3.65 is simulated and tested. In this figure, one roughing toolpath is represented as an example.

In the test, Al7075 workpiece material and a 6 mm diameter carbide cutting tool from are used. Simulated toolpath is generated in NX7.5 CAM software. In the toolpath, impeller hub surface is used as the drive surface and the orientation of the tool axis is set to be normal to the drive surface (normal to drive).

Spindle speed and the feedrate are selected as 5000 rpm and 750 mm/min (for constant feedrate case), respectively. Axial depth of cut varies approximately between 0 and 1.3 mm.

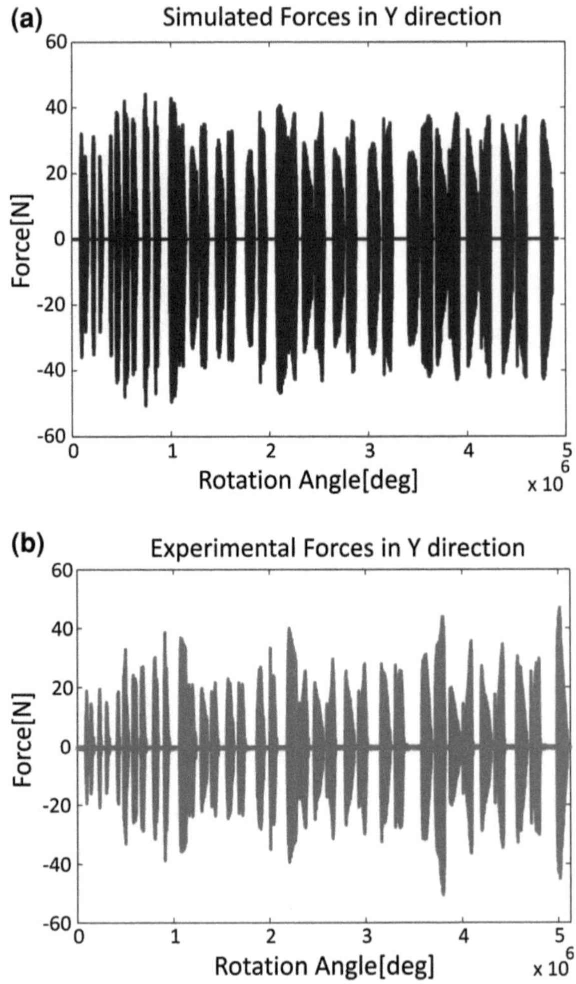


**Fig. 3.62** Comparison of simulated (**a**) and experimental (**b**) cutting forces for the impeller roughing toolpath in *Y* direction

The results of the predicted cutting forces with the experimental data are shown in Figs. 3.66 and 3.67.

As it can be found from Figs. 3.66 and 3.67, the cutting forces are not constant through the toolpath and by decreasing depth of cut and engagement angle at the end of the toolpath the cutting forces decrease. Therefore, by scheduling the feedrate value to increase during the forward movement of tool, it is possible to keep the cutting forces constant. In this case the threshold resultant cutting force in Eq. 3.40 was set to 190 N.

Figure 3.68 illustrates the result of feedrate scheduling validation test. In this figure the red line represents measured cutting forces during running scheduled feedrate values and the blue line represents measured cutting forces during constant feedrate, which is equal to 750 mm/min. According to Fig. 3.68 the cycling

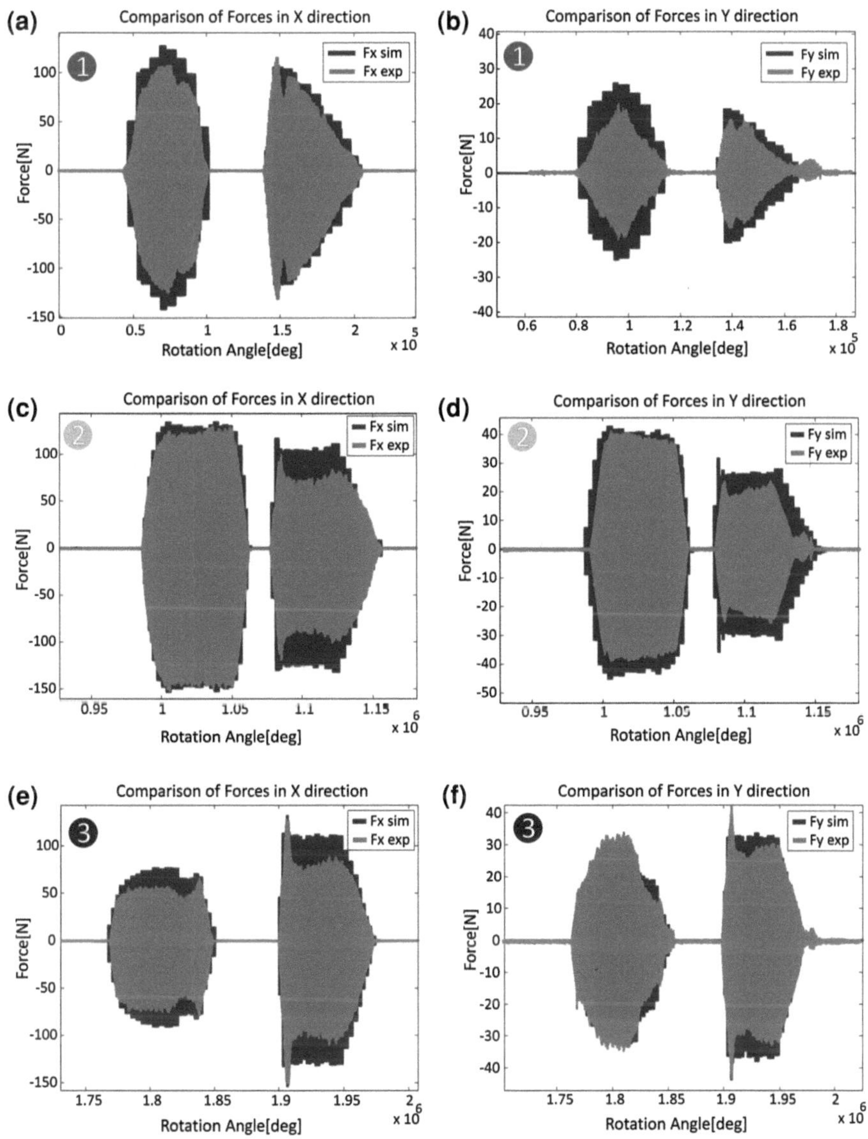


**Fig. 3.63** Comparison of cutting forces for regions 1–3. **a**, **c**, **e** Comparison of forces in *X* direction. **b**, **d**, **f** Comparison of forces in *Y* direction

time for machining of one path using constant feedrate is 4.561 s and using scheduled feedrate is 2.787 s. For manufacturing of a complete impeller there are 107520 CL point which take 202 min. Using the presented feedrate scheduling, cycling time was decreased 79 min, which is 39% of cycling time of the traditional constant feedrate machining.

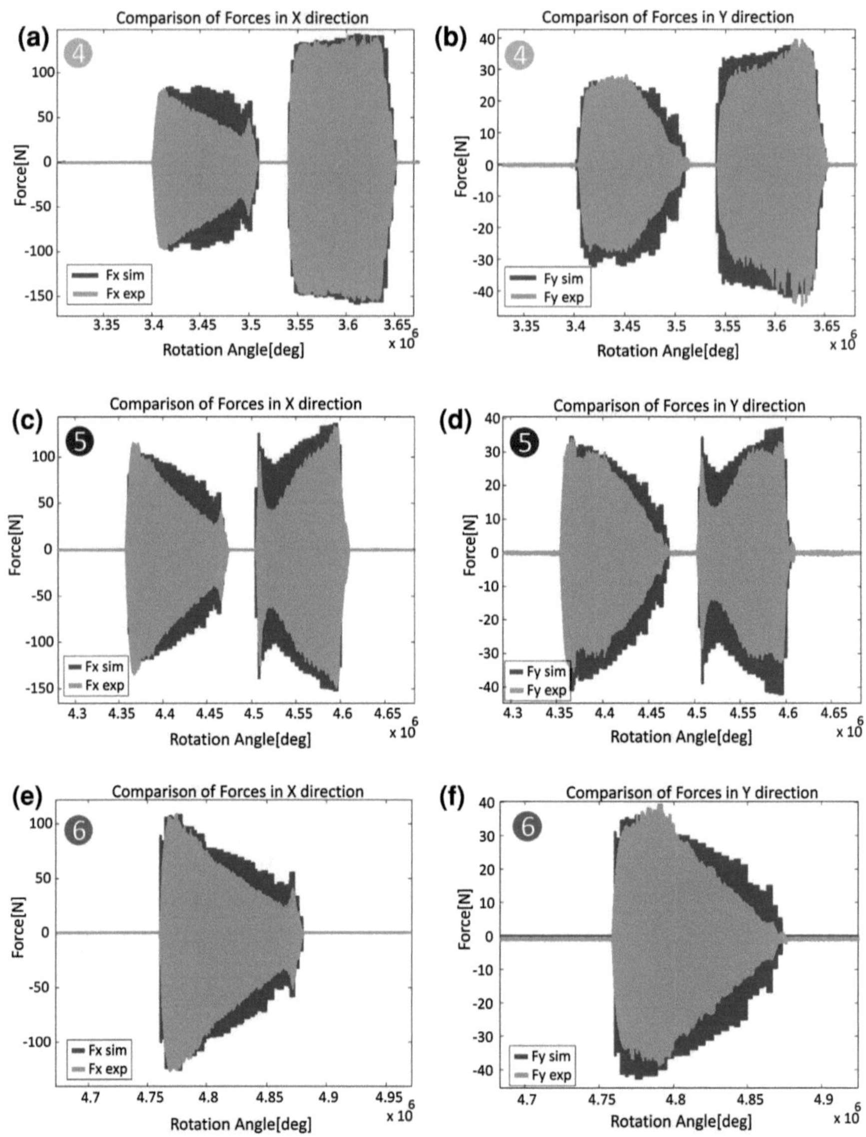


**Fig. 3.64** Comparison of cutting forces for regions 4–6. **a**, **c**, **e** Comparison of forces in *X* direction. **b**, **d**, **f** Comparison of forces in *Y* direction

Figure 3.69 illustrates the change of feedrate values in each cutter locations to keep the resultant cutting force constant. It should be mentioned that because of safety, in the feedrate scheduling algorithm the feedrate cannot bigger than 2,500 (mm/min).

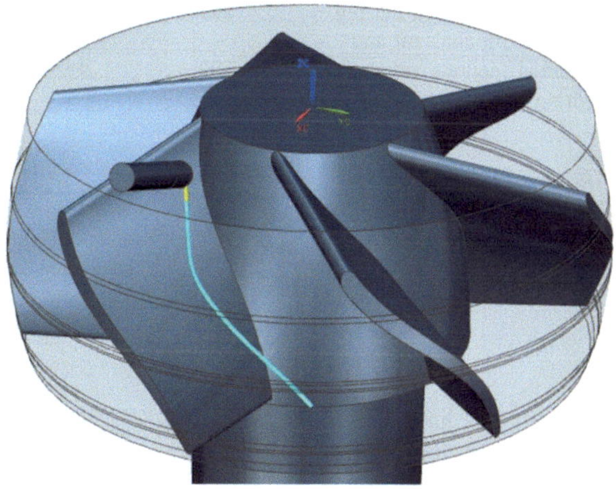

**Fig. 3.65** Simulated impeller roughing toolpath

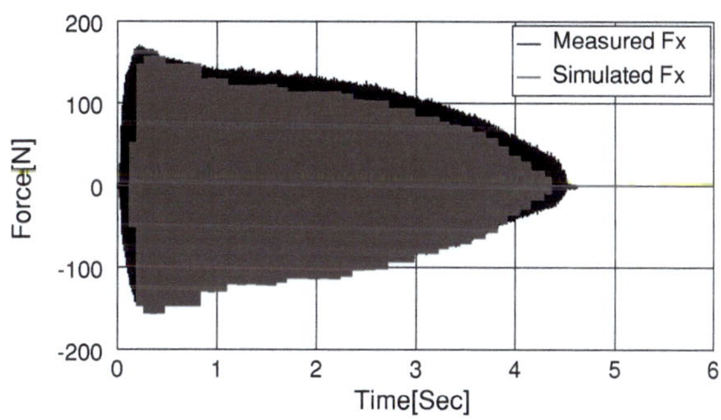


**Fig. 3.66** Measured and simulated cutting forces in *X* direction in the case of constant feedrate (750 mm/min)

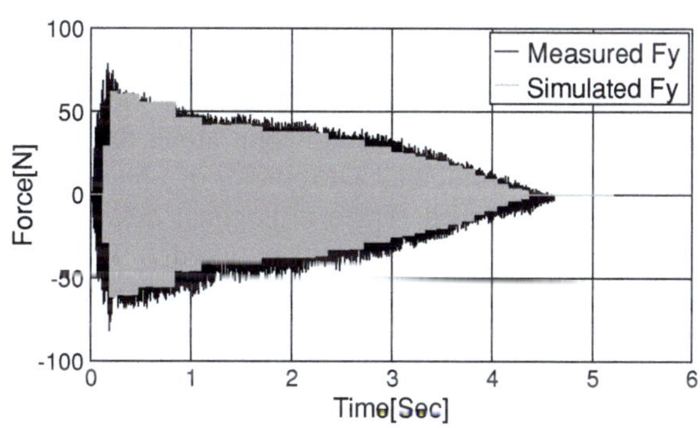


**Fig. 3.67** Measured and simulated cutting forces in *Y* direction in the case of constant feedrate (750 mm/min)

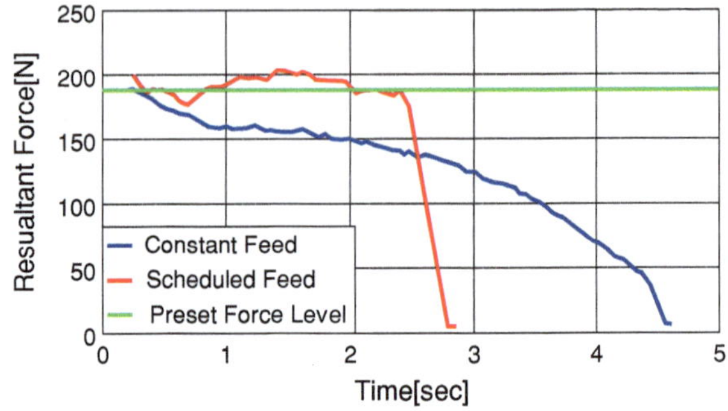


**Fig. 3.68** Measured cutting forces during constant and scheduled feedrate values for feedrate scheduling

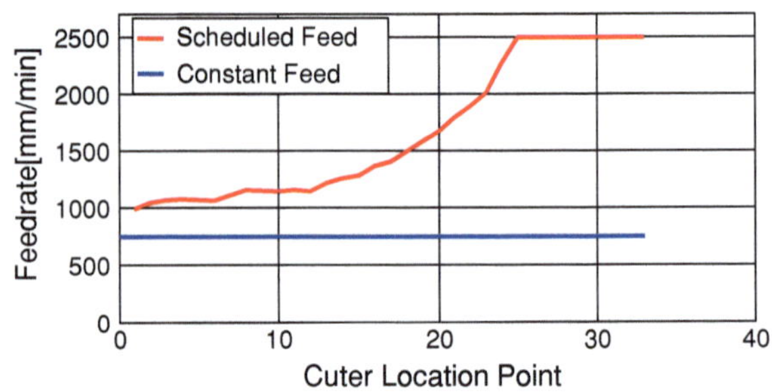


**Fig. 3.69** Scheduled feedrate values for each cutter location ($F_{\text{lim}} = 190$ N)

The variation of feedrate for each NC block is shown in Figs. 3.70 and 3.71.

The cutting force components for constant and scheduled feedrate are compared in Figs. 3.71 and 3.72.

## 3.15 Conclusion

This chapter presents a virtual machining system based on force modeling technique for 5-axis free-form surfaces.

Cutting forces in machining is determined by extracting the Cutter-Workpiece Engagement (CWE) from the in-process workpiece in the form of start and exit angles as a function of axial height along the tool axis. A novel discrete method, called Three-Orthogonal Dexelfield, of obtaining CWE maps for 5-axis ball-end milling is developed. Three-orthogonal dexelfield uses the depth buffer in three orthogonal directions. In other words, three-orthogonal dexelfield approach utilizes *Z*-map, *Y*-map and *X*-map simultaneously for improved accuracy. As a result, computation time is relatively long compared to conventional *Z*-map method. The simulation results of the CWE model showed that developed model can extract the CWE maps accurately but it takes longer time for simulation than the solid-modeler-based CWE method.

**Fig. 3.70** A generated NC program using feedrate scheduling method

```
G01 X-209.406 Y.342 Z132.313 F1100.
X-208.702 Y.177 Z132.013 F1100.
X-207.946 Y0.0 Z131.909 F1100.
X-207.387 Z131.414 B73.397 C27.185 F1100.
X-206.8 Z130.839 B73.556 C26.994 F1100.
X-206.138 Z130.378 B73.684 C26.802 F1100.8
X-205.508 Z129.867 B73.826 C26.61 F791.0
X-204.893 Z129.334 B73.975 C26.418 F783.5
X-204.252 Z128.839 B74.114 C26.226 F792.3
X-203.687 Z128.228 B74.285 C26.034 F843.2
X-203.06 Z127.711 B74.431 C25.842 F847.4
X-202.453 Z127.161 B74.586 C25.65 F843.5
X-201.852 Z126.603 B74.745 C25.459 F844.6
X-201.192 Z126.135 B74.878 C25.267 F844.5
X-200.615 Z125.537 B75.048 C25.075 F970.6
X-200.389 Z125.367 B75.097 C25.004 F965.8
X-199.6 Z124.676 B75.295 C24.854 F976.
X-198.732 Z124. B75.489 C24.882 F996.5
X-197.864 Z123.324 B75.685 C25.091 F1050.8
X-197.008 Z122.734 B75.856 C25.446 F1075.6
X-196.256 Z121.982 B76.075 C25.998 F1078.8
X-195.517 Z121.423 B76.238 C26.634 F1075.3
X-194.918 Z120.642 B76.467 C27.333 F1069.
X-194.483 Z119.6 B76.775 C28.088 F1117.2
X-194.195 Z118.317 B77.154 C28.873 F1158.1
X-194.187 Z116.563 B77.671 C29.712 F1157.
X-194.234 Z115.147 B78.089 C30.354 F1151.1
X-194.306 Z113.212 B78.66 C31.264 F1161.9
X-194.414 Z111.182 B79.258 C32.231 F1149.6
X-194.452 Z109.236 B79.832 C33.216 F1225.1
X-194.502 Z107.475 B80.35 C34.136 F1262.6
X-194.465 Z105.597 B80.904 C35.192 F1282.5
X-194.445 Z103.78 B81.439 C36.227 F1367.1
X-194.389 Z101.998 B81.964 C37.29 F1408.4
X-194.351 Z100.282 B82.47 C38.283 F1503.1
X-194.258 Z98.64 B82.954 C39.292 F1589.9
X-194.202 Z97.167 B83.388 C40.148 F1676.1
X-194.12 Z95.725 B83.814 C41.01 F1792.6
X-194.023 Z94.018 B84.318 C42.009 F1897.8
X-193.924 Z92.144 B84.872 C43.059 F1825.6
X-193.806 Z90.276 B85.424 C44.091 F2014.4
X-193.667 Z88.848 B85.846 C44.895 F2270.
X-193.484 Z86.904 B86.421 C45.94 F2500.
X-193.324 Z85.171 B86.935 C46.828 F2500.
X-193.033 Z83.391 B87.463 C47.803 F2500.
X-192.654 Z83.833 B87.331 C47.956 F2500.
X-192.123 Y.568 Z83.928 F2500.
X-191.628 Y1.099 Z84.222 F2500.
X-185.687 Y7.46 Z89.145 F2500.
G00 Z91.918
```

Mechanistic cutting coefficient calibration method is implemented for different cutting speeds, feedrates and tool geometry on Al7039 and Al7075 workpiece materials. Besides, cutting force calibration methodology in rotating coordinate frame is developed and implemented for the identification of cutting coefficients.

A cutting force prediction model for 5-axis ball-end milling is developed. Cutting force modeling is performed in the fixed coordinate frame (for table type dynamometer) and in the rotating coordinate frame (rotating coordinate dynamometer). Several validation tests for complex free-form surfaces are presented in the study. These validation tests are performed on Al7039 and Al7075 workpiece

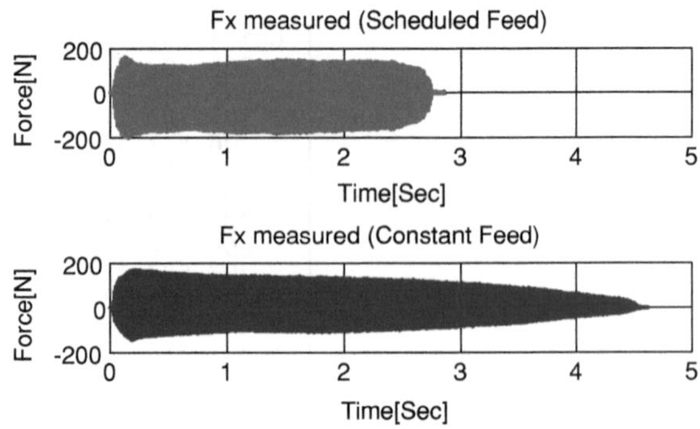


**Fig. 3.71** Compression of forces and machining time in the cases of constant feedrate and scheduled feedrate for Fx

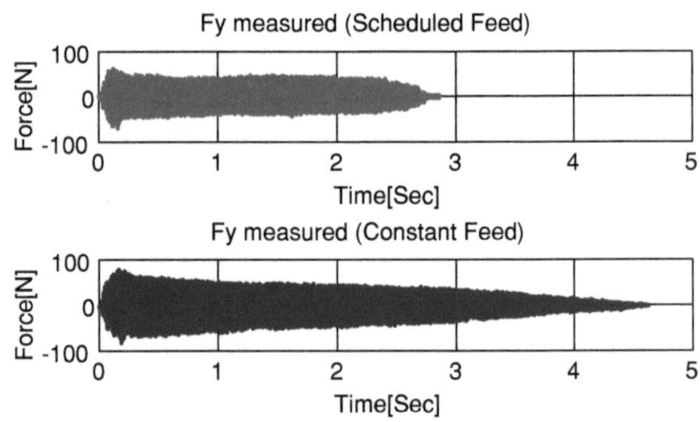


**Fig. 3.72** Compression of forces and machining time in the cases of constant feedrate and scheduled feedrate for Fy

materials with carbide cutting tools. These validation tests demonstrate that presented model is computationally efficient and force predictions are in good agreement with the measured data.

An enhanced Force-model-based Feedrate Scheduling (FFS) technique in 5-axis machining of parts with complex free-form surfaces is introduced that can be used in new generation of CAM software to schedule the feedrate in order to decrease the cycle time and increase the productivity.

Next generation CAM technologies such as force-based feedrate scheduling and toolpath generation demand the use of variable feedrate implementation along the toolpath. For this reason, a virtual machine simulation model, which is capable of simulating machine tool movements from the NC code, is presented.

**Acknowledgments** The authors acknowledge the Machine Tool Technologies Research Foundation (MTTRF), the Mori Seiki Co. and the DP Technology Corp for the Mori Seiki NMV 5000DCG CNC Machining Center and Esprit CAM software supports. The authors also acknowledge Sandvik Coromant Company for providing cutting tools for the research.

## References

1. Kim GM, Cho PJ, Chu CN (2000) Cutting force prediction of sculptured surface ball-end milling using Z-map. Int J Mach Tools Manuf 40:277–291
2. http://www.moduleworks.com/
3. Wang WP, Wang KK (1986) Geometric modeling for swept volume of moving solids. IEEE Comput Graph Appl 6:8–17
4. Du S, Surmann T, Webber O, Weinert K (2005) Formulating swept profiles for five-axis tool motions. Int J Mach Tools Manuf 45:849–861
5. She CH, Huang ZT (2008) Postprocessor development of a five-axis machine tool with nutating head and table configuration. Int J Adv Manuf Technol 38:728–740
6. Tunc L, Budak L (2009) Extraction of 5-axis milling conditions from CAM data for process simulation. Int J Adv Manuf Technol 43:538–550
7. Lazoglu I, Boz Y, Erdim H (2011) Five-axis milling mechanics for complex free from machining. CIRP Ann 60(1):117–120
8. Erdim H, Lazoglu I, Kaymakci M (2007) Free-form surface machining and comparing feedrate scheduling strategies. Mach Sci Technol Int J 11(1):117–133 (Taylor & Francis Group)
9. Lazoglu I (2003) Sculpture surface machining: a generalized model of ball-end milling force system. Int J Mach Tools Manuf 43:453–462
10. Lazoglu I, Liang SY (2000) Modeling of ball-end milling forces with cutter axis inclination. ASME J Manuf Sci Eng 122:3–11
11. Ozturk B, Lazoglu I (2006) Machining of free-form surfaces—Part I: analytical chip load. Int J Mach Tools Manuf 46:728–735
12. Ozturk B, Lazoglu I, Erdim H (2006) Machining of free-form surfaces—Part II: calibration and forces. Int J Mach Tools Manuf 46:736–746
13. Erdim H, Lazoglu I, Ozturk B (2006) Feedrate scheduling strategies for free-form surfaces. Int J Mach Tools Manuf 46:747–757
14. Budak E, Lazoglu I, Guzel BU (2004) Improving cycle time in sculptured surface machining sculptured surface machining through force modeling. CIRP Ann 53(1):103–106
15. Ferry WB, Altintas Y (2008) Virtual five-axis flank milling of jet engine impellers—Part I: mechanics of five-axis flank milling. J Manuf Sci Eng 130:011005-1–011005-11
16. Ferry WB, Altintas Y (2008) Virtual five-axis flank milling of jet engine impellers Part II: feed rate optimization of five-axis flank milling. J Manuf Sci Eng 130:011013-1–011013-13

# Chapter 4
# Milling Tool-Paths Generation in Adequacy with Machining Equipment Capabilities and Behavior

**Matthieu Rauch and Jean-Yves Hascoët**

This chapter discusses the challenges associated with tool-paths generation and control for complex surfaces machining. The increase of the technical and economic requirements associated with a diversification of the machining equipment makes it compulsory to implement machine-specific milling strategies to be efficient. Major technological evolutions within this context are identified. Then, advanced CNC programing methods are presented and discussed. Based upon innovative concepts and a new vision of the tool-path generation process, they enable to select the most suitable parameterization. In particular, virtual manufacturing can be of great interest.

## 4.1 Introduction

With the rise of high speed multi-axis milling, complex sculptured surfaces machining has become a very common practice. In parallel, the requirements related to industrial products have dramatically increased over the last decades especially concerning time-market and customization. To meet this request, the manufacturing equipment is now segmented in terms of capabilities.

In the same way, the use of these highly specialized machine tools calls for specific the CNC programing. Tool-path generation and control has become a

M. Rauch · J.-Y. Hascoët (✉)
Ecole Centrale Nantes Institut de Recherche en Communications et Cybernetique de Nantes (IRCCyN) UMR CNRS 6597,
1 rue de la Noe, BP92101 44321
Nantes Cedex 03, France
e-mail: jean-yves.hascoet@irccyn.ec-nantes.fr

M. Rauch
e-mail: matthieu.rauch@irccyn.ec-nantes.fr

J. P. Davim (ed.), *Machining of Complex Sculptured Surfaces*,
DOI: 10.1007/978-1-4471-2356-9_4, 

central issue. The universal CAM/CNC parameterization which provides optimal results for any part to produce on any machine tool does not exist anymore. In particular, it is necessary to generate milling tool-paths that are in adequacy with the machining equipment capabilities and its behavior to meet the challenges of today's manufacturing industry.

In a first part, this chapter focuses on the challenges related to multi-axis milling strategies. Emerging tool-paths geometries, whose efficiency has dramatically increased with technological progresses, are introduced as well. It is necessary to identify the characteristics of modern multi-axis equipment to propose efficient tool-path parameterizations. These aspects are discussed in the second part of the chapter. Then, in the last part, a synthesis of some advanced CNC programing methods developed to meet challenges of complex surfaces machining by using modern manufacturing equipment is introduced. Most of the proposed development base upon simulation approaches, as they are efficient to test several configurations and compute optimal parameterizations at low cost.

## 4.2 Machining Strategies and Tool-Paths Geometries

The machining of complex scultpured surfaces frequently requires 5-axis milling strategies. Compared to classical 3-axis machining, the two additional degrees of freedom provided by the rotary axes compound the issue. Much of the research work has been carried out to define the best tool position and tool orientation.

### *4.2.1 Multi-Axis Tool-Paths*

Most of the difficulties associated with multi-axis manufacturing are concentrated in the computation of explicit 5-axis strategies. The varying orientations of a tool enable machining of a large group of shapes independently from the shape of the tool: different cutter geometries may be used to cut the same geometry. The reason for this is that so-called cutter contact point (CC) can move along the cutting edge of the tool according to the tool orientation. There are two important parameters that define a tool's orientation. The lead angle is an angle measured in the plane that contains the local vector tangent to the tool-path and the local vector normal to the surface being machined. The title angle is defined as in the plane with the local vector tangent to the tool-path as its normal.

Most the of actual issues and related research works associated with 5-axis machining strategies have been reviewed by Rauch and Xu in [1]. There are basically two types of 5-axis tool-paths—end milling and flank milling—depending on whether its end or flank of the cutter is used for machining. Both types have their own preferred application fields. However, end milling has attracted more research efforts because it offers more industrial applications.

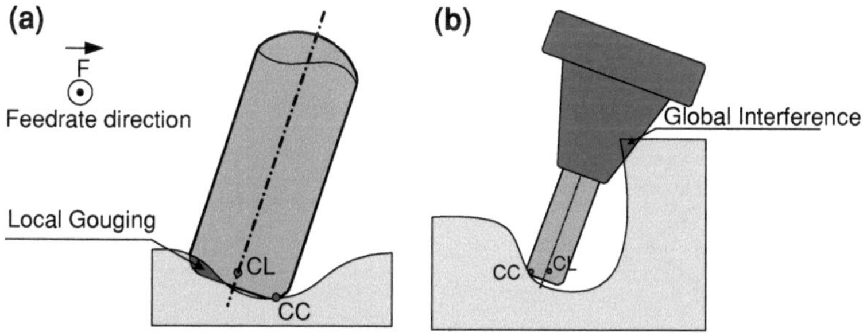


**Fig. 4.1** Collisions with a bull end cutter: **a** local gouging, **b** global interference

#### 4.2.1.1 End Milling Tool-Paths

For end milling strategies, CC is located near the tool tip. Depending on the tool shape, tool orientation can be modified within a certain range without modifying the machine surface. Three main issues are associated with end milling tool-paths: cutting tool positioning, tool orientation and tool-path generation.

Tool positioning is to define the exact position of CC. The major issue is to avoid local gouging. This phenomenon appears when a portion of the cutting edge of the cutter extends below the surface by more than the allowable surface tolerance and creates an overcut zone (Fig. 4.1).

Several tool positioning methods have been proposed. Computation is carried out either directly or in two stages, by putting the tool in the vicinity of final position before adjusting it to avoid gouges [2, 3].

The second issue is to define tool orientation. Most of the work is linked to global collision avoidance. Indeed, global interference avoidance is one of the key issues in 5-axis tool-path generation (Fig. 4.1). In practice, most of the proposed methods [4] are based on trial and error to find the collision interferences and it usually takes a long time to determine a final solution.

Some recently developed approaches take into account the kinematic and dynamic aspects of process and machine tool to set cutting tool orientation [5]. An original and efficient approach was proposed by Hascoet et al. in [6]. The purpose is to compute collision avoidance into the articular workspace of the machine tool. Another method introduced the concept of "Machining Surface", defined as a surface model for tool-path generation [7]. Two surfaces are considered: the guiding surface, which ensures tool positioning independently from the selected milling strategy, and the orientation surface, which controls the tool orientation (Fig. 4.2). From this surface, methods based on both topological and kinematical properties can be implemented [8].

Last issue is tool-path generation, which means the determination of six parameters for each controllable position of the cutter, three cutter location parameters and three cutter orientation parameters. Over the years, several tool-paths generation techniques have been proposed. The iso-parametric approach has been the most

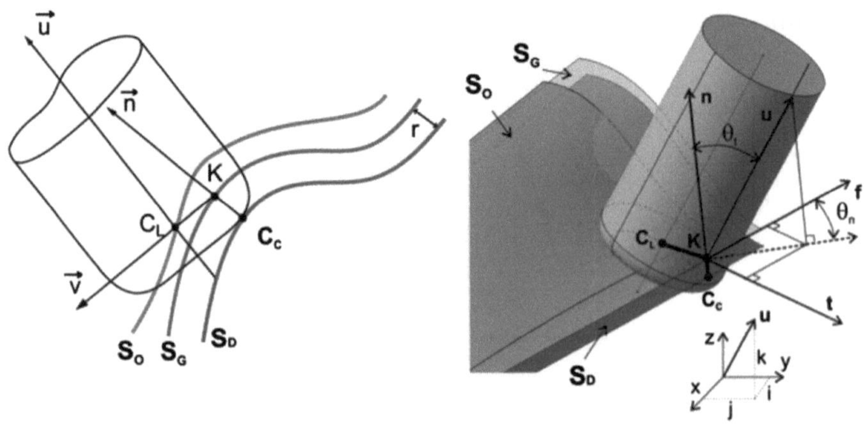


**Fig. 4.2** The machining surface concept (*source* [8])

**Fig. 4.3** Example of flank milling tool-path (*source* [12])

commonly employed method [9], thanks to the straightforwardness and simple computation associated. Optimal machined surface quality may not be achieved as scallop height cannot be precisely predicted. Optimization methods, such as the popular Space Filling Curve (SFC) approach [10], have been subsequently proposed.

#### 4.2.1.2 Flank Milling Tool-Paths

Flank milling operations are usually employed for machining ruled surfaces, the cutting tool following the surface ruling of the CAD model. Tool orientation consequently strongly affects part accuracy as the cutter flank generates the machined surface. Flank milling strategies are widely employed for machining complex sculptured surfaces such as impellers and turbine blades [11]. Removal rates are greater compared to end milling strategies but several constraints are assocociated with these tool-paths.

If the CAD surface is a developable ruled surface, the cutter can be positioned tangent to it over the complete rule and there is no major issue. If the CAD surface is not developable, the milling process often infers deviation errors. Hence, the definition of a tool-path that minimizes these errors becomes a central issue.

Most of the flank milling tool-path generations approaches rely on a first tool positioning at determined contact points followed by optimizations to reduce geometrical errors [13, 14].

Tool-path smoothness is essential as well. Kinematic focused methods have been introduced: the Domain of Admissible Orientation concept aims at defining a tool orientation that meets to both kinematical and functional requirements employed [15]; another approach, based on energy minimization applied to the machining surface concept [7] balances tool-path smoothness against geometrical deviations [16]. In order to control cutting forces along the tool-path Larue and Altintas [17] proposed a simulation tool dedicated to flank milling operations. Torque, power and static deflections can be maintained at safe levels. Another optimization direction is to focus on the tool profile rather than on the tool-path shape by employing half-barrel cutters [18].

### *4.2.2 Emergence of New Milling Strategies*

In addition to the spread of 5-axis machining, the performance of the latest machining equipments enabled innovative milling strategies, such as plunging tool-paths or trochoidal milling tool-paths, to become efficient for industrial application. The spectrum of available solutions for the CAM programer is consequently expanding: these strategies can be of great interest as long as they are implemented in adequacy with the machining equipment characteristics.

#### **4.2.2.1 Plunge Milling**

A plunge milling tool-path is obtained by making the cutter carry out several axial plunging trajectories (Fig. 4.4). Material is removed during the plunging phase. Plunge milling parameterization consists in determining the plunging points and the manner how the cutter goes from one to another. In contrast with usual radial tool-paths such as direction parallel and contour parallel which repeat one pattern at several levels, two patterns are necessary to define a plunge milling strategy. The plunging pattern defines the cutter motions at each plunging point. The guide curve defines the path between the plunging points.

Specific parameters need to be set to define a plunge milling tool-paths. They are given in Fig. 4.5 and can be separated into two groups. The first group gathers

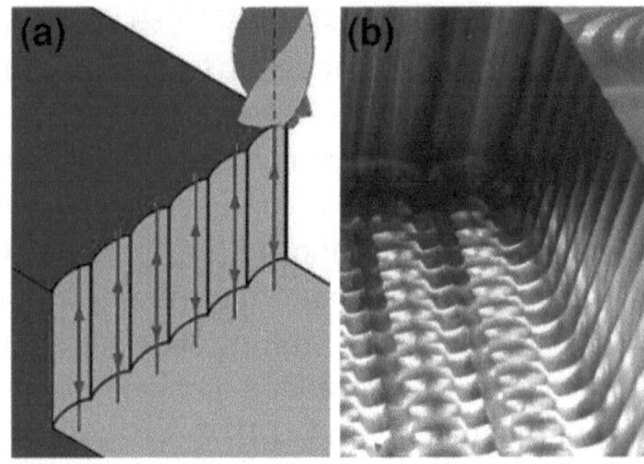


**Fig. 4.4** Plunge milling tool-paths: **a** example of tool-path, **b** machined surfaces

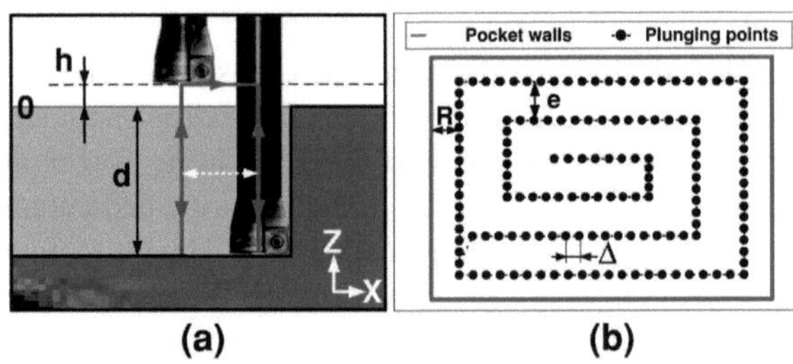


**Fig. 4.5** Plunge milling parameterization: **a** pattern, **b** guide curve

to plunging parameters: h, the height above the feature and d the feature height. The second gathers the guide curve parameters: Δ, the plunging step, e, the radial depth of cut and R, the tool radius. Guide curve type, which is often spiral or direction parallel, is also considered as a parameter.

Plunge milling tool-paths are increasingly employed for deep pockets and rough material machining [19]. The great potential associated with surface accuracy and machining times was showed by Wakaoka et al. [20]. Industrial concerns have shown a growing interest about plunge milling strategies and they are available in most CAM commercial software suites (Delcam Powermill, 3DS Catia, Missler Topcam …). It is however still difficult to circle the practical cases where plunge milling will be efficiently implemented. Feedback from the industry can provide general ideas but it is still difficult to predict whether these strategies are valuable for a specific combination of machine tool and part to machine.

Milling tests are a direct approach to compensate this lack of knowledge, but the associated times and costs are prohibitive. In contrast simulation is an effective way to identify the capability of plunge milling strategies in regards with process constraints. Unfortunately, the specificities of plunging tool-paths prevent the existing simulation software suites to propose efficient implementation configurations. Hence, it is still difficult for manufacturing experts neither to predict whether the implementation of plunge milling tool-paths will be profitable in regards with the available equipment nor to select the best tool-path parameterization. One can take advantage of specific simulation tool, based on Material Removal Rate (MRR) simulations.

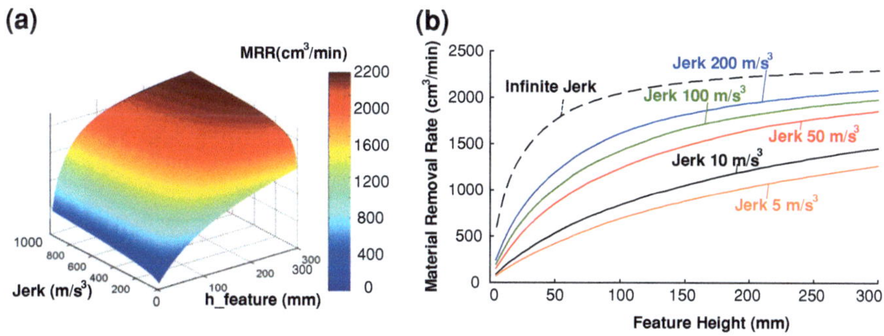


**Fig. 4.6** Plunge milling efficiency for a given machine tool: **a** complete MRR surface, **b** 2D projection

Typical use of this simulator is to determine the most efficient tool-path parameterization for a specific combination of machine tool and cutter. An example of result is proposed in Fig. 4.6. Great influence of the Jerk setting is illustrated too.

Another possible use case is the selection of the most suitable machine tool among a group to implement innovative tool-paths such as plunge milling. An illustration based on four high speed machining machines, two parallel kinematics machines (PKM) and two serial kinematics machines (SKM) is proposed here. MRR simulation results are given in Fig. 4.7.

Thanks to their high kinematics capabilities, PKM propose higher MRR. However, it would be wrong to conclude that plunge milling strategies shall not be implemented on SKM, looking on the results of SKM-1 for example. As a result, such simulation tool is highly useful to balance machine tool performances and tool-path programing with the product requirements.

In addition to that, the simulation tool can determine the kinematic properties to meet productivity requirements. When purchasing a new machine tool, each technical proposal can be evaluated and compared to the others, by using actual performances rather than nominal values. Tangible elements to decide whether the acquisition of new machining equipment associated with innovative machining strategies will be worthwhile are provided. The margins of progress linked to this technological investment are given as well, so that future needs and performances can be anticipated.

As a conclusion, plunge milling strategies can be of great interest as long as they are implemented in regards with part characteristics and machine tool behavior.

#### 4.2.2.2 Trochoidal Milling

Trochoidal milling tool-paths are 3-axis milling strategies mostly dedicated to roughing operations. The tool-path is defined as the combination of a uniform circular motion with a uniform linear motion.

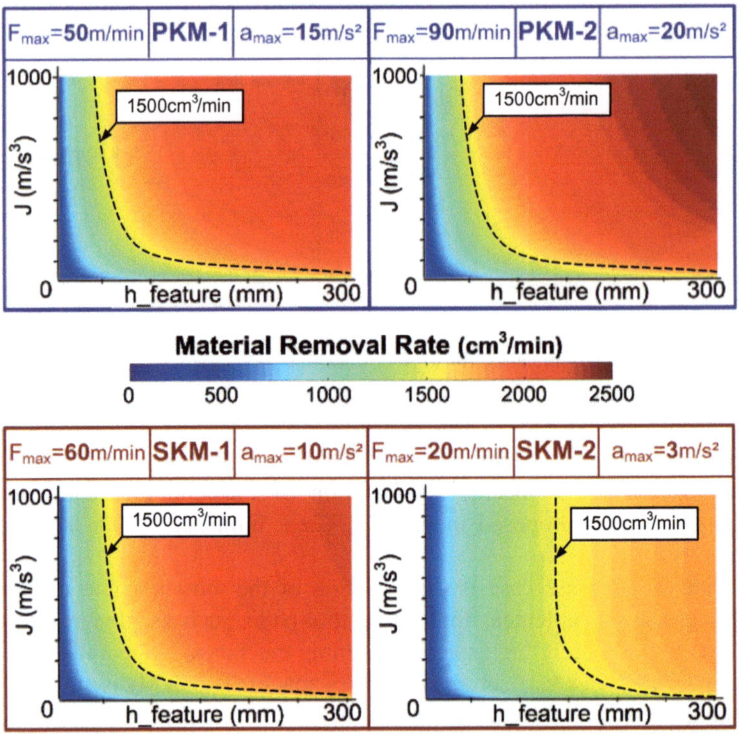


**Fig. 4.7** Comparison of a group of machine tools for plunge milling implementation

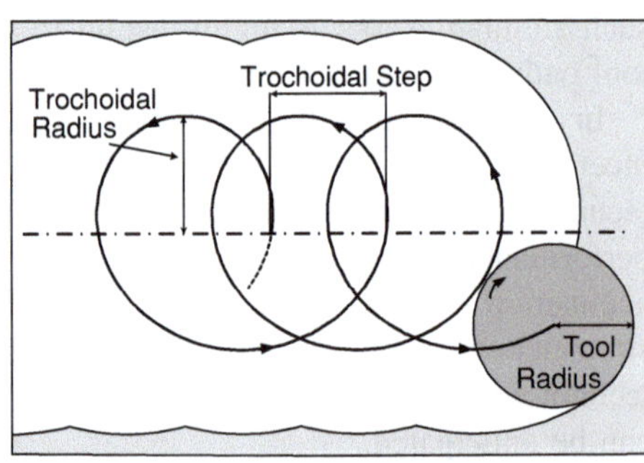


**Fig. 4.8** Example of trochoidal milling tool-path

As for plunge milling, trochoidal milling parameters can be sorted into two groups: trochoidal pattern parameters (trochoidal radius, tool radius and trochoidal step) which determine how material is removed and guide curve parameters which are linked to feature geometry. One great interest of trochoidal paths is consequently that process requirements (such as tool loads or chip thickness) are independent from the feature geometry.

Thanks to the properties of trochoidal curves, the generated tool-path is continuous in tangency and curvature. It creates favorable milling conditions in terms

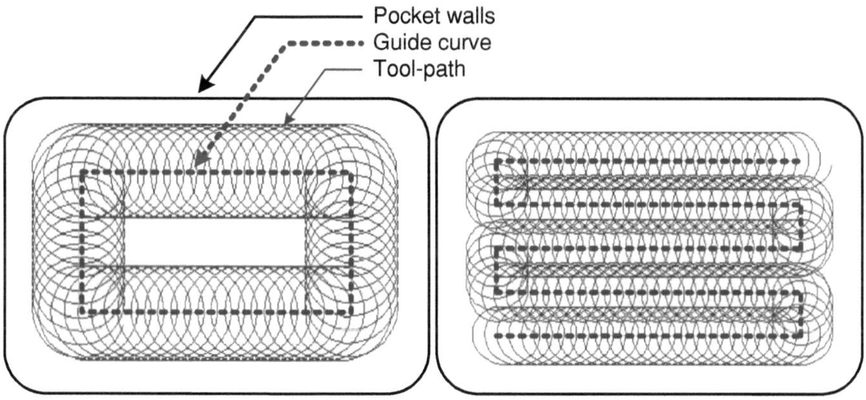


**Fig. 4.9** Example of guide curves for trochoidal pocket milling

of tool loads and kinematics. Furthermore, full-immersion milling configurations are avoided and it is possible to control the tool loads directly from trochoidal pattern parameters.

Nevertheless, the implementation of trochoidal strategies infers several requirements for the machine tool and its NC controller. Hence, the tool-path interpolation format has to reflect the continuity conditions of trochoids: spline formats have to be preferred to circular interpolation [19]. In addition to that, trochoidal milling generates many direction changes based upon small trochoidal radii. It is essential for the machine tool to dispose of high kinematic capabilities to respect programed feedrates [21].

Trochoidal milling tool-path length is much higher compared to standard tool-paths such as zigzag because large portions are outside the material. For most of the usual machining cases, the associated milling times are prohibitive. In contrast, trochoidal tool-paths are well adapted to complex milling cases, such as hard material roughing or pocket opening, for which process technological constraints decrease the efficiency of standard strategies.

These requirements explain why the idea of using trochoidal machining tool-paths is quite recent. However, their potential has already been detected by most CAM editors, who have included these strategies in their software suites to avoid high-load cuts. Indeed, the area of efficient implementation is growing with the on-going improvements of machine tools capabilities.

## 4.3 Technological Improvements Applied to Machine Tools

Implementing multi-axis and innovative tool-paths needs to know the characteristics of the manufacturing equipment to be carried out efficiently. Hence, this part focuses on some major improvements observed for the last years within the area of machine tools.

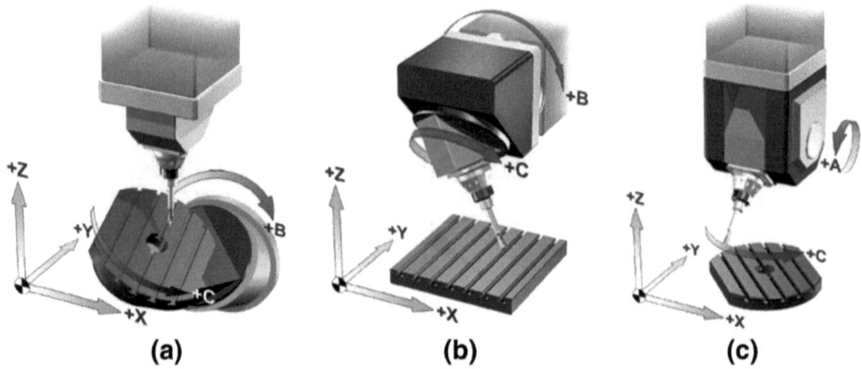


**Fig. 4.10** Example of 5-axis machines configurations: **a** spindle tilting, **b** table tilting, **c** spindle/table tilting (*source* [12])

## *4.3.1 Machines Tool Architectures*

Machine tools are the most employed industrial equipment to implement milling processes. Most of them are made with several motion axis organized according to a specific architecture, controlled by a NC controller. Most of them have a linear architecture, but since the middle of the nineties, parallel kinematics machines tools have been designed to carry out high speed milling operations.

### 4.3.1.1 Serial Machines Tools

Serial machine tools have their actuators positioned at each direction of the motion coordinate system. They move independently and are consequently stacked up: each actuator moves the whole downstream kinematic chain, which generated a large inertia.

Five-axis machine tools are usually configured to have three linear axes X, Y and Z plus two rotary axes. These rotary axis can be associated with the tool or workpiece:

- Tool-tilting or spindle-tilting machines have the tool spindle associated with both rotary axes (Fig. 4.10a)
- Table-tilting machines have the machine bed associated with both rotary axes (Fig. 4.10b).
- Spindle-table-tilting machines have one rotary axis associated with the tool and the other with the bed. These structures are however less common (Fig. 4.10c).

Selection of a machine tool is not a trivial task. It involves identification of machine's capabilities in relationship to the type of part to be machined. Indeed,

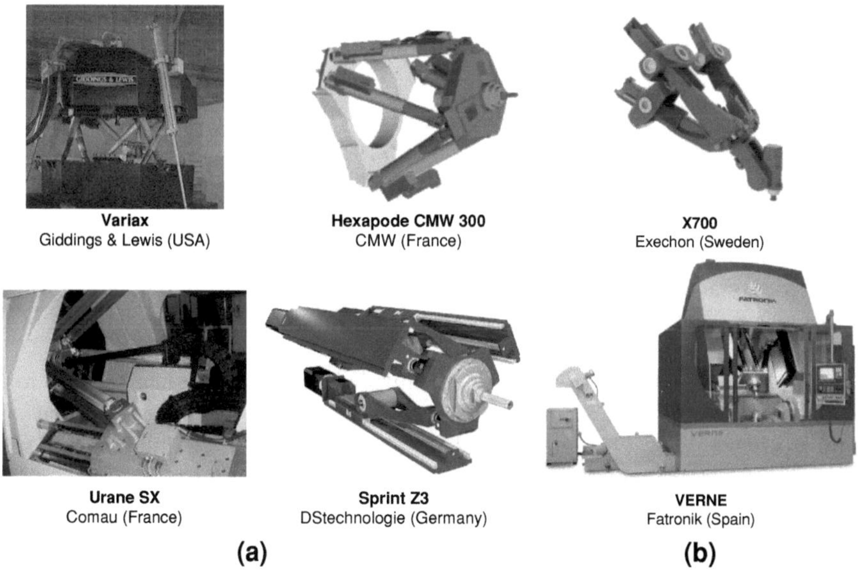


**Fig. 4.11** PKMs for high speed milling applications: **a** fully parallel mechanisms, **b** hybrid mechanisms

each configuration has its own forces and weaknesses. On these aspects, an interesting method to select optimal machine tool configuration and part setup has been developed by Risacher et al. [22]. In addition to that, machine tool models can have a great interest to be able to select the most appropriate one, by using simulation approaches for example.

#### 4.3.1.2 Parallel Machines Tools

The demand for high capabilities machining centers for the manufacturing of complex parts benefited from the progresses done in the field of robotics. Parallel kinematics machines (PKM) are a great illustration of this. Because of their singular architecture, these machines are doted with useful characteristics for multi-axis machining: high velocity, acceleration and jerk capabilities [23–25] and high rigidity [26] compared to SKMs.

Despite their heterogeneity in terms of motion capabilities and architecture, most of the PKM are able to achieve the whole panel of high speed machining. These machines can be classified into two families, the fully parallel machines and the hybrid machines, as shown in Fig. 4.11. The fully parallel machines are made of n degrees of freedom end-effectors connected to the base by n independent kinematics chains, each having an actuated kinematic joint. Within this family, three categories have been employed:

- Hexapods with variable-length struts and fixed joints,
- Glides, with fixed-length struts with moveable-based joints on fixed linear guide-ways,
- Delta, with extensible legs made of fixed-length components actuated by revolute joints.

Although they provide the best kinematic capabilities they suffer from some weaknesses such as the coupling between tool positioning and orientation or the size of the workzone. It is the reason the second family, hybrid machines, has emerged. By adding a serial module to the parallel mechanism, the idea is to get rid of these weaknesses. Indeed, most of the popular PKMs (i.e. Tricept [27], VERNE [28]) have an hybrid structure.

The nonlinear characteristics of the kinematics of a PKM can lead to variations of resolution, stiffness and mechanical properties inside the work-zone that are difficult to quantify and control [29]. They have consequently a different behavior compared to SKMs which make their optimal use points different.

It is a fact that industrial up-take of PKMs is quite slow and some myths exist among the machine tool users: PKMs are for example said to lack of precision. However, several comparative studies showed that PKMs can achieve the same precision as SKMs can [26, 30]. In addition to that, existing implementations of PKMs in automotive and aeronautics industries [23] have shown that the need for machining time reduction can be met by using high speed machining strategies such as plunge milling or trochoidal milling on parallel kinematics machines [19, 21].

## *4.3.2 Machine Tool Elements Technology*

In parallel to the rise of new architecture, machine tool elements have benefited to technological improvements. This paragraph centers on the progresses made in the field of motion axis and CNC controllers. The modeling of these elements can be carried out within machine tools models, which are presented afterward.

### 4.3.2.1 Displacement Axis Technologies

The architecture of a machine tool is designed to meet the requirements in regards with motions ranges, rigidity and precision. The technology and the assembly of the displacement axis have to be defined accordingly to guaranty these capabilities. These choices imply usually to balance conflicting constraints. Hence, to enable high acceleration and feedrate, inertia has to be as small as possible, which is in contradiction with the high rigidity required for machining.

For high speed machining applications, motion axis translation guides are usually made of preloaded ball screws (Fig. 4.12). They are adapted to work at high speed (over 50 m/min) but are vulnerable to collisions and impacts. Motions

**Fig. 4.12** Examples of translation guides: **a** preloaded ball screw drive and **b** linear drive (*source* [31])

are ensured by electric drives controlled in position and speed, which are connected to the ball screw directly or with a gear reduction.

Linear drives are a new technology that is increasingly employed by machine tool designers (Fig. 4.12). The linear motion is obtained directly from the electromagnetic forces in the drive, without any transformation mechanism. These actuators have several advantages: high kinematic properties (over 200 m/min feedrate, over 50 m/s$^2$ acceleration …), high precision and drastically simplified mechanics. The implementation of such equipment asks nevertheless for a high quality control system, a cooling system and a dedicated protective cover. Gantry assemblies are usually preferred to cope with the normal component of the electromagnetic force. In the same way, torque motors are growingly employed for rotative motion axis as they simplify the mechanical assemblies and dispose of high kinematics properties and precision.

As a result, the technology of the motions axis has been significantly modified since the 2000. Modern machine tools are doted with high kinematic properties and a new behavior. The proposal of efficient machining strategies has consequently to take this new situation into account.

#### 4.3.2.2 New CNC Controllers Capabilities

CNC controllers have a central role into the efficiency of the milling process as they interprete, interpolate and control the tool-paths.

CNC controller designers have taken advantage of the progresses of automatic and computer science to meet the requirements associated with the modern machining industry. Hence the computing capabilities have been strongly increased to face the increasing complexity of the tasks.

Finishing operations need tool-paths adapted to the geometry to be machined to produce accurate surfaces and adapted to the kinematic behavior of the machine tool to avoid perturbation phenomenon such as vibrations. As a result, usual linear and circular interpolations are not sufficient to meet the requirements, because they create significant acceleration and feedrate discontinuities, which penalize the machine tool kinematical behavior.

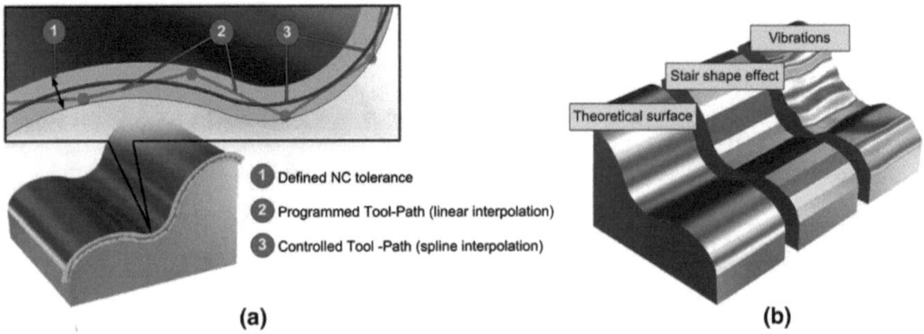


**Fig. 4.13** Tool-path implementation on NC controllers: **a** tool-path compression, **b** effects of linear interpolation on machined surfaces (*source* [12])

The use of high-level tool-path interpolation formats, such as B-splines polynomial curves or NURBS, has become consequently a common practice [32]. Even if linear interpolation is employed, the NC controller can use computation algorithms called "compressors" to interpret them as continuous 3D curves so that the kinematic behavior of the machine tool and the machined surface accuracy are improved by reducing chattering and "stair-shape" effects [12], as presented on Fig. 4.13. However, the use of these new interpolations formats makes the tool-paths generation more complex: for continuous 5 axis machining tool-paths, the cutter orientation transformation into articular coordinates strongly depends on the machine tool architecture.

Improvements have been made concerning the kinematical behavior of the machine tool during the milling process [33]. Axis motion functions aiming at crossing tool-path geometrical discontinuities, at redefining Jerk and acceleration settings or controlling the tool orientation in multi-axis milling have been developed.

As a result, recent NC controllers are doted with a wide spectrum of tool-path control algorithms and functions to adapt the machining process to the actual equipment and part to manufacture. However, it is still difficult to determine optimal configurations at once and optimization methods, such as process technological simulation are inescapable. The integration of network and communication technologies into NC controllers, such as the Ethernet protocol, highly simplified the availability of machine tools, by making them common elements of the enterprise network. Several improvements are made possible thanks to these evolutions, for process planning or maintenance for example.

In addition to that, NC kernel, CPU and actuators data (axis positions, spindle temperature, electrical power consumed) have become available: it is possible to make use of these process data to control the process and adapt milling tool-paths accordingly. In practice, these data are hardly employed for industrial applications and their use is mainly proposed by research works, such as the ICAM (Intelligent Computer Aided Manufacturing) method proposed in Sect. 4.4.1.2 of this chapter.

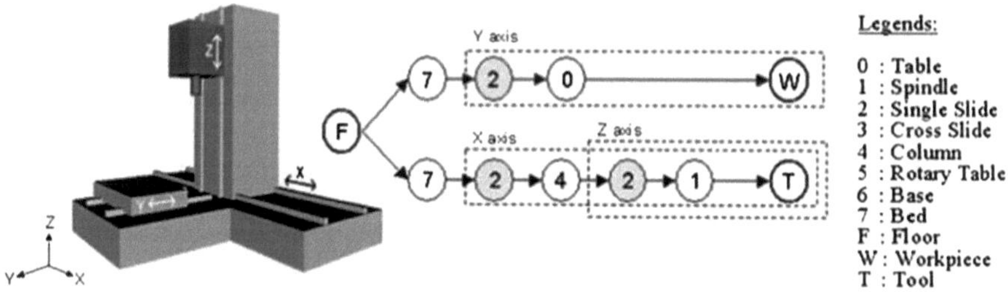


**Fig. 4.14** 3-axis machine tool modeling by using the model proposed by Suh et al. (*source* [34])

#### 4.3.2.3 Machine Tool Models

Machine tool models are information frameworks that encapsulate machine tool data, so that they can be, for example, used by simulation algorithms for virtual machining purposes.

In the field of complex surface machining, a first issue has been to predict the machine axis motions to detect and avoid collisions. Most of the machine tools models are consequently focusing on kinematics. Hence, the model proposed by Suh et al. in [34] aims at tool-path simulation and consist of a geometrical representation of a virtual machine tool. Direct and indirect kinematic transformations are made by following the kinematic chain from workpiece to tool.

Another machine tool kinematic model based on the EXPRESS language is proposed by Nassehi and Vichare in [35]. Mechanical machine elements, kinematic joint and axis of movement entities are defined. Each mechanical machine element is attached to a kinematic joint. Shape representation for virtual manufacturing applications is also provided.

A research effort driven by ISO group TC 184/SC1 aims to propose a STEP-NC machine tool model, which will be identified as ISO 14649-201. A part of the model is presented in Fig. 4.15. The purpose is to make machine tool data compliant with the process data to improve process planning and process parameterization. In contrast with the previous machine tool models, machine geometry is not the main focus. Most of encapsulated data are related to motion and dynamic capabilities and machine tool elements characteristics.

## 4.4 Tool-Paths Implementation and CNC Control Methods

Implementing milling tool-paths in regards with machining equipment capabilities and behavior implies to encapsulate and model the technical improvements. In addition to that, machining tool-paths are efficient only if both path geometry and process control meet the requirements, especially for complex sculptured

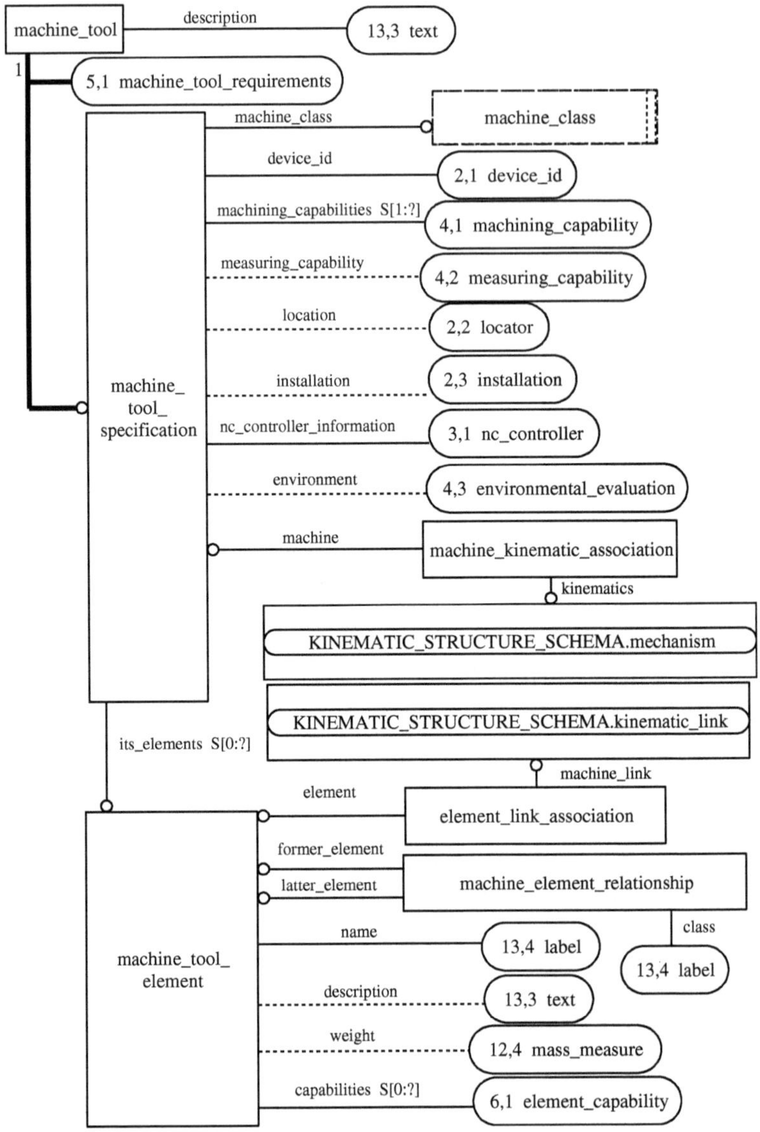


**Fig. 4.15** STEP-NC machine tool model developped by ISO TC 184/SC1 (*source* [36])

surfaces. Usual programing methods are not sufficient anymore. This part discusses the interest of several tool-paths implementation and control methods, which aim to optimize the machining process. Although these aspects are tightly independant in practice, they are presented here according to two categories: advanced CNC programing methods and virtual manufacturing.

## *4.4.1 Implementation of Advanced CNC Programing Methods*

This paragraph introduces three advanced programing approaches. Tool-path morphing, which aims to compensate tool deflexion during the machining process, ICAM approach, which bases on the use of process data to control the tool-path during the process and STEP-NC approach, which proposes a new efficient organization of CAD/CAM/CNC numerical chain.

### 4.4.1.1 Tool-Paths Morphing

The tool-path morphing method [37] focuses on tool deflection compensation. Main objective is to significantly decrease the effects of the tool deflection phenomenon associated with material removal processes. Initial tool-path is morphed to minimize the positionning error of the machined surface.

First, tool deflection is calculated within a dynamic analysis module, by using a cantilever beam model. The capability of such modeling has been successfully compared to FEM analysis approaches for this specific application. To decrease the computation times, the method focuses on the deflected position of the cutter center point rather than on the whole cutter envelope. Moreover, tool deflection is not calculated continuously along the whole tool-path but for defined sample positions. This frequency control factor makes the method applicable for any practical case as it adapts to NC controller computation capabilities. An example of this method is proposed in Fig. 4.16, where two cases are compared, one with and another without tool-path compensation.

This approach assesses the interest of using simulation results to improve the milling process. Theoretical tool positions are modified so that actual tool-paths generate a machined surface with a minimal positionning error.

### 4.4.1.2 Intelligent Computer Aided Manufacturing

The purpose of the Intelligent Computer Aided Manufacturing (ICAM) method is to make use of process data to compensate on line the machining tool-path [38]. Tool-path generation is shared by both CAM software and NC controller: basic tool-paths (strategy, depth of cut, …) are first generated at CAM level and then sent to NC controller. During the running of the program, these tool-paths are optimized online by uisng real time process data evaluation.

A major interest of ICAM method is the limited use of additional equipment neither to collect process data not to perform the adaptive tool-path control as these operations are carried out by the NC controller of the machine tool (Fig. 4.17). Another interest lies into the generic nature of ICAM method: it is flexible in terms of process (it has been successfully implemented for milling and incremental sheet forming [38]), of machine tool kinematic and of part characteristics (size, location within the working space…).

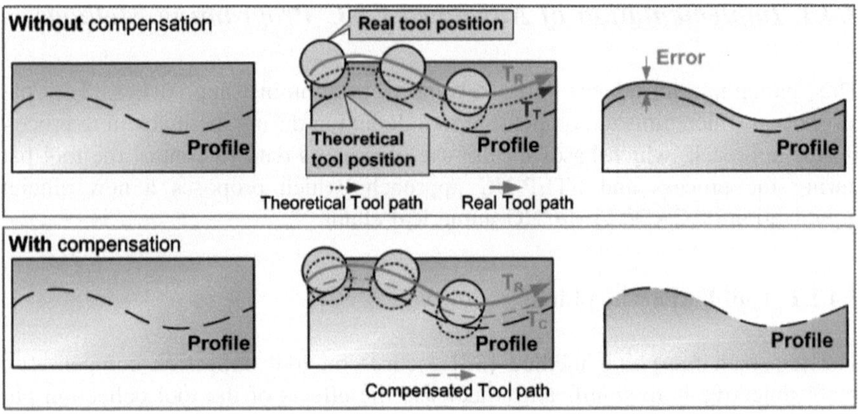


Fig. 4.16 Tool-path morphing method

Fig. 4.17 ICAM method for milling applications

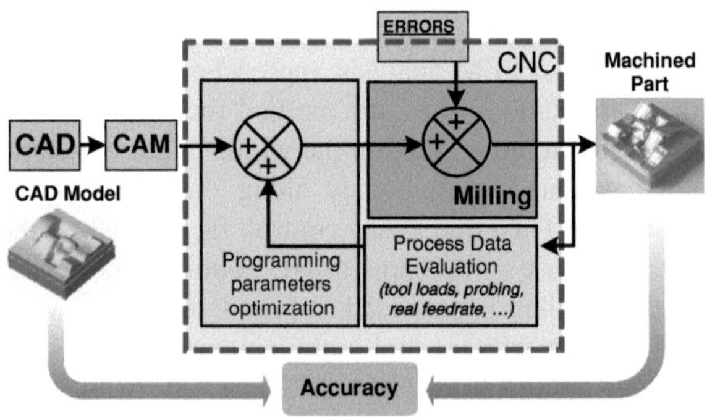


#### 4.4.1.3 STEP-NC Approach

Based on the STEP concept, STEP-NC aims at significantly modifying the CAD/CAM/CNC data chain by enhancing the amount and nature of exchanged data between the differents stages. It fits to modern CNC environments and simplifies the programing by being feature oriented. In this way, a direct consequence is that G-code is not necessary to transfer manufacturing data from CAM stage to CNC stage.

The manufacturing stage is fully included in the whole product development process: modification of a part geometry made for example on the shop floor can be saved and fed back to the design. Indeed, STEP-NC is fully STEP-compliant. Another important aspect lies into the use of a unique NC formalism for any machine tool software, as no post processing operations is needed. Communication language is high level: data relate the machining feature rather than simple axis coordinates control.

From a practical point of view, STEP-NC approaches help to shift the explicit tool-path generation to the shop-floor level. Intelligence and decision making are

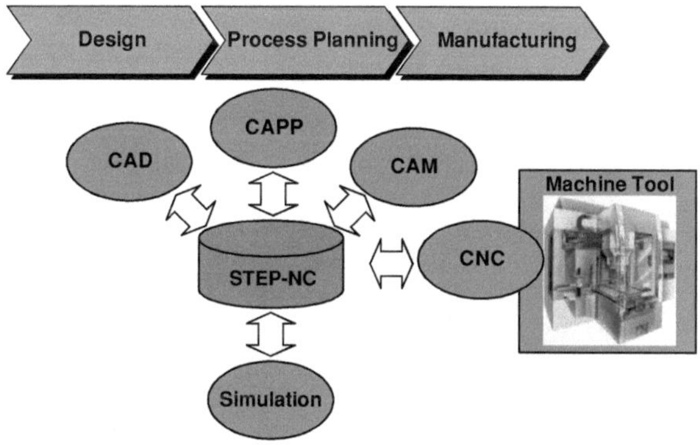


**Fig. 4.18** STEP-NC approach

transferred to the NC controller of the machine tool. Advanced CNC programing approaches can be proposed [39]. Interoperability and flexibility are increased too [40].

Eventually, by defining the CNC controller as any element of the CAD/CAM/CNC data chain, information feedback, tool-paths adaptations and "on the fly" modifications can be transferred [41]. The NC controller has consequently a central role by interpreting STEP-NC data and generating explicit tool-paths according to manufacturing feature characteristics and machine tool specificities for each manufacturing operation of the whole workplan (Fig. 4.18).

The SPAIM platform [39] is an example of the numerous intelligent programing possibilities offered by STEP-NC approach. This project aims at fulfilling two objectives. First it stands as a demonstrator to prove the benefits of STEP-NC on legacy industrial machines tools; secondly it serves as a development platform for future research and validations concerning the STEP-NC standard. SPAIM implementation has already been successfully carried out on both serial and parallel kinematics machines doted with commercial NC controllers (Fig. 4.19).

STEP-NC standard offers a range of new possibilities that need to reconsider the current practices of manufacturing processes. It is important to gradually deploy and integrate the STEP-NC data models into CNC controllers. The architecture of STEP-NC CAD/CAM/CNC data chain suggests that all practices involved in the present NC chain still have a role to play. STEP-NC is a data model standard that extends this knowledge. It calls for fundamental and even cultural changes. Yet, it still relies on existing capabilities and know-how in the manufacturing industry. Hence, it is really necessary to think out of the box to take fully advantage of this STEP-NC innovative programing approach.

### *4.4.2 Interest of Virtual Machining*

Virtual machining consists in providing simulation environments to ensure safety and accuracy of CAM-generated multi-axis tool-paths. Both geometric and graphic representation of machine tool elements are proposed in order to reduce the

**Fig. 4.19** SPAIM platform from IRCCyN

number of experimental tests, which are time-consuming and expensive, to a minimum.

#### 4.4.2.1 Generation of Collision Free Multi-Axis Tool-Paths

As explicited in a previous part of this chapter, the simulteaneaous motions of linear and rotative axes in multi-axis machining strongly increases the risk of collisions. What has been saved by in terms of workplan and setup simplication can be lost if the generation of a collision free milling strategy is a time and cost consuming process. To overtake this situation, a collision free tool-path generation method can be employed [6]. The specificity of such approach is to work carry out computations within the articular space of the machine tool.

The initial CAM-generated tool-path is first expressed as a sequence of sub-paths $t_i$ and $\tau_i$ (i = 1, n). $\tau_i$ paths are free of collision and $t_i$ paths have all their points in collision. Identification of collision is carried out with a simulation tool developed internally. Hence, it is enough to focus on $t_i$ paths to propose collision free tool-paths. $t_i$ paths are either milling either non-milling tool-paths.

It is important to notice that, within this context, the term "collision" applies to geometrical interferences, dynamic incompatibilities (such as feedrate limits) and carriage stops overrun.

Non-milling tool-paths are usually carried out at rapid feedrate and have no effect on the machined shape. They can be modified without affecting the

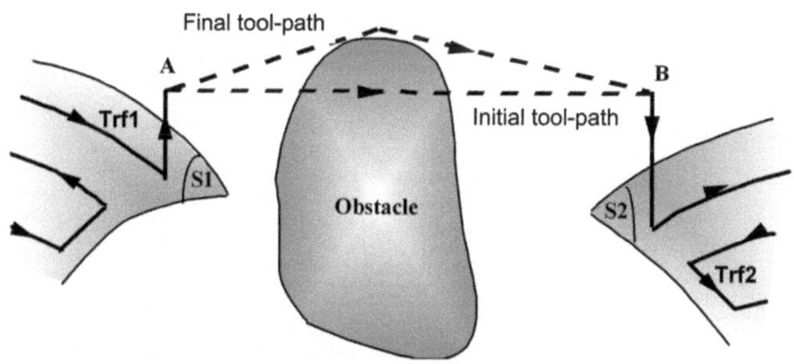


**Fig. 4.20** Example of collision avoidance for non-milling tool-path portions

machined part. The only requirement is to not modify the position of their ends, which are material entry and exit points. Modification of these tool-paths with the proposed approach bases upon an obstacle jump heuristic method derived from robotics (Fig. 4.20).

Multi-axis functional tool-paths need a different approach as several constraints have to be fulfilled in order to respect machined surfaces accuracy. Hence, they are corrected automatically by determining the tool orientations which reduce the volume in collision along the path into a null value. To do so, parametric curves such as Nurbs or B-splines are employed to describe tool-paths in order to control tool orientations with a limited number of parameters. Optimal orientation is then computed by resolving an optimization problem on the curve control polygon described in articular coordinates (Fig. 4.21). Last stage consists in generating the optimized tool-path by combining the optimized collision-free portions.

In addition to that, the simulation model is not limited to tool-path modification. A cutting conditions module enables to take process technology into account.

As any tool-path optimization method, this approach is efficient only if any element of the working space is precisely known: machine geometry, tool and tool holder characteristics, clamping fixtures ... However, this tool-path optimization approach is highly effective and fast to implement, as computation is carried out in the articular workspace. The interest of simulating the machine tool motions before the actual process is of great interest in multi-axis surface machining.

#### 4.4.2.2 Machine Tool Configuration Selection and Workpiece Orientation

Workpiece definition and positioning into the machine tool workspace has a great influence on both productivity and surface accuracy. However, these issues are usually not in the scope of CAM software tools and depend on programer know-how and expertise. The spread of 5-axis machines tools has made these tasks easier to carry out but for complex parts, several attempts may be necessary before finding a suitable solution. A method based on machine tool and part to machine

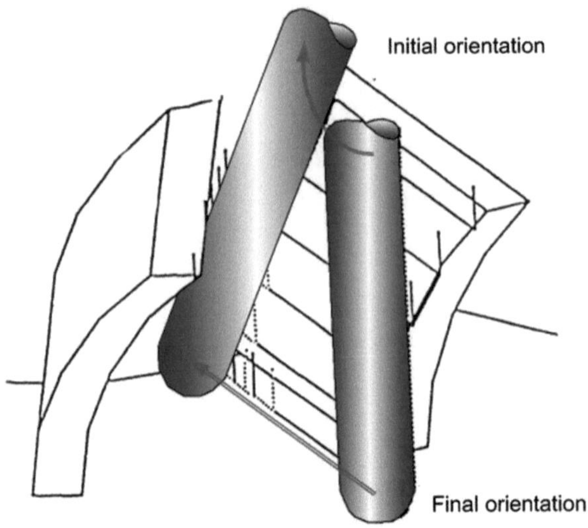


**Fig. 4.21** Example of a modification of tool orientation

visibilities has been proposed at IRCCyN to select the machine tool configuration and orient the workpiece into its workspace [22].

For the part, the visibility of each constitutive surface $S_i$ (this notion covers the area to be milled, independently from CAD features) is defined as a set $V_i$ of tool directions enabling the milling of the whole surface $S_i$, by considering collisions and technological constraints. A visibility is consequently a cone bounded by the extreme tool directions. In order to provide a more convenient representation, visibility cones can be represented as surfaces on the Gauss sphere: the borders of these surfaces are built from the intersections of the visibility cones with the Gauss sphere.

For the machine tool, the visibility is defined as the set of tool directions allowed by the machine kinematics. Hence, for 3-axis machines, the visibility is a single point on the Gauss sphere. For a 5-axis machine tool, the additional orientation axis makes the machine visibility as a spherical band or a cap on the Gauss sphere.

Then, the method proposed consists in searching the possible orientations (namely machine tool and surface visibility intersections on the Gauss sphere surface) for each surface of the part. Then, the solutions for the whole part are obtained by intersecting these sets of orientations. An empty intersection means that the part cannot be machined in a single setup on the selected machine tool. An example of implementation is proposed in Fig. 4.22.

Proposed approach is consequently of great interest when to select on which machine a piece can be produced, by testing successively different architectures. For suitable solutions, possible workpiece setups are defined as well. Tool-paths generation will then be made easier. In addition to that, the definition of the required architecture to meet specific needs in terms of part geometry can be carried out as well by using this setup definition method.

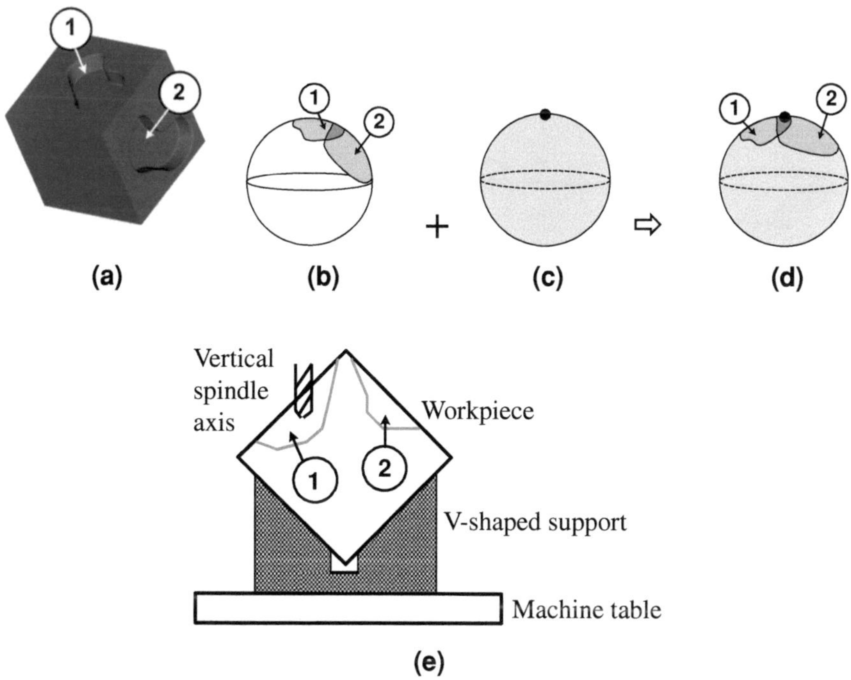


**Fig. 4.22** Example of setup for the machining of a cube with two pockets on a 3-axis machine tool: **a** Part to machine, **b** Part visibilities, **c** 3-axis machine visibility, **d** combination of part and machine visibilities, **e** Selected Setup

#### 4.4.2.3 Technological Simulation of the Machining Process

Simulation approaches have become inescapable and several machining simulation programs have been developed since the development of multi-axis high speed machining.

CNC simulation and validation at the tool-path planning stage are commonly available in CAM software tools (i.e. MachineWorks Simulation software). Standalone commercial simulation tools such as CGTech Vericut or Spring NC simul programs are mostly aiming to detect tool-path errors and inefficient motions. Feed rate optimization modules have been developed as well. Simulation tools and algorithms developed for research purposes model some specific aspects of the manufacturing system behavior [42, 43] (tracking errors, real feed rate, etc.) and analyze the cutting process [17, 44, 45] (tool deflection, chatter stability, etc.) in order to optimize the process offline.

When carrying out sculptured surface machining operations, surface properties and accuracy can suffer from the combined effects of errors related to the motions of the machine structure, the geometry of the defined tool-path and to the milling process itself (tool deflection, etc.). An enhanced machining simulator dedicated to

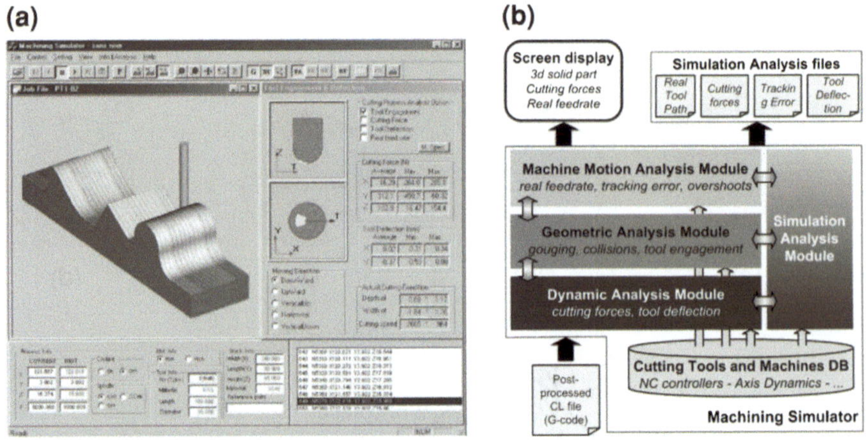


**Fig. 4.23** High speed machining simulator: **a** user interface, **b** architecture

**Fig. 4.24** Real feedrate simulation

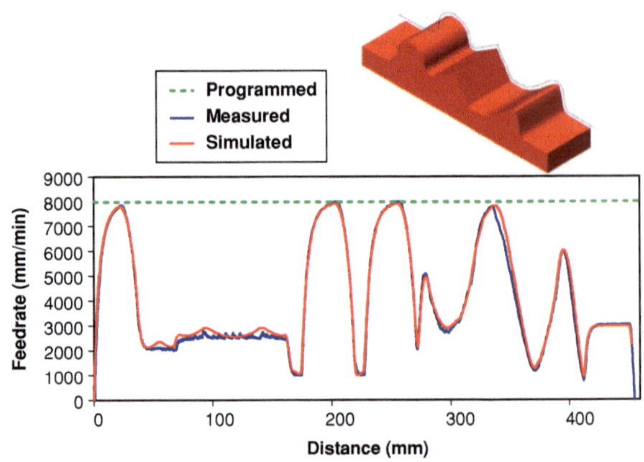


High Speed Milling (HSM) has been developed in the laboratory to analyze and estimate them [44]. Post-processed cutter location file, machine tool characteristics (structure and CNC controller) and cutting tool properties are used as inputs to carry out the simulations. Their results are displayed based on the 3D solid model of resulting part associated with the evolution of the process parameters, as presented in Fig. 4.23. The simulator is made of four modules, each of them focusing on one specific analysis.

The machine motion analyze module calculates real feedrates, tracking errors and overshoots by simulating the behavior of the machine dynamics (structure, actuator capabilities, etc.) associated to its NC controller. This module provides real feedrate computation by taking the dynamic capabilities of the machine into account, and the real tool-path, which is the tool-path resulting from the machine motions controlled by the CNC controller. An example of real feedrate result is

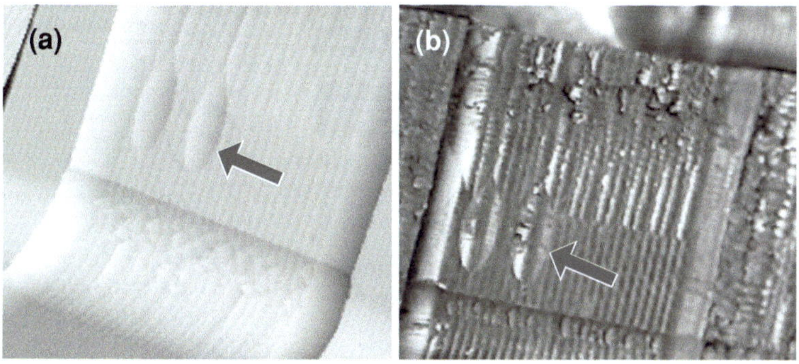


**Fig. 4.25** Adequacy of **a** simulation results and **b** machined part

**Fig. 4.26** Comparison of several milling strategies

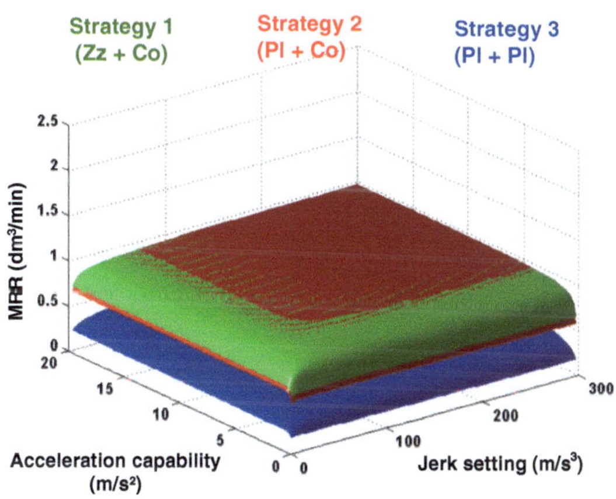


shown in Fig. 4.24. When the cutting feedrate varies during the process, surface integrity can be strongly affected. These simulation results identify one cause of surface defect. In addition to that, these outputs will be later used for cutting forces detection and collision detection, respectively. The geometrical analyze module computes gouging errors and collisions errors. Tool engagement is calculated too. The dynamic analyze module evaluates the causes of dynamic effects such as tool deflection or workpiece deformation and their effects on the obtained surfaces. The simulation analysis module compiles the results of the three other modules to evaluate their effects on the machined part. A comparison with the CAD model enables an accurate prediction of the defects localization and weight.

This HSM simulator is consequently a useful verification tool to assess the adequacy of a post-processed tool-path with a specific machine tool. Combined effects of CNC controller, cutting loads and tool deflection on machined surface are computed and displayed. The comparison of simulation results with the characteristics of real part shows the efficiency of this simulation tool to predict the final surfaces integrity as well, as illustrated in Fig. 4.25.

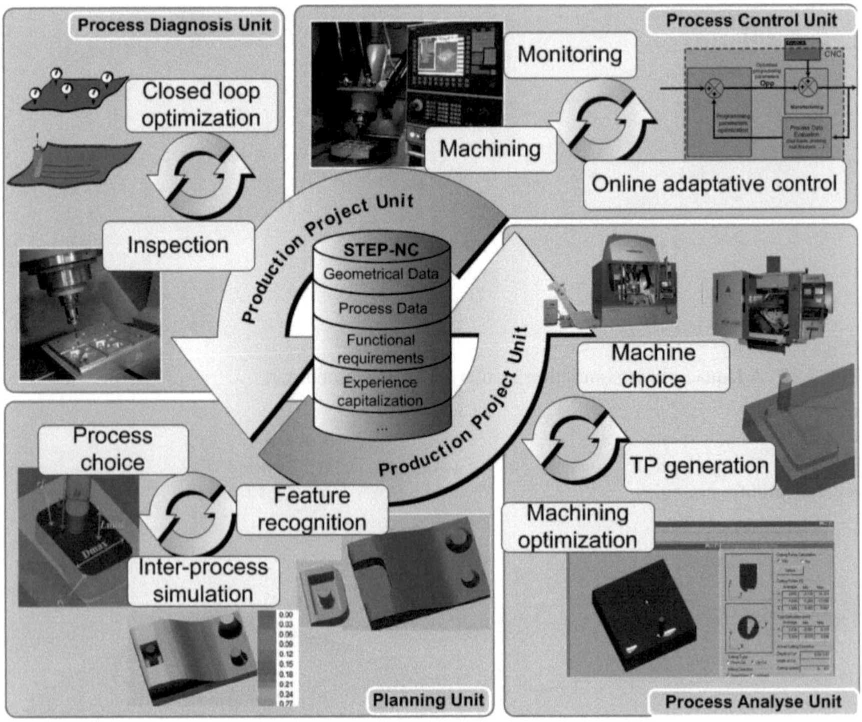


**Fig. 4.27** Integrated environment for advanced manufacturing

Technological simulation can be employed to compare several tool-paths parameterizations and select the most suitable one. A dedicated simulation tool has been developed to complete the existing environment by introducing innovative milling strategies such as plunge milling tool-paths [46].

Figure 4.26 provides the simulation results study concerning the evolution of the MRR according to the kinematics setting of the machine tool for several pocket milling strategies combinations: zigzag roughing and contour parallel finishing for strategy 1, plunging roughing and contour parallel finishing for strategy 2, and plunging for both roughing and finishing for strategy 3. With the selected machine tool, the results show that it is not viable to employ plunge milling as a finishing strategy: the machining times associated with an acceptable surface integrity are prohibitive. In contrast, plunge milling is an efficient choice for high speed milling cases where the kinematics settings have to be high.

Similar studies have been carried out for other parameters related to the machining equipment tool, such as the kinematics capabilities of the machine tool or the characteristics of the cutting tool, related to the part to machine, such as the feature heights, the material or the surface roughness or related to the milling strategy such as tool-path type or parameterization.

As a result, it is possible to get by simulation very accurate results to select the most effective process implementation and to measure the progress margins. Experimental campaigns can be run only for a limited number of tests, as a first selection can be achieved from the simulation results.

From a global point of view, these developments assess the interest of providing rich virtual machining environment to enhance the efficiency of manufacturing processes. When machining complex sculptured surfaces, they are particularly valuable as they enable to select the most appropriate parameterization.

## 4.5 Conclusions

With the technological progresses, the spread of multi-axis tool-paths combined with the emerging of new milling strategies dramatically modified the challenges associated with the machining of complex sculptured surfaces.

This chapter first discussed the actual state-of-play concerning multi-axis milling strategies and tool-paths geometries. As their control calls for a very good understanding of the actual machining equipment, major technological advances were introduced in the second part. Finally, advanced tool-paths generation and control methods were discussed. In particular, the interest of virtual machining has been demonstrated.

Future works will bring proposed methods together within an innovative advanced manufacturing environment. As presented in Fig. 4.27, the aim is to meet the requirements associated with CNC manufacturing by using a coherent and global approach based upon a coordination of optimization methods dedicated to each aspect of the manufacturing process.

## References

1. Rauch M, Xu X (2010) Five-axis machining: technologies and challenges. Int J Manuf Res 5(3):327–352
2. Gray PJ, Bedi S, Ismail F (2005) Arc-intersect method for 5-axis tool positioning. Comput Aided Des 37(7):663–674
3. Lauwers B, Dejonghe P, Kruth JP (2003) Optimal and collision free tool posture in five-axis machining through the tight integration of tool path generation and machine simulation. Comput Aided Des 35(5):421–432
4. Ilushin O, Elber G, Halperin D, Wein R, Kim M-S (2005) Precise global collision detection in multi-axis NC-machining. Comput Aided Des 37(9):909–920
5. Wang N, Tang K (2007) Automatic generation of gouge-free and angular-velocity-compliant five-axis toolpath. Comput Aided Des 39(10):841–852
6. Hascoët JY, Bennis F, Guerin F Optimization of tool trajectory for five axis control machine. In: IASTED robotics and manufacturing, Oxford (UK), 23–25 september 1993. pp 120–122
7. Duc E, Lartigue C, Tournier C, Bourdet P (1999) A new concept for the design and the manufacturing of free-form surfaces: the machining surface. CIRP Annals - Manuf Technol 48(1):103–106

8. Lavernhe S, Tournier C, Lartigue C (2008) Optimization of 5-axis high-speed machining using a surface based approach. Comput Aided Des 40(10–11):1015–1023
9. He W, Lei M, Bin H (2009) Iso-parametric CNC tool path optimization based on adaptive grid generation. Int J Adv Manuf Technol 41(5–6):538–548
10. Anotaipaiboon W, Makhanov SS (2005) Tool path generation for five-axis NC machining using adaptive space-filling curves. Int J Prod Res 43(8):1643–1665
11. Young HT, Chuang LC, Gerschwiler K, Kamps S (2004) A five-axis rough machining approach for a centrifugal impeller. Int J Adv Manuf Technol 23(3):233–239
12. Siemens AG (ed) (2009) Milling with SINUMERIK—5-axis machining Manual. 05/2009 edn. Siemens AG
13. Senatore J, Landon Y, Rubio W (2008) Analytical estimation of error in flank milling of ruled surfaces. Comput Aided Des 40(5):595–603
14. Li C, Mann S, Bedi S (2005) Error measurements for flank milling. Comput Aided Des 37(14):1459–1468
15. Castagnetti C, Duc E, Ray P (2008) The domain of admissible orientation concept: anew method for five-axis tool path optimisation. Comput Aided Des 40(9):938–950
16. Pechard P-Y, Tournier C, Lartigue C, Lugarini J-P (2009) Geometrical deviations versus smoothness in 5-axis high-speed flank milling. Int J Machine Tools Manuf 49(6):454–461
17. Larue A, Altintas Y (2005) Simulation of flank milling processes. Int J Machine Tools Manuf 45(4–5):549–559
18. Chaves-Jacob J, Poulachon G, Duc E (2009) New approach to 5-axis flank milling of free-form surfaces: Computation of adapted tool shape. Computer-Aided Design (in press), Accepted Manuscript
19. Rauch M, Hascoet JY (2007) Rough pocket milling with trocho and plunging strategies. Int J Machining Machinability Mater 2(2):161–175
20. Wakaoka S, Yamane Y, Sekiya K, Narutaki N (2002) High-speed and high-accuracy plunge cutting for vertical walls. J Mater Proces Technol 127(2):246–250
21. Rauch M, Duc E, Hascoet J-Y (2009) Improving trochoidal tool paths generation and implementation using process constraints modelling. Int J Machine Tools Manuf 49(5):375–383
22. Risacher P, Hascoët JY, Bennis F (1997) Workpiece shape and setup in milling. Paper presented at the intelligent manufacturing systems (IMS'97 - IFAC), Seoul (Korea)
23. Weck M, Staimer D (2002) Parallel kinematic machine tools—current state and future potentials. CIRP Ann Manuf Technol 51(2):671–683
24. Altuzarra O, Hernández A, San Martín Y, Larrañaga J (2009) Parallel Kinematics for Machine Tools. In: Lopez de Lacalle N, Lamikiz Mentxaka A (eds) Machine Tools for High Performance Machining. Springer, London, pp 335–368
25. Kanaan D, Wenger P, Chablat D (2009) Kinematic analysis of a serial-parallel machine tool: The VERNE machine. Mech Mach Theory 44(2):487–498
26. Terrier M, Dugas A, Hascoet JY (2004) Qualification of parallel kinematics machines in high-speed milling on free form surfaces. Int J Machine Tools Manuf 44(7–8):865–877
27. Xi F, Zhang D, Xu Z, Mechefske CM (2003) A comparative study on tripod units for machine tools. Int J Machine Tools Manuf 43(7):721–730
28. Terrier M, Gimenez M, Hascoet JY (2005) VERNE—A five-axis parallel kinematics milling machine. Proc Inst mech eng, Part B J Eng Manuf 219(3):327–336
29. Chanal H, Duc E, Ray P (2006) A study of the impact of machine tool structure on machining processes. Int J Machine Tools Manuf 46(2):98–106
30. Geldart M, Webb P, Larsson H, Backstrom M, Gindy N, Rask K (2003) A direct comparison of the machining performance of a variax 5 axis parallel kinetic machining centre with conventional 3 and 5 axis machine tools. Int J Machine Tools Manuf 43(11):1107–1116
31. Hascoet J-Y, Carabin G (2010) Système de génération du mouvement d'avance. In: Dunod (ed) Usinage à grande vitesse. Dunod
32. Langeron JM, Duc E, Lartigue C, Bourdet P (2004) A new format for 5-axis tool path computation, using Bspline curves. Comput Aided Des 36(12):1219–1229

33. Pateloup V, Duc E, Ray P (2004) Corner optimization for pocket machining. Int J Machine Tools Manuf 44(12–13):1343–1353
34. Suh S-H, Seo Y, Lee S-M, Choi T-H, Jeong G-S, Kim D-Y (2003) Modelling and Implementation of Internet-Based Virtual Machine Tools. Int J Adv Manuf Technol 21(7):516–522
35. Nassehi A, Vichar P (2009) A STEP-NC compliant methodology for modelling manufacturing resources. In: Xu X, Nee AYC (eds) Advanced design and manufacturing based on STEP. Springer, London, pp 261–281
36. ISO TC184/SC1/WG7 (2010) Cutting process machine tool schema. Paper presented at the 19th ISO TC184/SC1 Meeting, Nantes (France), 6 Oct
37. Hascoet J-Y, Dépincé P, Seo TI (1998) Compensation of tool deflection in ball-end millng simulation and experimental results. Paper presented at the CIRP International Seminar on Improving machine tool performance, San Sebastian (Spain), July 6–8
38. Rauch M, Hascoet J-Y, Hamann J-C, Plenel Y (2009) Tool path programming optimization for incremental sheet forming applications. Comput Aided Des 41(12):877–885
39. Rauch M, Laguionie R, Hascoet J-Y (2010) Achieving a STEP-NC enabled advanced NC programming environment. In: Xu X, Nee A (eds) Advanced design and manufacturing based on STEP. Springer series in advanced manufacturing. pp 197–214
40. Rauch M, Laguionie R, Hascoët JY, Xu X (2009) Enhancing CNC manufacturing interoperability with STEP-NC. J Machine Eng 9(4):26–37
41. Suh SH, Cheon SU (2002) A framework for an intelligent CNC and data model. Int J Adv Manuf Technol 19(10):727–735
42. Dugas A, Lee JJ, Terrier M, Hascoet JY (2003) Development of a machining simulator considering machine behaviour. Proc inst mech eng, Part B J Eng Manuf 217(9):1333–1339
43. Altintas Y (2000) Modeling approaches and software for predicting the performance of milling operations at MAL-UBC. Int J Machining Sci Technol 4(3):445–478
44. Dugas A, Lee J-J, Hascoët J-Y (2002) An enhanced machining simulator with tool deflection error analysis. J Manuf Syst 21(6):451–463
45. Merdol SD, Altintas Y (2008) Virtual cutting and optimization of three-axis milling processes. Int J Machine Tools Manuf 48(10):1063–1071
46. Rauch M, Hascoët J-Y (2011) Interest of multiphysics and multilevel simulation approaches to enhance the machining process. Adv Mater Res 223:891–899

# Chapter 5
# Intelligent Optimisation of 3-Axis Sculptured Surface Machining on Existing CAM Systems

**G.-C. Vosniakos, P. G. Benardos and A. Krimpenis**

This chapter discusses how the current practice for sculptured surface machining may be improved by embedding intelligent optimisation in the process planning steps for both roughing and finishing. Complexity and interdependencies between machining parameters are dealt with by stochastic evolutionary techniques whilst the need to take into account results of calculations lead to embedding meta-models of the machining process. Formulations are provided for the optimisation criteria including machining time, remaining material volume, distribution of remaining material volume, cutting force variation etc. and examples are given for considering machining constraints. Integration with existing CAM systems APIs is discussed and the possible future trends are presented.

## 5.1 Introduction

Machining parts with sculptured surfaces differ from machining other parts with regular surfaces. The reasons can be traced back to the very nature of the difference between sculptured and regular surfaces, i.e. the inherent freedom in designing a sculptured surface. Consequently, it is difficult to classify morphological features made of sculptured surfaces into very concrete categories, as in regular surface features which have concrete machining strategies and tools associated with them. Furthermore, morphological features made of sculptured surfaces are often machined in combination, whilst those with regular surfaces are normally machined individually. This often leads to usage of relatively few cutting

G.-C. Vosniakos (✉) · P. G. Benardos · A. Krimpenis
Department of Manufacturing Technology School of Mechanical Engineering,
National Technical University of Athens, Heroon Polytehneiou 9,
15780 Athens, Greece
e-mail: vosniak@central.ntua.gr

J. P. Davim (ed.), *Machining of Complex Sculptured Surfaces*,
DOI: 10.1007/978-1-4471-2356-9_5, 

tools when machining sculptured surface parts are compared to their regular surface analogues. Hence, this chapter deals with machining of complete sculptured surface regions without trying to divide them into subregions.

In more complex cases, four or five axes are needed for performing the machining task which creates the additional burden of determining the tool inclination to the machined surface. The latter task complicates any optimisation attempt with mathematical algorithms for multi-axis tool path planning in addition to machining strategy and tool selection; therefore it is left out in the course of this chapter, leading to consideration of sculptured surface machining optimisation in three axes, typically on vertical CNC machining centres.

Optimisation in this context refers to the best combination of both machining strategy, i.e. tool path pattern (including the variety of parameters defining the geometry of cutting passes), tool shape and size, as well as cutting parameters, typically feed and speed. The criteria against which any such combination is assessed refer to machining time and surface quality. The latter is possible to express either directly, e.g. as surface roughness, dimensional deviation etc. or indirectly through indicators, such as amplitude of vibration, variation of cutting forces etc. In addition, a combination is feasible subject to the satisfaction of some constraints often related to machine tool and to cutting tool limitations.

In practice, CNC programs for sculptured surfaces are created exclusively by employing CAM software systems. Very recently some CAM systems do claim optimum machining capabilities but without allowing the user to specify individual cost functions and constraints of interest. According to general practice, machining is performed by roughing first and finishing last (possibly with intermediate semi-finishing) within which tool and machining strategy selection is recognized to be heavily experience based. Once these selections are made, CAM systems offer an accurate enough prediction of machining time. Notwithstanding the inability of CAM systems to predict surface finish and dimensional accuracy, a number of 'what-if' CAM simulation runs are usually employed, so as to narrow down tool and strategy alternatives before concluding with the most appropriate choice.

Use of a CAM system in preparing a part program for sculptured surfaces conceals the real nature of the relevant tasks which is that of process planning. The relevant complexity and interdependency of process planning decisions will be dealt with in Sect. 5.2 in order to explain that machining simulation runs on a CAM system are time-consuming and hardly affordable in the industry, even if they are conducted in the framework of a systematic 'experimental' program, which, however, still does not guarantee optimality of solutions. Therefore, an 'automated' optimisation system would make sense, so much so if some sort of intelligence were built into it. The particulars of intelligent solutions offered are presented in Sects. 5.3 and 5.5 for roughing and finishing, respectively, whilst their implementation issues of interest are discussed in Sect. 5.5.

## 5.2 Problem Analysis

### 5.2.1 Current Machining Practice Based on CAM Systems

In the general case, sculptured surface machining practically needs to be programmed on a CAM system, since programming languages are by and large an obsolete practice. A typical CAM system works with solid/surface models of the parts to be machined and of the stock, should the latter be a casting. The user is to decide how to position and clamp the stock on the machine in one or more consecutive setups. In each setup he/she is required to mentally decompose the difference between the stock and the final part into regions and assign to each one of them machining strategies including basic type of tool path, the applicable tool and the values of the corresponding machining parameters. Approximate solid models of tools such as end mills, drills and milling cutters are usually available to choose from. The system calculates tool paths according to the values of the machining parameters and gives the user the possibility to edit them, divide them, apply various transformations or even combine two or more of them together. The cutter location file produced like this is, then, augmented by a post processor with machine tool-specific information (axis configuration, syntax peculiarities etc). Machining simulation is primarily visual, including collisions with the tool holder etc., as well as rest material that was not possible to remove with the designated machining strategy. It also includes capabilities which include calculation of material volume that was removed as well as length of the tool paths and corresponding machining time.

In practice, CAM software supports a 'trial-and-error' process starting with definition of the regions to be machined and the applicable order, proceeding with selection of the type of tool path on each region and finally, by transitioning the tool from one cutting pass to the next, especially between regions, see Fig. 5.1. This process involves a lot of recurrent loops among its distinct phases but has remained the same since the beginnings of CAM usage. Overall, CAM software fragments 3-axis (as well as 4- and 5-axis) capability into various dedicated functions essentially corresponding to specific types of tool paths.

It is worth noting that rough machining and finish machining are treated separately, although the results of the former are the starting point of the latter. In the rare case where the user might run out of options in specifying the finish machining strategy due to unsatisfactory results, he/she might need to go back to the roughing stage and specify a different strategy there, see Fig. 5.2

### 5.2.2 Combinatorial Complexity of Detailed Process Planning

In roughing, the most often used machining strategies are raster and offset, in both of which the cutting tool operates on plane slices of stock (2.5D-based roughing),

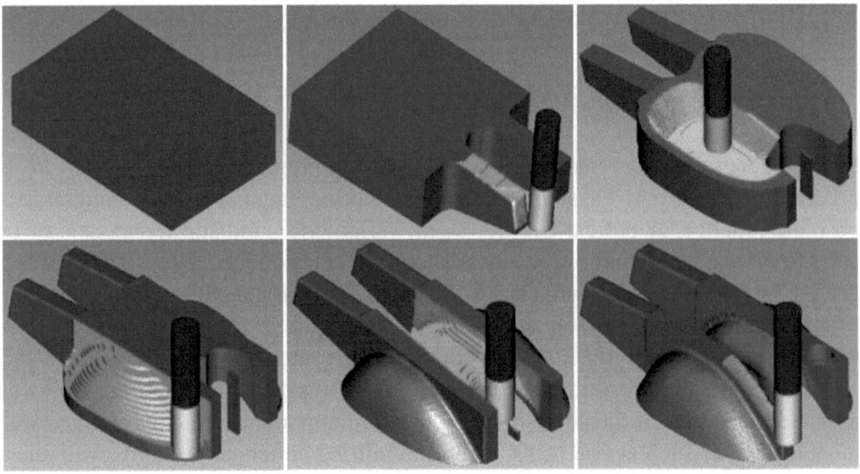

**Fig. 5.1** Examples of sculptured surface roughing by region. Intermediate finishing by region is not shown

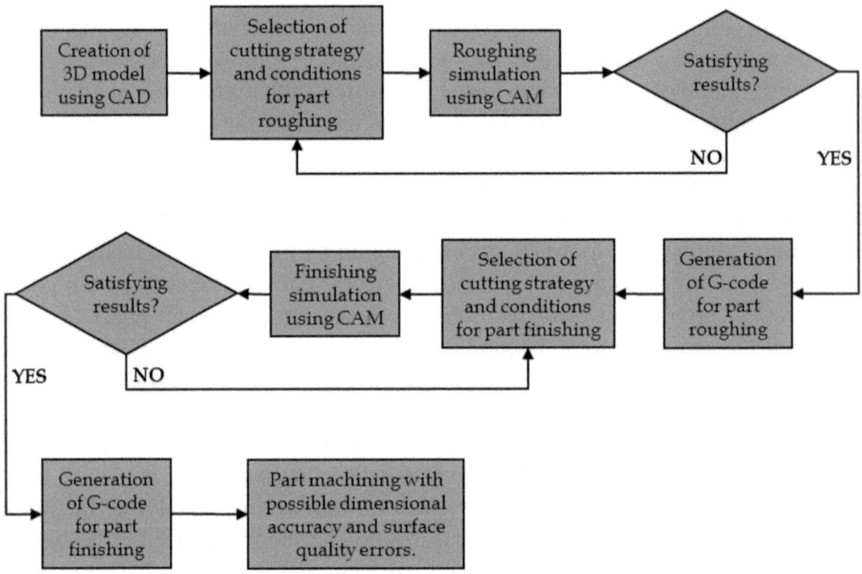


**Fig. 5.2** Typical CNC machining process planning flow

see Fig. 5.3. Less often, a profiling strategy is used, in which the cutting tool operates on surfaces being successive 3D offsets of the final part surface and is most common in 4- or 5-axis machining.

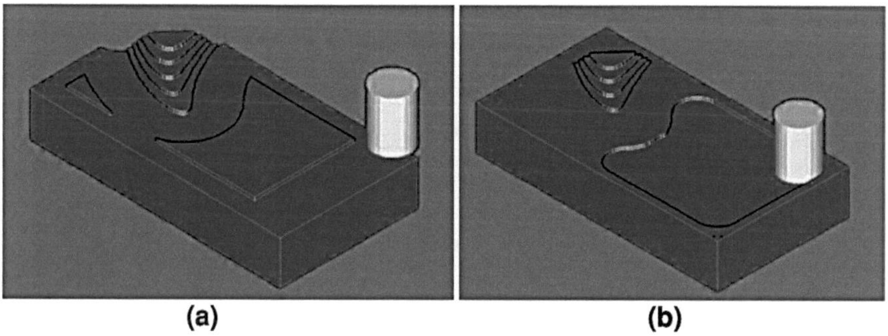


Fig. 5.3 2.5D-based roughing strategies, **a** Raster **b** Offset

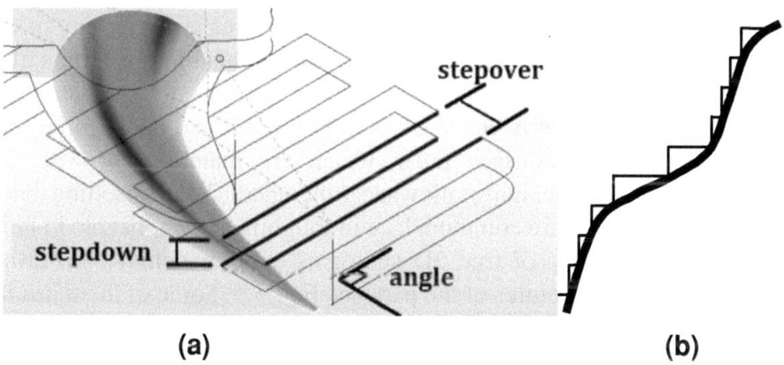


Fig. 5.4 Raster roughing strategy. **a** angle, stepover, stepdown **b** cusps

In 2D-based roughing strategies a number of planar slices are defined denoting successive roughing planes, their distance (stepdown) being the same or different as the user wishes, depending on the slope of the boundary surface and the allowable cusp height. In the raster (hatch) strategy the tool path consists of linear parallel segments at a distance from each other that needs to be defined by the user (stepover) and at an angle to the X-axis that needs to be defined by the user, too, see Fig. 5.4. In addition, there are two alternative directions of cut that lead to up-milling or down-milling. The way in which the tool enters the stock slice to be cut, e.g. plunging, ramping, side cutting etc. is also a matter of choice. Furthermore, in some cases rest machining is performed, in order to clear material that was left over from a previous operation. Regions in which the tool can move are also to be designated, both in cutting and fast moving mode, e.g. at a specific distance around the stock.

The choice of each of the above values, as well as their combination, in relation to the tool type and diameter choice affects the cutting force, the remaining material on the part that needs to be removed in finishing or semi-finishing stage, the dynamic performance of the machining system and, of course, machining time.

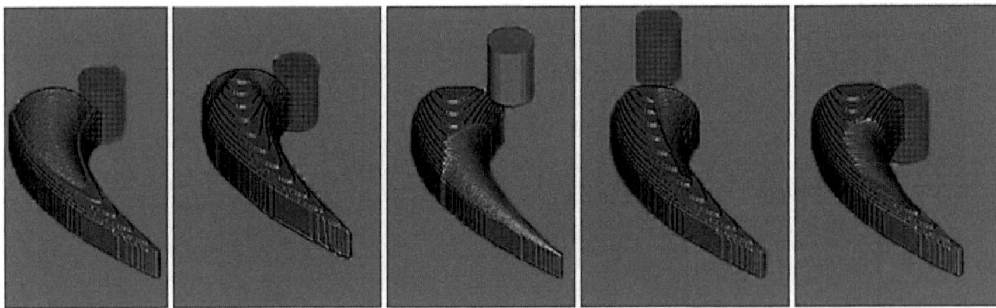


**Fig. 5.5** Finishing, **a** offset **b** 3D profiling **c** linear **d** radial **e** spiral

A similar approach applies to offset (contour) strategy the main difference being that the tool path consists of successively offset closed splines rather than parallel lines. It is also possible that the tool performs a profiling cut after any of the two roughing strategies in 2D, in order to further reduce remaining material still at the roughing stage.

In finishing, a small part of the stock is only present and needs to be removed under the constraints of surface finish which are mainly dealt with through appropriate choice of machining allowance/tolerance. 2.5D offsetting but especially 3D profiling which are commonplace in roughing are applicable in finishing, too. In addition, a variety of true 3D strategies may be defined in finishing to exploit particular shape features of the part, see Fig. 5.5. Some of them just impose a constraint on 3D profiling regarding the direction of the tool path lines, e.g. spiral, radial, depending on the shape being machined. Some other strategies result from a projection of basic planar shapes (line, rectangle, circle etc.) onto the final part surface coupled again with specific tool path directions. In addition, material that is left over locally, typically due to the part radius of curvature being locally smaller than that of the cutting tools, is removed by local finishing with small tools, which is commonly termed pencilling. Each of these new strategies is defined in detail by several parameters, which affect surface finish, machining time, as well as machining force variation and machine dynamics especially in high-speed milling. For instance: in finishing by radial pattern, stepover, angle and circle radius need to be assigned values; in finishing by 2.5D offsetting just a minimum and a maximum stepdown need to be defined so that the actual stepdown is defined through tolerance constraint; in 3D profiling just the stepover needs to be given to enable the CAM system to calculate individually variable stepdown value for each new offset tool path profile derived.

The multitude of parameters involved makes it very difficult to optimise process planning decisions for sculptured surface machining in a both holistic and detailed manner. In addition, there is no unique correlation between the process parameters and the result of the process, meaning that there is no one combination of values that satisfies the requirements set. Furthermore, there is always the possibility for the process parameters to constantly change along the tool path due to the complex geometry of the sculptured surface parts. Finally, the particularities of the equipment used, especially of the machine tool—cutting tool—workpiece system, must be taken into account to yield realistic solutions.

It is no surprise that CAM systems only recently have started making an attempt to consider optimisation issues, whilst the research literature indicates that all optimisation attempts are confined to a subset of parameters. Tool paths may be generated by sub-dividing a complex sculptured surface into a number of easy-to-machine surface patches and by identifying the favourable machining setup/orientation for each patch [1]. Roughing decomposition is aimed at multiple tools, whilst for the finishing operation it may be based on multiple tool path patterns, depending on optimal decomposition values (tool diameters, surface slopes) that minimize machining time [2]. Heuristic-based tool type selection and a mathematical optimisation model approach for tool combination often exploits face loops in the solid model representation of the part [3]. The Taguchi design of experiments method and ANOVA have been proven to be a good tool to determine suboptimum values for all parameters involved in each milling strategy as well as the most significant of those parameters [4]. Cutting force calculations to take into account in the optimisation criteria was advocated [5], as well as uniformity of distribution of the remaining material on part surface after roughing [6]. Overall, despite substantial efforts, a comprehensive and practically feasible methodology for the optimal determination of the process parameter values has not been established until today and therefore process planning decisions for sculptured surface machining mostly remain experience based.

### 5.2.3 Optimisation Formulations

Optimisation criteria and constraints in machining processes always evolve around three major axes: quality, cost and productivity. Besides, these are the cornerstones in the modern industry. The constantly posed question is how to achieve optimal products at the highest rate with minimum cost. Researchers worldwide perceive this question in a different manner. For some, quality of the produced part is undertaken as a technological fact; it is strictly contained within predefined bounds that come from both a part's functional properties and the manufacturing cell it is produced at. This is common for industries with vast knowledge and experience on manufacturing a certain part category and it partially derives from the confidence in the embedded technology and equipment. However, since machining parts may vary from simple prismatic to highly sophisticated with sculptured surfaces, other researchers disagree tending to embrace the notion of continuously producing part with even better quality, especially when demand for new parts arises.

It is common ground that production and machining cost should be as low as possible, hopefully with no compromise to products' quality. Thus, in an optimisation procedure cost is always regarded as a criterion that should be minimal. Cost of manufacturing processes can be attributed and distributed into a number of components, such as actual machining time, length of rapid moves, number of tool changes, loading and unloading the CNC machine tool, part setup, energy consumption, tool usage and wear, work-hour and level of trained employees, associated secondary processes on the part and in general anything that adds value to the part. Productivity refers to the efficient usage of all involved resources in the manufacturing chain per time unit. It comprises personnel's work productivity, machine tool productivity, tool productivity etc. An optimisation procedure may be formulated to include as many as possible or it can be focused in a specific part of the manufacturing chain.

In abstract mathematical terms, a holistic treatment of machining processes optimisation would need to find:

$$\vec{f}(a_1, a_2, \ldots, a_n) = \text{optimal}\left(\vec{Q}, \vec{C}, \vec{P}\right) \tag{5.1}$$

where

$a_1, a_2, \ldots, a_n$: manufacturing process parameters, e.g. feedrates, spindle speed, depth and width of cut, tool, machine tool, etc.

$(\vec{Q}, \vec{C}, \vec{P})$: a vector with values of Quality, Cost and Productivity that should be optimised; maximum Quality, minimum Cost and maximum Productivity are addressed.

$\vec{Q}$: a vector with components such as mean surface roughness, operational behaviour, total remaining volume, distribution of remaining volume, etc.

$\vec{C}$: a vector with components such as machining time, total production time, etc.

$\vec{P}$: a vector with components such as number of produced parts per time unit, scrap material per number of products, machining cell efficiency, etc.

Implementing Eq. 5.1 in machining optimisation problems inherently demand for high computing times owing to the vast solution space. It is therefore unpractical to do so without taking under consideration specific constraints imposed both by the manufacturing cell technical specifications and by managerial decisions or priorities. Constraints can be divided into two major categories: (a) technological constraints (TC) and (b) quality constraints (QC). The former constraint category concerns technical limitations of a manufacturing cell, proper usage of machining equipment and elimination of uncontrollable factors during machining. The latter category is a set of limits (lower, upper or both) applied to optimisation objective values and/or to process parameters, as long as they do not supplant any technical constraints, and are external to specific manufacturing processes, but internal to the overall manufacturing procedure and to the enterprise's quality/performance policy.

In the relevant literature, researchers have considered a number of constraints depending on the solved problem's nature [7–10], such as:

a. Maximum available spindle power, maximum cutting force, maximum allowed loading of cutting edge, maximum tool deflection.
b. Maximum spindle torque, maximum spindle speed, maximum feedrates.
c. Parameter value limits.

d. Tool geometry.
e. Tool wear and tool life, heat during cutting.
f. Geometric and technical limits, stiffness of CNC machine.
g. Surface quality specifications, machining errors in parts.

Constraints in (e) could be characterized either as TCs or as QCs. The rest constraints fall in the TC category with the exception of (g), for which QC category is more appropriate. Spindle speeds, feedrates, machine power and torque, parameter value limits, available tool properties, CNC machine tool size and stiffness constitute technical properties of the available equipment and cannot be neglected in an optimisation procedure.

In order to quantize the effect of constraints in an optimisation problem, a mathematical formulation is necessary. So, let $TC_i$ be a technological constraint in a machining optimisation problem. Then for any $TC_i$ it should be:

$$\min TC_i < TC_i < \max TC_i \tag{5.2}$$

where $i = 1, \ldots, l$, l is the number of TCs, min $TC_i$ and max $TC_i$ are the lower and upper limits of each $TC_i$ respectively.

As before, let $QC_j$ be a quality constraint in the same optimisation. Then for any $QC_j$ it should be:

$$\min QC_j < QC_j < \max QC_j \tag{5.3}$$

where $j = 1, \ldots, m$, m is the number of QCs, min $QC_j$ and max $QC_j$ are the lower and upper limits of each $QC_j$ respectively.

Expressed in natural language, a suitable optimisation algorithm is called to solve Eq. 5.1 taking into account constraints described by (5.2) and (5.3). The number of optimisation criteria, as well as the number of independent variables, defines the complexity of solved problems, thus the sophistication of the optimisation algorithm that should be used. It is very difficult, if not impossible, to approximate an analytical function that correlates machining process variables to the objectives. Even if this was possible, this function would involve discontinuities and corner points. Hence, classical optimisation algorithms, which mainly work with deviation and local extremes of continuous functions, tend to perform poorly or have a large response time. Artificial Intelligence Algorithms, however, are capable of managing discontinuous functions and do not need deviation information while they converge to global optima.

### *5.2.4 Intelligent Optimisation Methods in Machining*

The multitude of parameters and the span of their values makes it difficult to apply classic optimisation in machining due to their high computational cost, even by current hardware standards. This is aggravated by the multitude of criteria that may be used for evaluating a possible solution. Modern optimisation approaches have instead been applied in the field, such as (along with characteristic references): Simulated annealing and genetic simulated annealing [10], genetic and evolutionary algorithms [11], particle swarm optimisation [9], tribes method [8], heuristics in GAs [12], GAs with ANNs [13], memetic algorithms [14] and ant colony optimisation [15].

These optimisation algorithms have been proven efficient at a varying calculation cost taking into account a limited number of process parameters, usually feedrate, spindle speed, depth and width of cut. However, when it comes to more detailed process parameters, such as the ones given in CAM software or if the machining strategy has a variant width or depth of cut for example, there is need for an even more sophisticated method in order to find the global optima. Apart from real valued process parameters, there are some that operate as on–off switches, for example the cutting tool that performs final contour pass or not, and others that take integer values, for example a cutting tool or a CNC machine tool. Besides, if the machining strategy involves variant width or depth of cut, then it is important to appropriately model the machining problem and, consequently, the solving algorithm. In other words, the sophistication of optimisation algorithm is directly driven by a problem's complexity.

A logical way to create a more sophisticated optimisation algorithm is to couple two or even more optimisation algorithms to form a new hybrid method. One common approach is to utilize artificial neural networks (ANNs) as surrogate metamodels that operate as prediction functions of the objectives after proper training. This notion has drastically improved response times of optimisation algorithms, but still needs to be implemented with care, since ANNs ought to generalize well after training with large data sets. Although the ANN approach minimizes calculation cost, it does not cope with issues regarding problem complexity. The most efficient way to both minimize calculation cost and accurately solve a complex machining of sculptured parts is to combine Genetic or Evolutionary Algorithms with another optimisation algorithm. This has been reflected in the following trends.

- Implementing genetic operators in another algorithm, e.g. Genetic Simulated Annealing [16]
- Implementing another method's operators in an Evolutionary Algorithm, e.g. Heuristics [12]
- Formulating Games that combine two or more Evolutionary Algorithms (EAs), typically in a distributed [17] or hierarchical [6] manner.

## 5.3 Rough Milling Optimisation

### *5.3.1 Criteria*

The main objective of rough milling operations on sculptured surface parts is to remove as much material as possible from an original raw material block in the least amount of time or, equivalently, with the highest possible rate. This is to be carried out with respect to cutting conditions during machining operations that follow roughing, such as semi-finishing or finishing. An appealing notion is to utilize material removal rate (MRR) or mean material removal rate as an optimisation criterion, since it combines machining time and removed material implicitly. Although MRR is a means to measure roughing productivity, it does not offer a direct indication of machining time or of the material that must be removed during the rest machining processes.

Machining time, which includes rapid moves and tool changes, directly measures a roughing operation's cost as it allows comparison among different alternatives. For example, it is easy to compare the cost of two different machining strategies when 1,000 and 10,000 s are respectively needed. It is widely accepted that time is cost as far as energy consumption, human and other resources are concerned. Besides, occasional cost of resource usage is also implied. However, considering cost as an objective while neglecting quality is a serious mistake.

In roughing operations, part quality cannot be measured by surface roughness. The latter is meaningful when measured on the final part, as it defines the final part's operational behaviour, ergonomics and/or aesthetics. More correctly, a roughed part's quality is thus measured by the remaining volume that is to be machined during finishing or semi-finishing. The amount of remaining volume drives finish machining time as well as finish cutting conditions. Variation of cutting forces during finishing primarily imposes the finished part quality; the less the variation the smaller the surface roughness in critical sections of the final part. It has been proven that cutting forces variation is an immediate result of removed volume variation. In other words, roughing remaining volume topology affects finishing quality. Hence, both the amount of remaining volume and its topology must be introduced as rough machining objectives in an optimisation procedure. Volume topology can be measured by the remaining volume's standard deviation.

If the generalized Eq. 5.1 is to be applied to roughing processes optimisation, it can be formulated as follows:

$$\vec{f}(a_1, a_2, \ldots, a_n) = \min(\text{MT, RV, SD}) \tag{5.4}$$

where

$\alpha_1, \alpha_2, \ldots, \alpha_n$: rough milling parameters, e.g. CNC machine tool, tool, feedrate, spindle speed, depth of cut, etc. depending on applied machining strategy.

MT: rough milling machining time.
RV: remaining volume after roughing.
SD: standard deviation of remaining volume.

The optimisation procedure should find the global minimum of all three objectives. Note that normalization of the three objectives is not necessary, if multi-objective optimisation is applied; it should be carried out if single-optimisation optimisation is considered, though, as it will be explained next. Also note that productivity is implicitly calculated making use of machining time and remaining volume, or removed volume, and therefore there is no need to include it in the optimisation scheme.

### 5.3.2 Constraints

As mentioned in Sect. 5.2.3, TCs do not concern just machining parameters, which constitute the optimisation problem's independent variables, but other technical quantities that result from their combination, such as power, torque, cutting edge loading etc. It is possible to include the CNC machine tool as a machining parameter in the optimisation procedure, if there is more than one available machine tool that can efficiently machine a particular sculptured surface part. Tools are also considered as machining parameters, since there is always an available set that can be implemented for the current machining strategy. Obviously, these two parameters have integer values, for instance 1–10, reflecting the ID of a certain CNC machine tool or tool respectively. Depending on the machine tool and the tool used, parameters such as feedrates and spindle speed are confined within specific technical limits coming from considered equipment. Moreover, tool manufacturers propose specific limits for values of feedrates, spindle speed, depth and width of cut, so as to maximize tool life; these limits should also be applied as constraints.

Thus, each machining parameter takes values that lie within limits:

$$a_{i\,\min} \leq a_i \leq a_{i\,\max} \tag{5.5}$$

where

$\alpha_i$: machining parameter $i, i = 1, \ldots, n$

$\alpha_{imin}$, $\alpha_{imax}$: minimum and maximum values of parameter $\alpha_i$

Cutting power measures feasibility of a machining scenario (Eq. 3.6). It is limited by the maximum available power of the CNC machine tool (Eq. 3.7).

$$P_c = \frac{a_p \times a_e \times v_f}{60,000,000 \times \eta} \times k_c \tag{5.6}$$

$$0 < P_c < P_{\max} \tag{5.7}$$

where:

$P_C$: cutting power in k Watts.
$P_{\max}$: maximum power of spindle motor in kW.
$\alpha_P$: depth of cut in mm; it is one of the machining parameters $\alpha_i$
$\alpha_e$: width of cut in mm; it is one of the machining parameters $\alpha_i$
$v_f$: cutting feedrate in mm/min; it is one of the machining parameters $\alpha_i$
$\eta$: efficiency coefficient of spindle motor.
$k_C$: fraction stiffness in N/mm$^2$, which depends on the part material.

Apart from the TCs, a set of QCs should be applied in order to facilitate convergence of the optimisation procedure. Experience on manufacturing processes imposes limits to the rough machining objectives of sculptured surface parts. In practice, machining time should always be a small as possible, but with an upper bound, so as to avoid excessive amounts of time spent on the first machining phase (Eq. 5.8). Besides, zero machining time means that no machining has been performed on the raw material, which is inadmissible. The same rule applies for the remaining volume. It should be at least larger than the final part's volume, but smaller than a predefined value, e.g. less than 110% of the total ideally finished part volume (Eq. 5.9). It is obvious that standard deviation of a roughed part's remaining volume should be less than a maximum value, since after this threshold vibrations on the tool during finishing would lead to a distorted final sculptured surface part. However, considering a lower bound to standard deviation's value is meaningful as there is a value under which no measurable tool vibrations appear; calculation cost is thus kept within reasonable limits (Eq. 5.10).

$$\mathrm{MT}_{\min} < \mathrm{MT} \leq \mathrm{MT}_{\max} \tag{5.8}$$

$$RV_{\min} < RV \leq RV_{\max} \tag{5.9}$$

$$SD_{\min} < SD \leq SD_{\max} \tag{5.10}$$

where

$\mathrm{MT}_{\min}$, $\mathrm{MT}_{\max}$: lower and upper limit of machining time respectively.
$\mathrm{RV}_{\min}$, $\mathrm{RV}_{\max}$: lower and upper limit of remaining volume respectively.
$\mathrm{SD}_{\min}$, $\mathrm{SD}_{\max}$: lower and upper limit of standard deviation respectively.

### *5.3.3 Single and Multi-Objective Optimisation*

Before applying a certain optimisation method, researchers need to decide on the type of philosophy for the rough milling optimisation problem; either single-objective or multi-objective optimisation can be considered. Single-objective optimisation converges to a single optimal solution that is directly applicable. On the other hand, multi-objective optimisation exploits Pareto optimality and offers a set of optimal solutions out of which the user is expected to choose and apply to the manufacturing system.

Since there are more than one physical objectives in rough machining processes as described by Eq. 5.4, one would wonder how it is possible to formulate a single-objective optimisation. This can be overcome by implementing weighing factors $w_i$ and formulating Eq. 5.11.

$$O(a_1, a_2, \ldots, a_n) = w_1 \cdot \mathrm{MT} + w_2 \cdot \mathrm{RV} + w_3 \cdot \mathrm{SD} \tag{5.11}$$

where
$O(\alpha_1,\alpha_2,\ldots,\alpha_n)$: the objective function with machining parameters $\alpha_1,\alpha_2,\ldots,\alpha_n$
$w_1,w_2,w_3$: real valued weighing factors of MT, RV, SD respectively, taking values in [0, 1] and summing up to 1.

Degrading the objective function's complexity with Eq. 5.11, thus the optimisation method's also, is feasible, but needs normalization of values prior to using them in the equation. This is carried out easily by simple numerical transformations of the objectives' values, such as dividing MT with a large MT value or dividing RV with the 3D part model volume. If normalization is not applied, optimal solutions are not reliable, since values of the three physical objectives are of different degrees. Machining Time is measured in seconds, varying from several hundred to several hundred thousand seconds according to the sculptured surface part complexity. Remaining volume varies from several hundred thousand to several hundred million cubic millimeters, while Standard deviation of remaining volume varies from microns to a few millimeters. Thus, summing up values of these three physical objectives is a priori biased towards the physical objective with the larger values of the three.

For multi-objective optimisation, Eq. 5.4 is transformed into Eq. 5.12. In this equation, $\vec{O}(a_1,a_2,\ldots,a_n)$ is a vector function. Although normalization of physical objectives' values is not necessary, researchers are advised to do so because uniformity of objective components contributes to robustness of the optimisation algorithm.

$$\vec{O}(a_1,a_2,\ldots,a_n) = (\text{MT,RV,SD}) \tag{5.12}$$

## 5.3.4 Examples

### 5.3.4.1 Optimisation Using a GA Coupled with ANN for Objective Prediction

Feedforward back-propagation ANNs can be trained to predict objective values. ANN training is performed using data sets obtained by systematic data gathering, such as by Design of Experiments. Here, two different ANNs are considered; the first predicts machining time values and the second predicts the remaining volume values after roughing simulation in the CAM software.

Single-objective optimisation is carried out by a simple GA, see Fig. 5.6 for the sculptured surface part depicted in Fig. 5.7a starting off from a rectangular block and employing a flat endmill. The values of objectives, as offered by the ANNs, are normalized, weighed and summed according to the process described in Sect. 5.3.3. For this case, the two objectives are considered equally important, thus $w_1 = w_2 = 0.5$. Raster roughing strategy is applied and Table 5.1 shows the optimal machining parameter values when the GA converged.

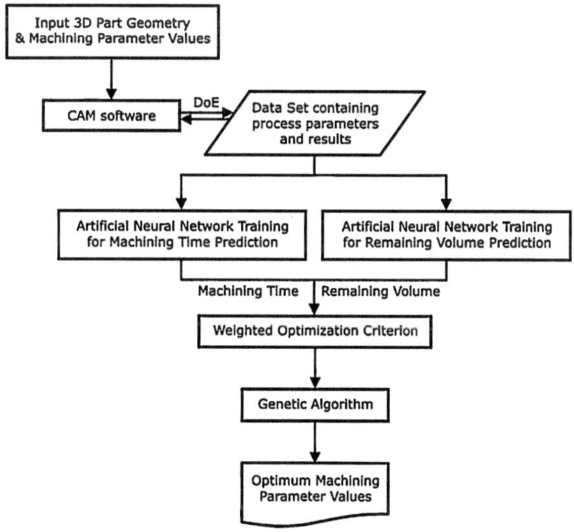


**Fig. 5.6** GA optimisation with ANN as predictors of objective values

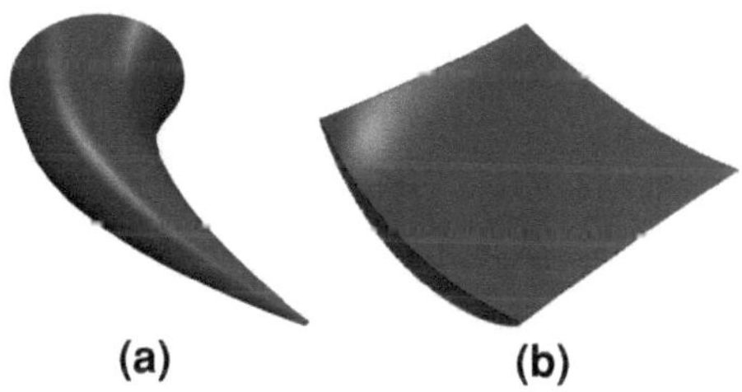


**Fig. 5.7** Target sculptured surface parts

#### 5.3.4.2 Optimisation Using a Genetic Nash Game

In Games, parameters and objectives are distributed among the players. According to Game theory, a game reaches "Nash equilibrium" at which if any player changes his strategy, no player gains more profit. The formulated Genetic Nash Game evolves in generations, as any GA. In each generation, half of the problem variables are appointed to one player, who optimises one objective, e.g. Machining Time, while it considers the rest of the variables constant. The rest of the variables are appointed to the other player, who optimises the other objective, e.g. Remaining Volume after roughing. A player's optimal variable values are passed on to the other player as constants for his next generation.

Figure 5.8 depicts the swapping of variable values in each generation. This process continues until convergence. Prior to the objective values extraction through CAM software, the technological and quality constraints are applied to a candidate solution, as seen in the flowchart of Fig. 5.9.

In this paradigm, feedrate and spindle speed were also included as process parameters and they were distributed between the two players as seen in Table 5.2. The 2-player Genetic Nash Game utilizes Pareto optimality for the solution set.

**Table 5.1** Optimisation results for GA coupled with ANN method

| | |
|---|---|
| Tool diameter | 25 mm |
| Stepover | 19.8 mm |
| Stepdown | 5 mm |
| Thickness | 0.55 mm |
| Allowance | on (automatic) |
| Profiling | During |
| Raster angle | 90° |
| Pass joining range | on (infinite) |
| Machining time | 345 s |
| Remaining volume | 10167.7 mm$^3$ |

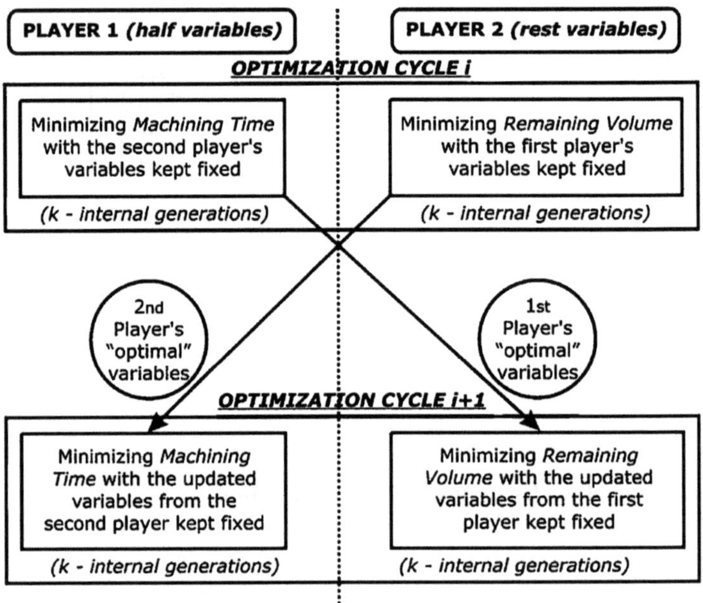


**Fig. 5.8** Two-player Genetic Nash Game

The results of the algorithm application on a sculptured surface part shown in Fig. 5.7b has shown that this method is about twice as fast compared to a simple EA, but it is not always as efficient as the EA.

#### 5.3.4.3 Optimisation Using Coupled EA and $\mu$-GA in a Stackelberg Game

In the direction of genetic games, a novel methodology is implemented, as seen in Fig. 5.10. This is different from the previous one in the fact that it couples an EA with a Micro-Genetic Algorithm ($\mu$-Ga) in a hierarchical game. A $\mu$-GA is in

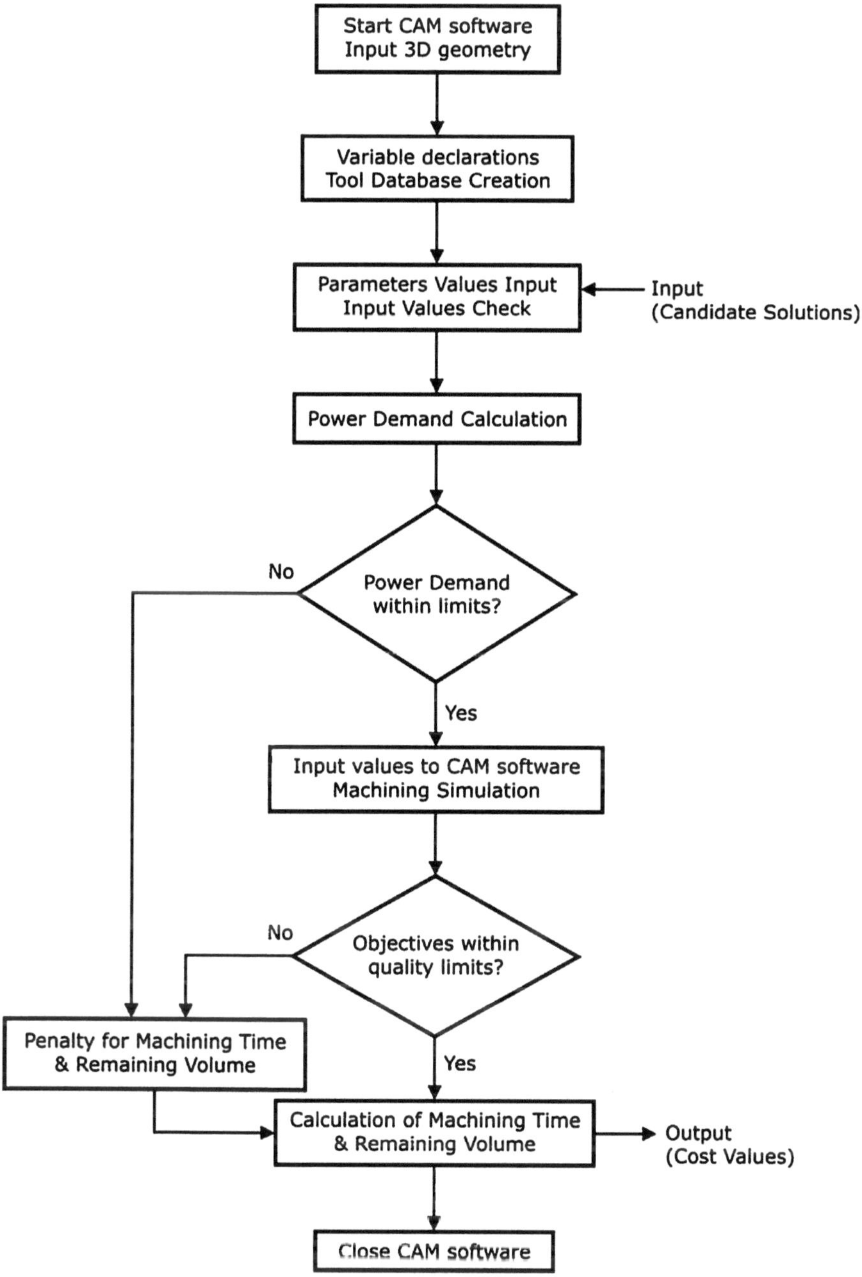


**Fig. 5.9** Flowchart of the evaluator

**Table 5.2** Variable limits and distribution in the genetic nash game

| No. | Variable | Lower limit | Upper limit | Player 1 | Player 2 |
|---|---|---|---|---|---|
| 1 | Tool ID | 1 | 8 | X | |
| 2 | Stepover (% D) | 0.1 | 99.9 | | X |
| 3 | Thickness (mm) | 0.5 | 3.0 | | X |
| 4 | Stepdown (mm) | 0.1 | 6.0 | X | |
| 5 | Profiling | 1 | 4 | X | |
| 6 | Raster angle (degrees) | 0.1 | 90.0 | X | |
| 7 | Allowance (mm) | 0.1 | 2.7 | | X |
| 8 | Joining range (on–off) | 0 | 1 | | X |
| 9 | Feedrate (% max F) | 0.0 | 1.0 | | X |
| 10 | Spindle speed (rpm) | 10.0 | 6,000.0 | X | |

general a GA with small population size. The EA is considered as the main GA, working as the leader in the game, while the $\mu$-GA function as the followers. The main GA optimises process parameters as seen in Table 5.2 with the difference that Variable 4 (Stepdown) takes integer values that represent the number of stepdowns that will be performed during the roughing operations. Then, $\mu$-GA variables become the stepdown heights and are optimised according to three objective functions: (a) machining time (b) remaining volume and (c) standard deviation of remaining volume. For every candidate solution of the leader, a different follower is created, in accordance with the value of variable four after a specific number of generations.

The Main GA performs multi-objective optimisation utilizing Pareto optimality, while the $\mu$-GAs perform single-objective optimisation, with the same three "natural" objectives in a properly weighed sum. Evolution of the $\mu$-GA is carried out in steps, e.g. of five generations per step, so as to minimize evaluation cost for candidate solutions with poor objective values. Each time the same value of Variable 4 appears in the main GA, the $\mu$-GA picks up the last best parameter values and evolves for another step.

Figure 5.11 depicts the flowchart of $\mu$-GA evolution. Schematically, it resembles any typical EA. However, $\mu$-GAs demand high elitist behaviour and strong mutation schemes. So, some of the genetic operators are designed from scratch; for example the ranking function boosts survival rate of optimal individuals of the population and the mutation scheme starts off with high rate, diminishing in relation to the number of generations.

Application of this algorithm on sculptured surface parts, such as the ones depicted in Fig. 5.7, leads to obtaining optimal rough machining parameter values with high efficiency and in reasonable time, considering the problem complexity. Moreover, the algorithm calculates the optimal stepdown height distribution. Choosing a solution out of a Pareto front, at which this algorithm converges, can be done based on specific demands regarding process, CNC machine tool, tool or manufacturing cell performance. Both the normalized and the real values of the three objectives are presented in Table 5.3.

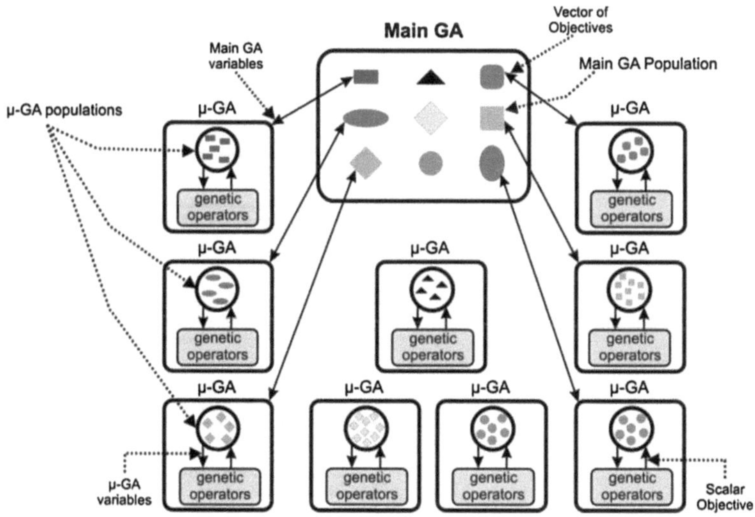


**Fig. 5.10** Optimisation using Stackelberg's hierarchical genetic game

The proposed methodology takes advantage of the high exploration capability of the EA and the high exploitation capability of the $\mu$-GA. Thus, the solution space is searched both quickly and efficiently as far as the global optimal is concerned. It is noted that this algorithm performs well irrespective of the type and size of sculptured surface parts.

## 5.4 Finish Milling Optimisation

### *5.4.1 Criteria*

The main goal of the finishing stage is to satisfy all the morphological and qualitative requirements that will render the final part in accordance with its technical specifications. Commonly used morphological requirements are the geometrical tolerances and dimensional deviations of the final part. The most common qualitative requirement is surface roughness.

Following the formulation of Sect. 5.2, these are the components of Quality, see Eq. 5.1. It is therefore desirable to maximize quality, i.e. to use the above magnitudes as criteria in the optimisation procedure. Of course, Cost, see Eq. 5.1, is always being taken into consideration and in the case of the finishing stage, time is usually the criterion of choice due to the low feeds and a very large number of cutting tool movements employed. While time can be easily determined through simulations on CAM systems, calculating or even estimating the values of the

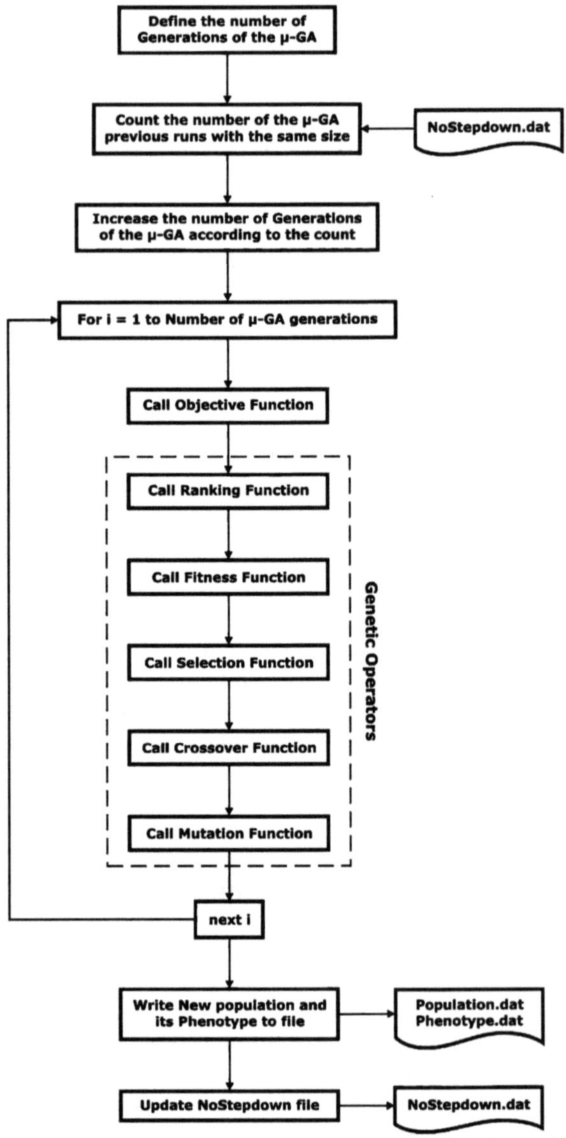


**Fig. 5.11** Flowchart of $\mu$-GA optimisation

Quality criteria is very difficult due to the large number of factors that affect them as well as the interactions between these factors [18].

The complexity of the problem, see Fig. 5.12 is further increased due to the following:

- There is no unique correlation between these factors and the result of the finishing stage, meaning that there is not only one combination of values that satisfies the accuracy requirements.

**Table 5.3** Minimum and maximum values of the three objectives for the studied sculptured surface parts

| Part | MT | | RV | | SD | |
|---|---|---|---|---|---|---|
| | Min (s) | Max (s) | Min ($mm^3$) | Max ($mm^3$) | Min (mm) | Max (mm) |
| Fig. 5.7a | 0.0075 (81) | 0.3642 (3,933) | 0.0705 (5,659) | 0.4337 (34,809) | 0.1879 (1.429) | 0.4184 (3.182) |
| Fig. 5.7b | 0.0364 (393) | 0.1539 (1,662) | 0.4002 (1,034,394) | 0.4789 (1,237,899) | 0.0911 (4.527) | 0.1764 (8.768) |

*MT* machining time, *RV* remaining volume after roughing, *SD* standard deviation of remaining volume in normalized values
Numbers inside parentheses are the real values of the objectives

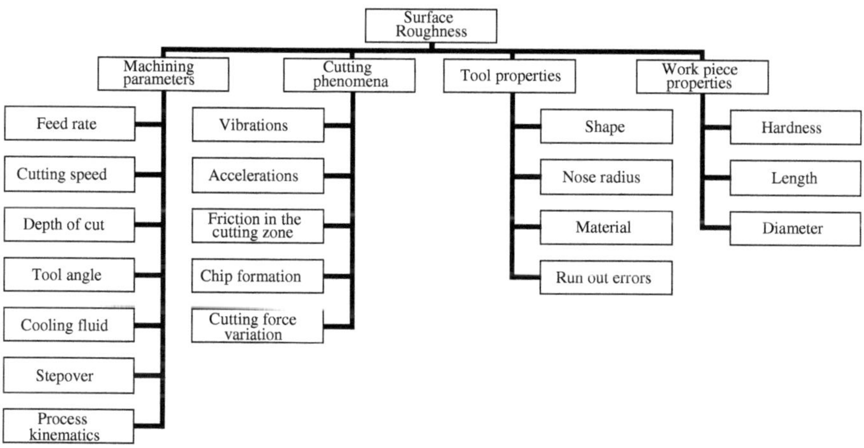


**Fig. 5.12** Factors that affect surface roughness in machining

- Due to the complex geometry of the sculptured surface parts, cutting conditions are almost constantly changing along the cutting tool path.
- Each machine tool—cutting tool—workpiece system is characterized by different properties and particularities (for example stiffness).

To overcome these problems, two modelling approaches are typically adopted. The first one develops models that attempt to correlate the process' controllable parameters (cutting speed, feed, depth of cut etc.) and the dimensional accuracy and surface roughness of the final part [19–21]. The second one tries to achieve the same goal but in an indirect way by correlating the process' controllable parameters and the developed cutting forces, since these are linked with the characteristics of the generated surface [22–24].

Despite the fact that generally cutting forces are low during the finishing stage, their variations can be high enough to be responsible for dimensional deviations and poor surface quality in the final part. Consequently, the problem of determining the best solution in regard to the values of $Q$ in Eq. 5.1, is transformed to determine the best solution in regard to the developed cutting force values $F_C$. The latter also largely depend on other parameters such as roughed part geometry, tool path strategy etc. making it therefore necessary to include them in the overall optimisation procedure. The material removed per cutting tool revolution is often a good representation of such parameters because it directly quantifies the engagement of the cutting tool with the roughed part.

Taking into consideration all of the above, in the case of finishing stage optimisation Eq. 5.1 can be formulated as follows:

$$\vec{f}(a_1, a_2, \ldots, a_n) = \min(\mathrm{MT}, F_C) \tag{5.13}$$

where
$\alpha_1, \alpha_2, \ldots, \alpha_n$: finish machining process parameters, e.g. feedrate, spindle speed, geometry and tool path parameters, e.g. material removed per cutting tool revolution and others.

MT: Finishing stage time.
$F_C$: Cutting force during finishing.

### 5.4.2 Constraints

The discourse in Sect. 5.2.3 regarding the Technological Constraints (TCs) for the optimisation of the roughing stage is also valid for the optimisation of the finishing stage, so Eqs. 5.5–5.7 still apply.

Regarding time (MT), it is typically very large during finishing of sculptured surface parts due to the low pick feeds used and the very large number of cutting tool movements. In practice, very conservative cutting conditions are chosen to deal with the uncertainty concerning the quality of the final part. As a result, the time needed to finish machining the roughed part can vary greatly depending on selected strategy and available resources (selected machine tool, cutting tools etc.). Therefore, MT can only be lower bounded (Eq. 5.14) mainly to avoid cases of very short finishing that would result in a final part of poor quality. In this sense (MT) is used as a criterion in order to choose between solutions that are otherwise equivalent.

$$\mathrm{MT}_{\min} < \mathrm{MT} \tag{5.14}$$

where
$\mathrm{MT}_{min}$: lower limit of machining time.

Regarding cutting forces ($F_C$), a rough estimation can be made beforehand using analytical or even empirical formulas in order to establish upper and lower limits for $F_C$. However, it is advisable to leave $F_C$ unbounded because there are cases where it is significantly different than expected. For example, at the time of the cutting tool's entrance to the roughed part, $F_C$ is many times higher compared to when the cutting tool remains in contact with the roughed part. Similarly, when executing pencil cuts, $F_C$ can be almost negligible. An inaccurate bounding of $F_C$, has the possibility to dismiss solutions that would in fact be the optimum ones.

### 5.4.3 Optimisation Approaches

Due to the important role of SSM, its optimisation has gathered increased research interest and as a result there are a lot of different approaches adopted. Especially for the finishing stage these approaches range from analytical and numerical methods to artificial intelligence techniques. While an exhaustive review of the relevant literature is beyond the scope of this chapter, several representative approaches will be described and their advantages and disadvantages highlighted.

Experimental methods attempt to optimise a response variable, which in the case of finishing is usually the surface roughness. This is done by selecting appropriate control and noise parameters and executing a number of experiments in order to analyse the data and determine the control parameters' optimum values. The Taguchi experimental design method can be used in this context [25]. This method is accurate and can maximize the use of available resources. It can also take into consideration the available equipment and actual cutting conditions through the conducted experiments. However, the determined values cannot be generally applied.

Analytical methods based on geometric analysis or mechanistic force models have also been employed extensively. These approaches aim to either deal with cutter positioning and/or tool path selection problems or cutting condition value optimisation by examining the developed cutting forces [5]). The developed algorithms can easily be used in various optimisation schemes and are fairly accurate. On the other hand, they are usually computationally expensive and despite the rigorous mathematical background, they still involve several simplifications about the developed cutting phenomena.

To cater for increased computational cost while simultaneously embedding process knowledge in the optimisation procedure, several artificial intelligence approaches have been adopted. Various tools are used such as artificial neural networks, genetic and evolutionary algorithms, simulated annealing as well as combinations resulting in hybrid methods [26]. Intelligent-based optimisation exhibits good computational efficiency, increased accuracy and ease of integration both in online and offline schemes. The main disadvantage lies in the difficulty in modelling the desired phenomena.

### 5.4.4 Example

In order to illustrate the implementation of intelligent optimisation in the finishing stage, a selected approach using an ANN cutting force model in combination with

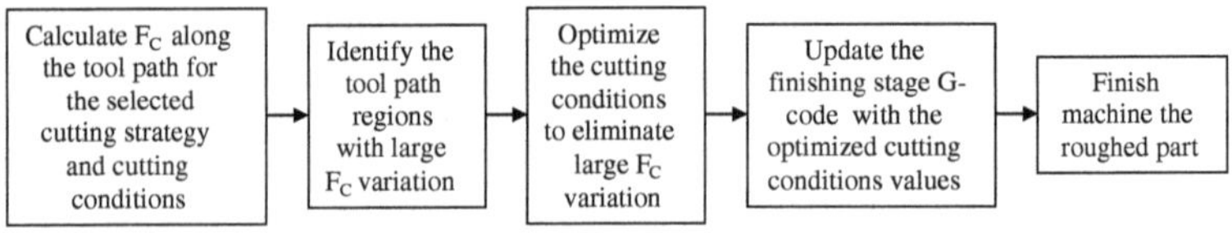


**Fig. 5.13** Philosophy of finishing stage optimisation

a GA is presented step by step. Focus is given on reasoning for each of the choices made, the way that they are fused together and on the resulting benefits. The philosophy of the optimisation approach is given in Fig. 5.13.

The optimisation process is subsequently divided into two stages. The first stage involves an ANN model for the prediction of $F_C$ based on the cutting conditions and part geometry. The second stage employs a GA that incorporates the ANN model in the objective function and that is responsible for optimising the $F_C$ in terms of cutting conditions.

#### 5.4.4.1 ANN Model Development

The selection of the ANN modelling procedure over other alternatives was due to the number of factors that affect the cutting forces, the complex interdependencies among them as well as the fact that ANN modelling can capture the machine tool—cutting tool—workpiece system properties through supervised training with experimental data.

A feedforward ANN is selected, with the model input parameters including the cutting speed, feedrate and material volume removed per cutting tool revolution. These parameters are selected because both cutting speed and feedrate directly affect the $F_C$ and at the same time they are the only controllable cutting conditions during sculptured surface finishing. Furthermore, as already mentioned, the material volume removed per cutting tool revolution directly quantifies the engagement of the cutting tool with the roughed part. Given the finishing strategy and the fact that for selected regions the cutting conditions remain constant, the only thing that affects this magnitude is the roughed part geometry. The model has a single output parameter, the predicted $F_C$. The overall ANN model architecture is shown in Fig. 5.14.

In order to train the model, a series of cutting experiments must be conducted to gather data that relate the cutting conditions and removed volume to the developed $F_C$. Continuing the optimum use of resources trend throughout the procedure, the Design of Experiments technique can be followed in order to minimize the number of required experiments. Once the data are gathered, the ANN can be trained to determine the most appropriate architecture and connection weights. There is a variety of training methods and algorithms but the only systematic procedure that also takes into account the generalization performance of the trained ANN [26].

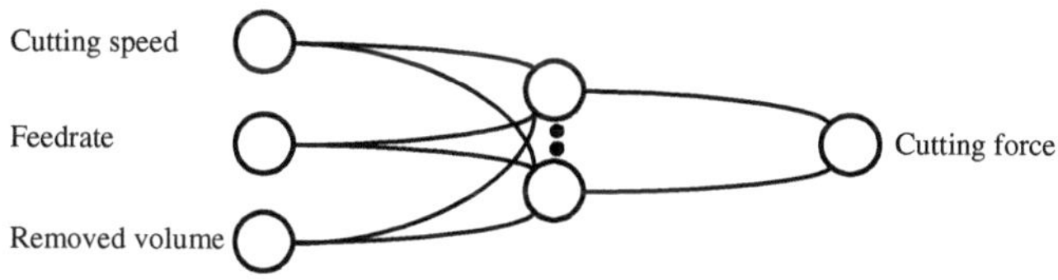


**Fig. 5.14** Architecture of ANN cutting force model

**Fig. 5.15** 3D models of cutting tool types

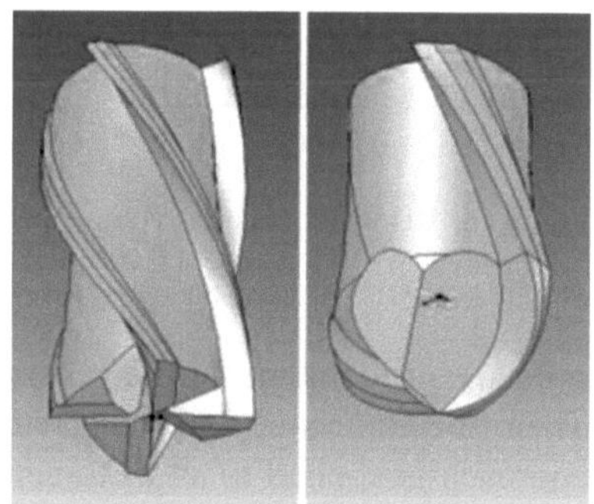

After training is completed, the ANN model can be used to accurately predict the $F_C$ for any point along the cutting tool path by feeding into it the corresponding values of the input parameters. Initial values must be selected for the cutting conditions. In order to calculate the removed material volume, accurate 3D models of the cutting tool and the roughed part are used (Fig. 5.15).

By using the cutting tool path coordinates and the cutting conditions values, the cutting tool's motion can be discretized and for each intermediate point Boolean operations between the cutting tool and roughed part are performed. The resulting benefits are that the removed material is accurately calculated and at the same time the 3D model of the roughed part is updated. The procedure's flowchart is described in Fig. 5.16 and can be implemented using almost any existing CAD software.

#### 5.4.4.2 GA Optimisation

The flowchart of the GA employed for the optimisation of the $F_C$ is depicted in Fig. 5.17.

Each chromosome is broken down into two values that correspond to the values of the cutting speed and feedrate. The type of coding determines the type of values that can be used. If binary coding is used, then only selected discrete values can be represented by the chromosomes in prespecified intervals (for example, feedrates of 200–225–250–275–300 mm/min). If real coding is used then any value can be represented (similarly, any feedrate between 200 and 300 mm/min). There is no

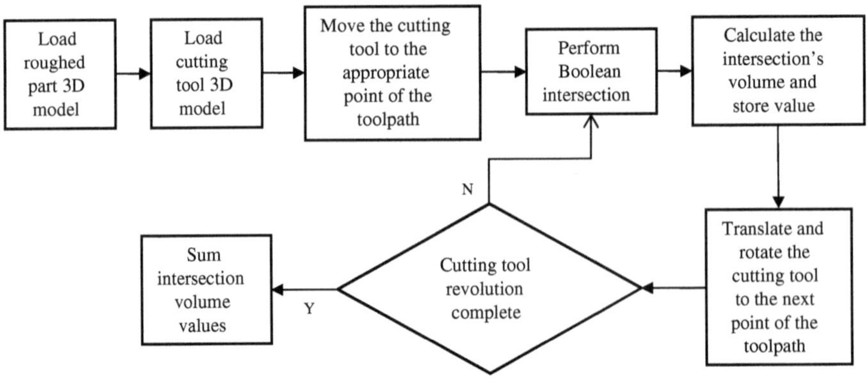


**Fig. 5.16** Removed material volume per cutting tool revolution calculation

special consideration for the genetic operators (selection, crossover and mutation). Their specific values are selected according to each individual case study the approach is applied to so as to cater for avoidance of local minima and speed of convergence.

The final step towards optimising the $F_C$ with the use of the GA is to formulate the objective function. As previously mentioned, the optimisation needs to minimize $F_C$ variation. Equivalently, since optimisation occurs at any single point of the cutting tool path, it needs to minimize the difference between the developed $F_C$ and another value (Eq. 5.15).

$$\text{ObjValue} = \min(F_C - F_{\text{des}}) \tag{5.15}$$

where
$F_C$: cutting force value as calculated by the ANN model
$F_{\text{des}}$: desirable cutting force value

This value can be the $F_C$ of the previous tool path point or the mean $F_C$ of a selected tool path region or in general any desired value. The latter effectively means that desirable $F_C$ patterns can be created and the optimisation procedure will attempt to calculate the cutting conditions values for these patterns. This is made possible by the fact that the ANN force model is included in the objective function formulation, allowing for a variety of different optimisation scenarios.

To exemplify the above, a scenario involving the finishing of the suction side of an airfoil (Fig. 5.18) is presented. An offset profile cutting strategy and a 6 mm ball endmill are selected. A variable stepover is used such that the scallop height in the final part will not be greater than 0.010 mm. Initial spindle speed is set at 3,000 RPM and initial feedrate is 200 mm/min.

Figure 5.19, shows the $F_C$ as calculated by the ANN model (Force). Two regions can be identified, the first corresponding to the initial engagement between the cutting tool and the roughed part and the second corresponding to the subsequent parallel passes of the cutting tool. In this scenario, the goal is to minimize

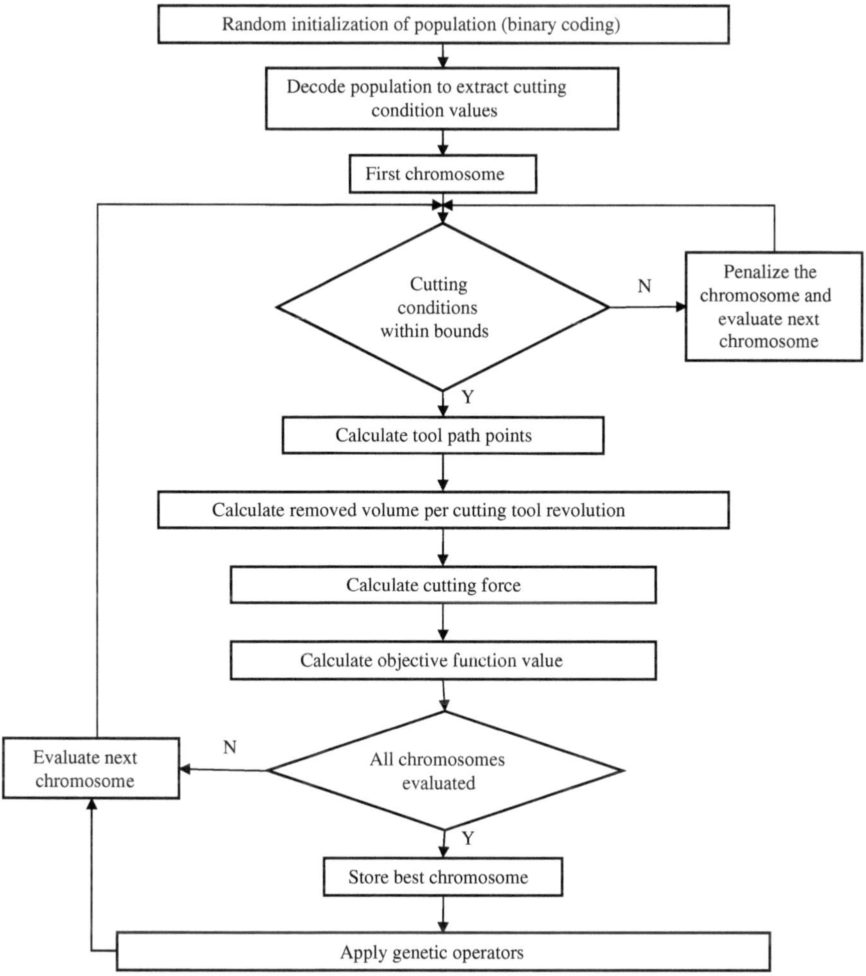


**Fig. 5.17** Flowchart of the GA-based optimisation procedure

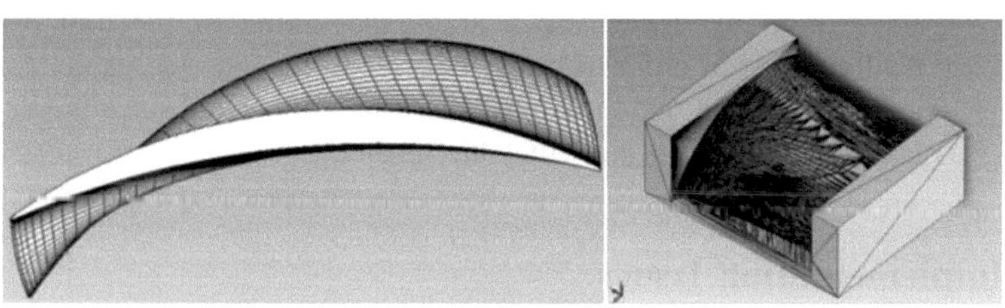

**Fig. 5.18** Airfoil design surface (*left*) and roughed part (*right*)

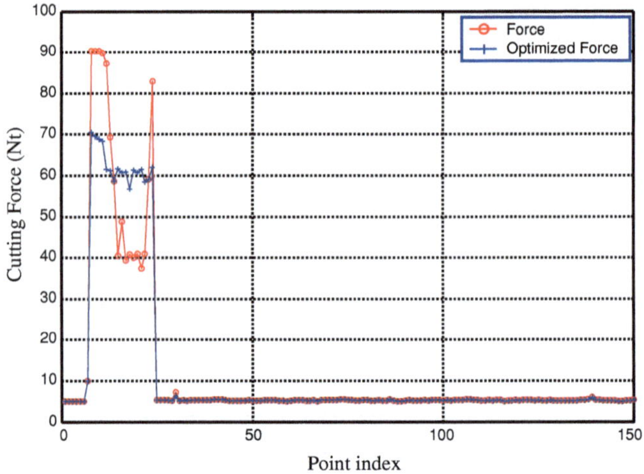


**Fig. 5.19** Non-optimised (*red*) and optimised (*blue*) cutting force

**Table 5.4** Comparison between non-optimised and optimised cutting force

| Cutting force (N) | Non-optimised | Optimised |
|---|---|---|
| Average | 61.63 | 62.52 |
| Standard deviation | 22.22 | 4.19 |

the variation of $F_C$, by minimizing the difference between the actual cutting force and the average cutting force of each of the two regions. So Eq. 5.15 is formulated as follows:

$$\text{ObjValue} = \min(F_C - \bar{F}_i) \tag{5.16}$$

where
$F_C$: cutting force value as calculated by the ANN model
$F_i$: non-optimised cutting force of the ith tool path region (the bar denoting average).

The results of the optimisation procedure are shown in Fig. 5.19. It is obvious that for the first region of the tool path the optimisation procedure has modified the $F_C$ as expected, both by decreasing it at the beginning of the tool path and increasing it afterwards in order to bring it closer to the average value. It is also obvious that for the second region, no optimisation was needed since the variation in $F_C$ was minimal. Table 5.4 quantifies the results.

## 5.5 Implementation Issues

All modern CAM software includes an application programming interface (API), usually in Microsoft's Visual Basic or macros in the native programming language, in order to maximize software customization. Developing applications that interact with CAM software helps in extracting critical machining results and statistics. Since CAM software offers a lot of information regarding the machining process, it can substitute the actual machining process in iterative calculation algorithms. This is done with no material loss and without the need to measure optimisation objectives on the machine tool or on the part.

Thus, an application must be built and be incorporated in the optimisation procedure as the "evaluator" of objective values that interacts with both the CAM and the optimisation algorithm. Practically, the application inputs parameter values and all relative choices for the machining strategy to CAM and extracts machining results, such as the machining time, the machined part, number of tool lifts and changes, tool path length etc.

As an example, the approach described in Sect. 5.4 involves the following tools and methods, closely linked with each other in order to calculate the optimum values of the cutting conditions:

- An ANN model is used to calculate the developed cutting force at any desirable point along the cutting tool path.
- In order to perform this calculation, the removed material per cutting tool revolution is needed (it is an input parameter in the ANN model), which is in turn calculated through CAD/CAM software and accurate 3D models of the cutting tool and roughed part.
- Finally, the ANN model is incorporated in the objective function of a GA responsible for optimising the cutting condition values.

Consequently, the GA is coded as the main routine performing the actual optimisation, whereas the ANN and CAD/CAM software are coded as subroutines, called to provide data to the GA. The main concern regarding the integration of the above is data exchange. While this depends on each chosen platform, two general methodologies can be identified. If the CAM software has an open API, then a single application can be developed and parameter values are simply passed as function arguments. In this case, custom code for the GA and ANN execution would have to be created. If the CAM software API cannot be used as the common programming language, then data exchange can be achieved by using file I/O operations (reads and writes of parameter values to predetermined files). The same is also true in case there are preferred development platforms for the GA and/or ANN. The latter methodology was adopted in our example. A common programming platform for the development of the GA and ANN was chosen as well as a CAD/CAM software with an extensive API based on Visual Basic, so as to maximize the use of its features. The conceptual diagram in Fig. 5.20 presents the flow of information between the different platforms.

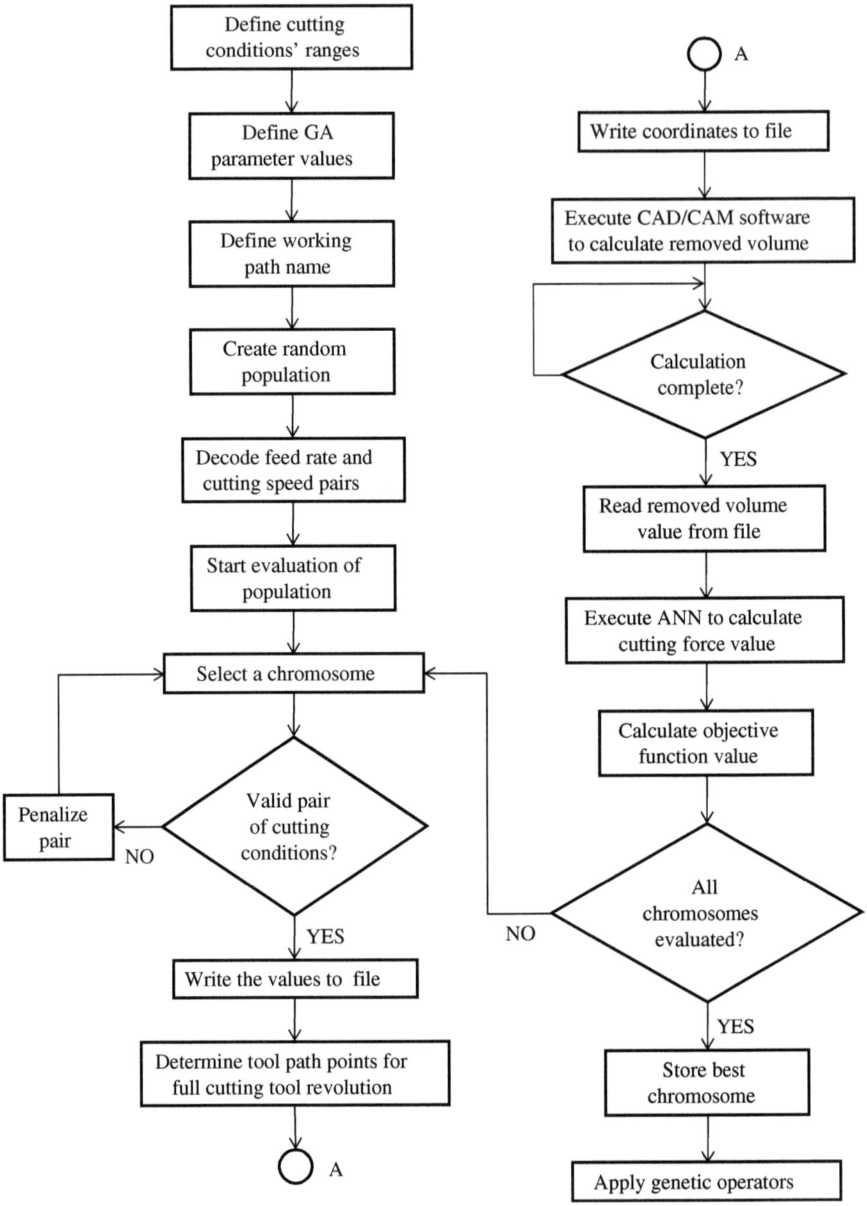


**Fig. 5.20** Integration flowchart between GA, ANN and CAD/CAM

## 5.6 Conclusion

CAM systems are a key point in the optimisation chain of milling operations. High experimentation times and relative costs on actual CNC machine tools can be kept to a minimum by exploiting CAM software. Besides, G-codes for sculptured surface parts machining are strictly produced by postprocessors embedded in CAM software, as they comprise more than some thousands of blocks. Needless to say, these G-codes are almost impossible to write by hand.

Due to the multitude of parameters that are akin to any of the machining strategies used in sculptured surface machining as well as to the variety of strategies as such human-driven experimentation with them on a CAM system is virtually impossible, although an experienced process planner might be able to approach relatively good combinations. However, the quality of results can be assessed by the human only in one dimension (criterion), although at least two dimensions are always present, i.e. machining time and surface quality.

Intelligent optimisation can simultaneously consider all process parameters and find its way through the search space towards optimal or suboptimal combinations using single or multiple criteria, even defining additional criteria as necessary. Furthermore, calculations regarding notably cutting force can easily be included in the optimisation in the form of meta-models that can be built from experiments that capture the idiosyncratic nature of the particular equipment used.

The resulting benefits of the described approach are:

- It is of general application, meaning it can be applied for the optimisation of the roughing and finishing stage of any sculptured surface part, regardless of cutting strategy.
- It is easily modifiable and expandable in order to adapt to different requirements and targets as well as specific cases.
- It incorporates knowledge about the real production floor conditions as well as the cutting mechanisms' interactions, included in the experimental data used to train metamodels.

In order to narrow the search space a rough-cut process plan may be made, based on practical experience of machinists, serving as guidance to the user of a commercially available CAM system. Expert, neural and fuzzy systems are tools towards this end. For instance, an expert system may advise on tool specifications, work holding method, type of milling operation, direction of feed and offset values. A human-assisted fuzzy logical reasoning knowledge base may determine the ranges of operational parameter values (feed, cutting velocity, etc.) to ensure quality specifications, such as surface roughness, are kept within the stipulated range. Finally, a neuro-fuzzy model that uses the concept of 'feature manufacturability' may recognise individual sculptured surface part features and evaluate their manufacturing capability.

**Acknowledgments** Most of the work presented in this chapter was funded as project number 01EΔ131 of the PENED2001 action funded at 75% by the European Union—European Social Fund and 25% the Greek State—Ministry of Development, General Secretariat for Research and Technology and the Private Sector (companies Axon Engineering and VIORAL) under Measure 8.3 of the Operational Programme "Competitiveness" 2000–2006.

## References

1. Chen ZC, Dong Z, Vickers GW (2003) Automated surface subdivision and tool path generation for 3-axis CNC machining of sculptured parts. Comput Ind 50(3):319–331
2. Sun G, Sequin CH, Wright PK (2001) Operation decomposition for freeform surface features in process planning. Comput Aided Des 33(9):621–636
3. Wang Y, Ma H-J, Gao C-H, Xu H-G, Zhou X-H (2005) A computer aided tool selection system for 3D die/mould-cavity NC machining using both a heuristic and analytical approach. Int J Comput Integr Manuf 18(8):686–701
4. Krimpenis A, Fousekis A, Vosniakos G–C (2005) Assessment of sculptured surface milling strategies using design of experiments. Int J Adv Manuf Technol 25(5–6):444–453
5. López de Lacalle LN, Lamikiz A, Sánchez JA, Salgado MA (2007) Toolpath selection based on the minimum deflection cutting forces in the programming of complex surfaces milling. Int J Machine Tools Manuf 47(2):388–400
6. Krimpenis A, Vosniakos G-C (2008) Rough milling optimisation for parts with sculptured surfaces using genetic algorithms in a Stackelberg game. J Int Manuf 20(4):447–461
7. Cus F, Milfelner M, Balic J (2006) An intelligent system for monitoring and optimisation of ball-end milling process. J Mater Process Technol 175:90–97
8. Onwubolu GC (2006) Performance-based optimisation of multi-pass face milling operations using Tribes. Int J Mach Tools Manuf 46:717–727
9. Tandon V, El-Mounayri H, Kishawy H (2002) NC end milling optimisation using evolutionary computation. Int J Adv Manuf Technol 42:595–605
10. Wang ZG, Rahman M, Wong YS, Sun J (2005) Optimisation of multi-pass milling using parallel genetic algorithm and parallel genetic simulated annealing. Int J Mach Tools Manuf 45:1726–1734
11. Agrawal RK, Pratihar DK, Choudhury AR (2006) Optimisation of CNC isoscallop free form surface machining using a genetic algorithm. Int J Mach Tools Manuf 46:811–819
12. Akturk MS, Ghosh JB, Kayan RK (2007) Scheduling with tool changes to minimize total completion time under controllable machining conditions. Comput Oper Res 34(7):2130–2146
13. Tansel IN, Ozcelik B, Bao WY, Chen P, Rincon D, Yang SY, Yenilmez A (2006) Selection of optimal cutting conditions by using GONNS. Int J MachTools Manuf 46:26–35
14. Baskar N, Asokan P, Saravanan R, Prabhaharan G (2006) Selection of optimal machining parameters for multi-tool milling operations using a memetic algorithm. J Mater Process Technol 174(1–3):239–249
15. Baskar N, Saravanan R, Asokan P, Prabhaharan G (2004) Ant Colony algorithm approach for multi-objective optimisation and surface grinding operation. Int J Adv Manuf Technol 23:311–317
16. Wang ZG, Wong YS, Rahman M (2004) Optimisation of multi-pass milling using genetic algorithm and genetic simulated annealing. Int J Adv Manuf Technol 24:727–732
17. Kampolis IC, Giannakoglou KC (2009) Distributed evolutionary algorithms with hierarchical evaluation. Eng Optim 41(11):1037–1049
18. Benardos PG, Vosniakos G-C (2003) Predicting surface roughness in machining: a review. Int J Mach Tools Manuf 43(8):833–844
19. Baek DK, Ko TJ, Kim HS (2001) Optimisation of feedrate in a face milling operation using a surface roughness model. Int J Mach Tools Manuf 41:451–462

20. Benardos PG, Vosniakos GC (2002) Prediction of surface roughness in CNC face milling using neural networks and Taguchi's design of experiments. Robot Comput Integr Manuf 18:343–354
21. Suresh PVS, Venkateswara Rao P, Deshmukh SG (2002) A genetic algorithmic approach for optimisation of surface roughness prediction model. Int J Mach Tools Manuf 42:675–680
22. Bae S-H, Ko K, Kin BH, Choi BK (2003) Automatic feedrate adjustment for pocket machining. Comput Aided Des 35:495–500
23. Guzel BU, Lazoglu I (2004) Increasing productivity in sculptured surface machining via off-line piecewise variable feedrate scheduling based on the force system method. Int J Mach Tools Manuf 44:21–28
24. Kim S-J, Lee H-U, Cho D-W (2006) Feedrate scheduling for indexable end milling process based on an improved cutting force model. Int J Mach Tools Manuf 46:1589–1597
25. Zhang JZ, Chenb JC, Kirby ED (2007) Surface roughness optimisation in an end-milling operation using the Taguchi design method. J Mater Process Technol 184:233–239
26. Benardos PG, Vosniakos G-C (2007) Optimising Feedforward Artificial Neural Network Architecture. Eng Appl Artif Intell 20(3):365–382

# Chapter 6
# Process Planning for 5-Axis Milling of Sculptured Surfaces Based on Cutter's Accessibility Analysis

**L. Geng and Y. F. Zhang**

In recent years, industries such as aerospace, shipbuilding, automotive and even wood carpentering have seen a fast growing use of sculptured surfaces to meet functional or esthetic needs. In this chapter, an automated process planning system for 5-axis finish milling is introduced. The whole system is built on an accessibility analysis algorithm that produces a cutter's accessible posture range to a surface point in the form of A-maps.

## 6.1 Introduction

### *6.1.1 Sculptured Surfaces and 5-Axis Milling*

These surfaces usually have non-uniform curvature distribution and are typically manufactured by end-milling using computer numerical control (CNC) machines. The traditional 3-axis CNC machines can only produce translational movements of the cutter, which means the cutter will remain in a fixed orientation relative to the workpiece in a setup. This greatly confines the cutter's accessibility to the workpiece, resulting in more step-ups, poor material removal rate and finally, lowered machining efficiency. Only for the so-called 'open' sculptured surfaces with simple geometry can 3-axis CNC machining deliver satisfactory performance.

---

L. Geng · Y. F. Zhang (✉)
Department of Mechanical Engineering, National University of Singapore,
9 Engineering Drive 1, Singapore 117576, Singapore
e-mail: mpezyf@nus.edu.sg

J. P. Davim (ed.), *Machining of Complex Sculptured Surfaces*,
DOI: 10.1007/978-1-4471-2356-9_6, © Springer-Verlag London Limited 2012

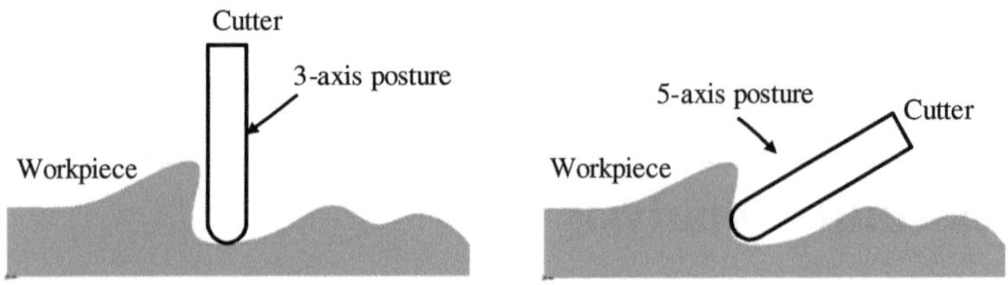


**Fig. 6.1** Accessibility comparison between 3-axis and 5-axis machining

Unlike their 3-axis counterparts, 5-axis machines can produce 5-DOF movements, enabling the cutter to be placed at any point in any orientation within the machine joint limits. Such a feature gives 5-axis machines many advantages over 3-axis machines. First, with the two revolute joints, the cutter's accessibility is greatly enhanced to deal with geometrically and topologically complex surfaces. In one setup on a 3-axis machine, only those regions visible from a particular direction can be milled and inaccessible regions need to be milled by reconfiguring the workpiece setup. However, on 5-axis machines, the orientation of the cutter can change dynamically to access all areas of the surface in a single setup, producing increased machining efficiency and accuracy (see Fig. 6.1). Second, 5-axis machines can also achieve increased material removal rate by orienting the cutter to closely match the local surface shape. To achieve the same level of shape error for machining a given sculptured surface, 5-axis machining needs much shorter tool paths than 3-axis machining.

Although 5-axis machining provides many advantages over 3-axis machining, the flexibility of 5-axis machines comes at a price. Process planning for 5-axis machines is much more complicated than their 3-axis counterparts, as programmers need to provide both location and orientation for the cutter at each cutter contact point. Possible interferences of the cutter with machining/non-machining surfaces also add to the gravity of this task. Most of the currently available commercial CAM software takes a conservative trial-and-error approach, which relies heavily on human intervention and cannot fully explore the capabilities of 5-axis machines. In this chapter, an automated process planning system aimed at selecting an optimal cutter set and generating the interference-free 5-axis tool-paths with maximized machining efficiency will be introduced.

### *6.1.2 The Proposed Process Planning System Structure*

There are two distinctive steps in tool-path generation for 5-axis milling of sculptured surfaces, i.e., generation of cutter location (CL) data and post-processing. CL data describe the cutter's positions and orientations in the workpiece frame and are machine-independent. After CL data have been generated, post-

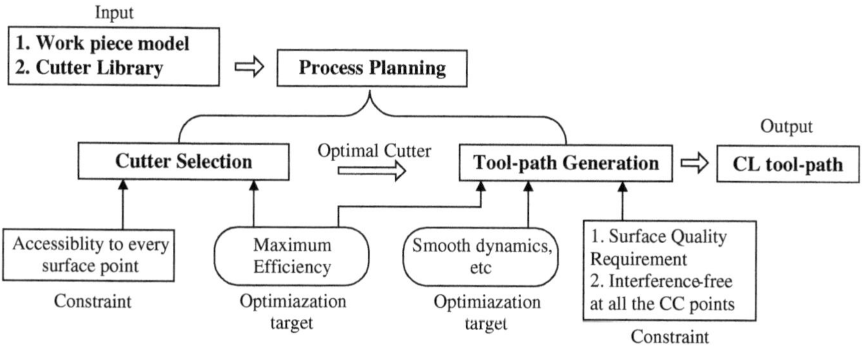


**Fig. 6.2** The proposed process planning system for 5-axis finish machining

processing will be carried out to convert them into joint commands in the form of NC codes. This process heavily relies on machine structures and controller design, and will not be discussed here. Process planning to be discussed in this chapter takes CL data as the final output. Besides, during process planning, various geometric (cutter size and design, tool-path topology and distribution, etc.) and non-geometric (cutting force, cutting dynamics, etc.) factors should be considered. In this chapter, only geometric aspects are considered. Finally, the tool paths generated are intended for 5-axis finish machining, whose target is to produce the final part with satisfactory quality in one pass.

Process planning tasks for 5-axis milling of sculptured surfaces include *cutter selection* and *tool-path generation*. The former refers to the selection of one or a set of cutters that can traverse the whole machining surface without any interference. The optimization criterion is cutter size, as large cutters produce large material removal rate, hence better machining efficiency. The latter selects a tool-path pattern, generates the cutter contact (CC) points based on surface accuracy requirement and determines the interference-free posture (orientation) at each CC point. In the proposed system, these two tasks can be carried out automatically with minimal human intervention. A layout of the proposed system structure is given in Fig. 6.2.

### *6.1.3 The Cutter and Workpiece Models*

The proposed system takes the workpiece model and cutter library as inputs. For the workpiece, the machining and non-machining surfaces all come in the form of NURBS patches. NURBS is chosen as the format because of its wide application in the CAD/CAM industry. Along with the workpiece comes the workpiece coordinate system $\mathbf{O}_G\text{–}\mathbf{X}_G\mathbf{Y}_G\mathbf{Z}_G$, as shown in Fig. 6.3a. The cutters used in this study are fillet-end cutters, which also cover ball-end and flat-end cutters when their minor

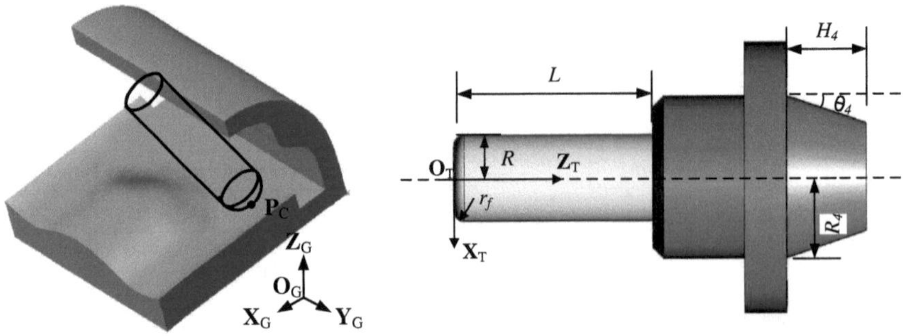


**Fig. 6.3** The models of the workpiece, cutter, and holder

radii are at extremities. A fillet-end cutter is described by its major radius $R$, minor radius $r_f$ and length $L$. The holder of the cutter is also considered during interference checking, which is modeled as a series of cylinders or truncated cones (see Fig. 6.3b). Each segment $m$ in the holder is described by its height $H_m$, bottom radius $R_m$ and angle $\theta_m$. The cutter center refers to the center of the cutter's bottom circle. A tool coordinate system $\mathbf{O_T}$–$\mathbf{X_TY_TZ_T}$ is assigned to the cutter with $\mathbf{O_T}$ coincident with cutter center and $\mathbf{Z_T}$ along the cutter axis (see Fig. 6.3b).

## 6.2 Cutter Accessibility at a Point on the Machining Surface

For 5-axis machining, the range of feasible postures a cutter can take at a point on the surface is confined by many factors. The travel limits on the two revolute joints construct the first boundary for the feasible posture range of a cutter. Other than this, interference avoidance is the biggest issue. Due to the complicated cutter movement, it is common for the cutter to have interference with the machining/non-machining surfaces. It is therefore highly desirable to find the accessible posture range of the cutter before a particular posture is selected. In this section, an algorithm has been developed to find the interference-free posture range for a cutter at any surface point. This algorithm forms the foundation for the automated cutter selection and tool-path generation functions to be introduced in the following sections.

Generally, there are three kinds of machining interferences, i.e., *local gouging*, *rear gouging*, and *global collision*. For each type of interference, there is a corresponding interference-free posture range for a cutter at a surface point. The intersection of these three ranges and the one determined by joint limits, if not empty, gives the accessible posture range for a cutter at a surface point, named *accessibility map* (A-map). Before introducing the algorithms for constructing the A-map, we need to bring in a new local coordinate system $\mathbf{O_L}$–$\mathbf{X_LY_LZ_L}$ (see

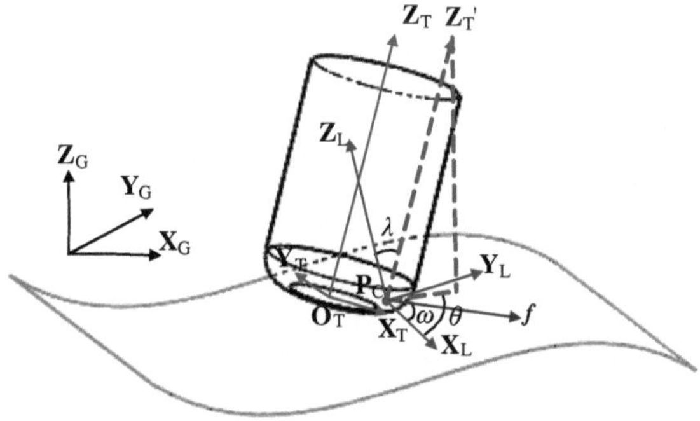


**Fig. 6.4** The local and tool coordinate systems for accessibility analysis

Fig. 6.4). The origin $\mathbf{O}_L$ is coincident with the surface point $\mathbf{P}_C$, with $\mathbf{Z}_L$ coincident with the surface normal, $\mathbf{X}_L$ along the maximum principal direction and $\mathbf{Y}_L$ along the minimum principal direction at the point respectively. As mentioned earlier, a tool coordinate system $\mathbf{O}_T$–$\mathbf{X}_T\mathbf{Y}_T\mathbf{Z}_T$ has been formed with its origin at the center of the cutter bottom and $\mathbf{Z}_T$ along the cutter axis. For a clearer definition, we also make $\mathbf{X}_T$ stays in the plane determined by $\mathbf{P}_C$ and $\mathbf{Z}_T$, pointing towards $\mathbf{P}_C$. In the local coordinate system, a cutter's posture is defined by an angle pair ($\lambda$, $\theta$). $\lambda$, known as the inclination angle, is defined as the angle between $\mathbf{Z}_L$ and $\mathbf{Z}_T$ (or $\mathbf{Z}_T'$, which passes through $\mathbf{P}_C$ and is parallel to $\mathbf{Z}_T$, see Fig. 6.4). Meanwhile, $\theta$ is the rotational angle that describes how much the cutter rotates about surface normal $\mathbf{Z}_L$. $\theta$ equals the angle between $\mathbf{X}_L$ and the projection of $\mathbf{Z}_T'$ on plane $\mathbf{X}_L$–$\mathbf{Y}_L$.

## *6.2.1 Accessible Range for Local Gouging Avoidance*

At a surface point $\mathbf{P}_C$, local gouging occurs when the curvature of the cutter's local surface is smaller than that of the part surface such that the cutter cuts excess material. Therefore, given a posture ($\lambda$, $\theta$) of the cutter, the normal curvatures of the cutter and the part surface in every possible direction need to be compared to detect local gouging. Suppose $\boldsymbol{x}_\omega$ is a unit vector on plane $\mathbf{X}_L$–$\mathbf{Y}_L$ (the surface's tangent plane at $\mathbf{P}_C$). The angle between $\mathbf{X}_L$ and $\boldsymbol{x}_\omega$ is given by $\omega$ ($0 \le \omega \le 2\pi$). As shown in Fig. 6.5, the surface curvature at $\mathbf{P}_C$ on plane $\boldsymbol{x}_\omega$–$\mathbf{Z}_L$ is given as:

$$\kappa_{s\omega} = \kappa_{\max} \cos^2 \omega + \kappa_{\min} \sin^2 \omega \tag{6.1}$$

where $\kappa_{\max}$ and $\kappa_{\min}$ are the maximum and minimum principal curvatures of local part surface at $\mathbf{P}_C$, respectively. On the same plane, the curvature of the cutter is given as [1]:

$$\kappa_{t\omega} = \frac{\cos^2(\omega - \theta)}{r_f} + \frac{\sin^2(\omega - \theta)}{\dfrac{(R - r_f)}{\sin \lambda} + r_f} \tag{6.2}$$

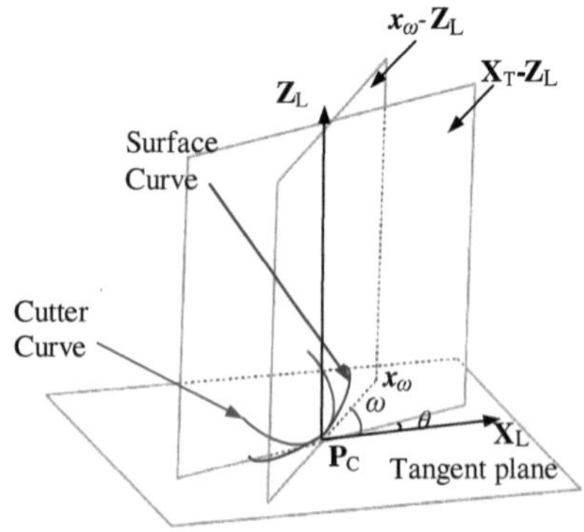


**Fig. 6.5** Intersection curves of the cutter and the surface with $x_\omega$–$\mathbf{Z}_\mathrm{L}$

To make sure the cutter is free of local gouging, for any $\omega$, we must have $\kappa_{t\omega} - \kappa_{s\omega} > 0$. Combining Eqs. 6.1 and 6.2, we obtain the following two inequalities that are independent of $\omega$:

$$\sin\lambda > \frac{r_1(r_f\kappa_{\max} - \cos^2\theta)}{r_f(1 - r_f\kappa_{\max})} \tag{6.3}$$

$$\sin\lambda > \frac{r_1\kappa_{\max}(1 - r_f\kappa_{\min}) - r_1(\kappa_{\max} - \kappa_{\min})\cos^2\theta}{(1 - r_f\kappa_{\min})(1 - r_f\kappa_{\max})} \tag{6.4}$$

where $r_1 = R - r_f$. Given a $\theta$, two minimum values, $\lambda_{\min-1}$ and $\lambda_{\min-2}$, if there is any, can be obtained. The accessible range is thus $[\lambda_{\theta-lg}, 90°]$, where $\lambda_{\theta-lg} = \max(\lambda_{\min-1}, \lambda_{\min-2})$.

## 6.2.2 Accessible Range for Rear Gouging Avoidance

At a surface point $\mathbf{P}_\mathrm{C}$, rear gouging refers to the removal of excess material due to the intrusion of the cutter bottom surface into the part surface. The detection of rear gouging employs a point-based approach. The machining surface is sampled into a high-density point set. For each value of $\theta$, regarding each sampled point $\mathbf{P}_i | i = 1, \ldots, n$, an accessible range $(\lambda_{\theta-rg1-i}, \lambda_{\theta-rg2-i})$ can be obtained at $\mathbf{P}_\mathrm{C}$. The intersection of all the $(\lambda_{\theta-rg1-i}, \lambda_{\theta-rg2-i}) | i = 1, \ldots, n$, is taken as $(\lambda_{\theta-rg1}, \lambda_{\theta-rg2})$, which is the cutter's rear-gouging free posture range at $\theta$. As shown in Fig. 6.6a, for a given $\theta$, when $\lambda$ changes, the cutter rotates around $\mathbf{Y}_\mathrm{T}$ with point $\mathbf{O}$ as the pivot point (the point along surface normal with distance $r_f$ from $\mathbf{P}_\mathrm{C}$). Based on this, the following three conditions are established for a sampled point $\mathbf{P}(x_T, y_T, z_T)$ to be a rear-gouging prone candidate point (note that the point coordinate is expressed in tool frame): (1) $|\mathbf{OP}| \le 2R - r_f$; (2) $\mathbf{P}_\mathrm{C}\mathbf{P} \cdot \mathbf{Z}_\mathrm{L} > 0$; (3) $-R \le y_T \le R$.

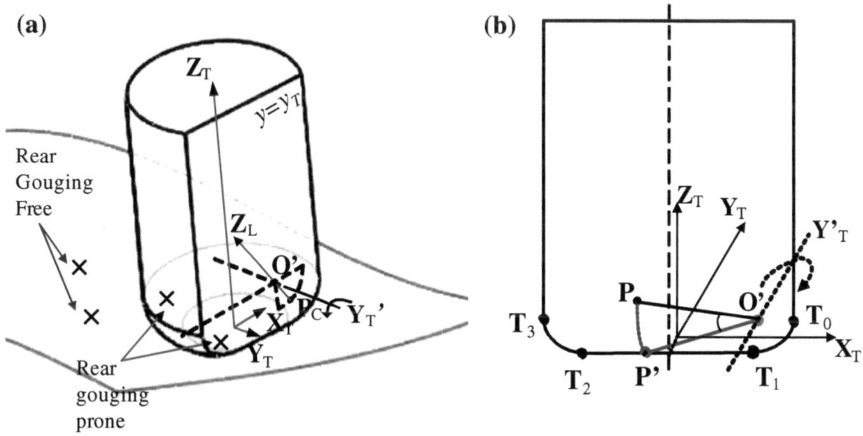


**Fig. 6.6** Rear gouging free posture range. **a** Goughing prone point and the cutter; **b** section curve on the cutter

As shown in Fig. 6.6b, when we set $\lambda$ as 0 and use a plane $y = y_T$ to section the cutter bottom, a section curve of 3 segments is produced: two arcs $\mathbf{T}_0\mathbf{T}_1$ and $\mathbf{T}_2\mathbf{T}_3$ (fillet part) and one horizontal line $\mathbf{T}_1\mathbf{T}_2$ (cutter bottom). If $\mathbf{P}$ is above the section curve, rear gouging occurs. If we increase $\lambda$ by rotating the cutter about $\mathbf{Y}'_{\mathrm{T}}$, $\mathbf{P}$ tends to move towards underneath the section curve. Therefore, we need to find the minimum $\lambda$ to move $\mathbf{P}$ to the cutter outer surface at position $\mathbf{P}'$. Depending on the segment $\mathbf{P}'$ falls onto, calculation of the increment $\Delta\lambda$ that moves $\mathbf{P}$ to $\mathbf{P}'$ is different. We define $d$ as the distance between $\mathbf{P}$ and axis $\mathbf{Y}'_{\mathrm{T}}$, and use $d_0$, $d_1$, $d_2$, and $d_3$ to represent the distances from $\mathbf{T}_0$, $\mathbf{T}_1$, $\mathbf{T}_2$, and $\mathbf{T}_3$ to $\mathbf{Y}'_{\mathrm{T}}$, respectively. The calculation of $\Delta\lambda$ is given as follows:

(1) When $d_1 \leq d \leq d_2$, $\mathbf{P}'$ falls between $\mathbf{T}_1$ and $\mathbf{T}_2$, and its coordinates in the tool frame are given as $\mathbf{P}'(x_T, y_T, z_T) = \left(r_1 - \sqrt{d^2 - r_f^2}, y_T, 0\right)$ and $\Delta\lambda$ is calculated as:

$$\Delta\lambda = \cos^{-1}\left(\frac{r_f - z_T}{d}\right) - \cos^{-1}\left(\frac{r_f}{d}\right) \tag{6.5}$$

(2) When $d_0 \leq d \leq d_1$ or $d_2 \leq d \leq d_3$, $\mathbf{P}'$ falls between $\mathbf{T}_0$ and $\mathbf{T}_1$, $\mathbf{T}_2$ and $\mathbf{T}_3$, respectively. Its coordinates in the tool frame is given as:

$$\mathbf{P}'(x_T, y_T, z_T) = \left(a + \frac{a^2 + y_T^2 - r_1^2}{2(r_1 - a)}, y_T, r_f\left(1 - \sqrt{1 - \left(\frac{a^2 + y_T^2 - r_1^2}{2r_f(r_1 - a)}\right)^2}\right)\right) \tag{6.6}$$

where $a = \frac{2r_1^2 + r_f^2 - d^2 - y_T^2}{2r}$. $\Delta\lambda$ is then calculated as:

$$\Delta\lambda = \sin^{-1}\left(\frac{z_T - r_f}{d}\right) + \sin^{-1}\left(\frac{r_f - z_T'}{d}\right) \quad (6.7)$$

Regarding point **P**, the rear gouging free posture range at $\mathbf{P}_C$ with rotational angle set as $\theta$ is $[\Delta\lambda, 90°]$. After applying the algorithm to all the candidate points, the accessible region for rear gouging avoidance at $\theta$ is $[\theta, (\lambda_{\theta-rg}, 90°)]$, where $\lambda_{\theta-rg} = \max\{\Delta\lambda_i | i = 1, \ldots, n\}$. However, if $\lambda_{\theta-rg} > 90°$, rear gouging is inevitable for the cutter at $\mathbf{P}_C$.

### 6.2.3 Accessible Range of Global Collision Avoidance

Global collision happens when the non-cutting parts of the cutter (cutter shaft, holder) intersect with the machining surface or the non-machining surfaces. To find the accessible posture range for global collision avoidance, a similar approach as that used in rear gouging is taken. Let **n** denote the surface normal at any point $\mathbf{P}(x_T, y_T, z_T)$ on the machining/non-machining surface. For **P** to be a candidate point for global collision, the following conditions should be met: (1) $\mathbf{n} \cdot \mathbf{P}_C\mathbf{P} < 0$ (see Fig. 6.7a). (2) Let $R(z_T)$ denote the radius of the cutter at the height of $z_T$, $-R(z_T) < y_T < R(z_T)$.

Similarly, we use the plane $y = y_T$ to section the cutter. For a certain $\lambda$ ($\theta$ is fixed), if **P** falls inside the section curve, global collision happens. Unlike accessibility analysis for rear gouging, the upper limit for $\lambda$ is no longer 90°. So we need to find both the minimum $\Delta\lambda$ the cutter should be rotated to avoid global collision and also the minimum $\Delta\lambda$ the cutter should be rotated to induce collision. The relative positional relationship between **P** and the section curve can be categorized into the following 5 scenarios:

(1) $z_T < r_f$, **P** is collision-free and its accessibility range is [0, 90°].
(2) $x_T < -R(z_T)$, and $\mathbf{n} \cdot \boldsymbol{X}_T > 0$ $(z_T \geq r_f)$, **P** is collision-free.
(3) $x_T \geq -R(z_T)$, and $\mathbf{n} \cdot \boldsymbol{X}_T > 0$ $(z_T \geq r_f)$, global collision exists (see Fig. 6.7b). The minimum $\Delta\lambda$ that the cutter must be rotated clockwise to avoid global collision is:

$$\Delta\lambda = \cos^{-1}\frac{r_1 - x_T}{d} - \cos^{-1}\frac{r_1 + \sqrt{R(z_T)^2 - y_T^2}}{d} \quad (6.8)$$

where $d$ is the distance from **P** to $\mathbf{O}'$. The accessibility range is $[\Delta\lambda, 90°]$.
(4) $x_T < r$ and $\mathbf{n} \cdot \boldsymbol{X}_T < 0$ $(z_T \geq r_f)$, accessibility range is NULL.

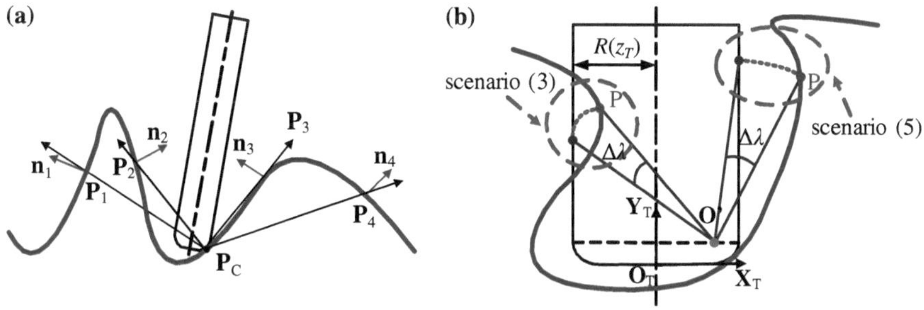


**Fig. 6.7** Rear gouging free posture range. **a** Collision-free and collision-prone points; **b** section curve on the cutter

(5) $x_T \geq r$, and $\mathbf{n} \cdot \mathbf{X}_T < 0$ ($z_T \geq r_f$), **P** is collision-free (see Fig. 6.7b). The minimum $\Delta\lambda$ that the cutter must be rotated clockwise to cause global collision is given as:

$$\Delta\lambda = \cos^{-1}\frac{r_1 - x_T}{d} - \cos^{-1}\frac{r_1 + \sqrt{R(z_T)^2 - y_T^2}}{d} \tag{6.9}$$

The accessibility range is $[0, \Delta\lambda]$.

Using the above method, the accessible ranges for all global-collision prone points can be obtained as $[\lambda_{\theta-gc1-i}, \lambda_{\theta-gc2-i}] \mid i = 1, \ldots, n$. The overall accessible range for global-collision avoidance is $[\theta, (\lambda_{\theta-gc1}, \lambda_{\theta-gc2})]$, where $\lambda_{\theta-gc1} = \max\{\lambda_{\theta-gc1-i} \mid i = 1, \ldots, n\}$ and $\lambda_{\theta-gc2} = \min\{\lambda_{\theta-gc2-i} \mid i = 1, \ldots, n\}$. However, if $\lambda_{\theta-gc1} > \lambda_{\theta-gc2}$, the accessible range for the cutter at $\theta$ is null.

### *6.2.4 Finding the A-Map*

Using the methods introduced in Sects. 6.2.1, 6.2.2, 6.2.3, for a certain $\theta$, the accessible range of the given cutter in terms of $\lambda$ at $\mathbf{P}_C$ can be obtained as the intersection of the three accessible ranges. To obtain the cutter's A-map at $\mathbf{P}_C$, we sample $\theta$ evenly in the range of $[\theta_{min}, \theta_{max}]$. By connecting the maximum and minimum values of $\lambda$ obtained at all the sampled $\theta$, a numerical approximation of the accessible range can be obtained, which is called the A-map. The algorithm is given in the following:

**Algorithm: Finding the A-map of a cutter at a CC point $\mathbf{P}_C$**
**Input**: (a) *Sampled point set* $\{\mathbf{P}_k, k = 1, 2, \ldots, m\}$ *and CC point* $\mathbf{P}_C$
(b) *A filler-end cutter*
(c) *Titling angle range* $[\lambda_{min}, \lambda_{max}]$, *rotational angle range* $[\theta_{min}, \theta_{max}]$
**Output**: *Accessibility map and the accessibility of the cutter at* $\mathbf{P}_C$

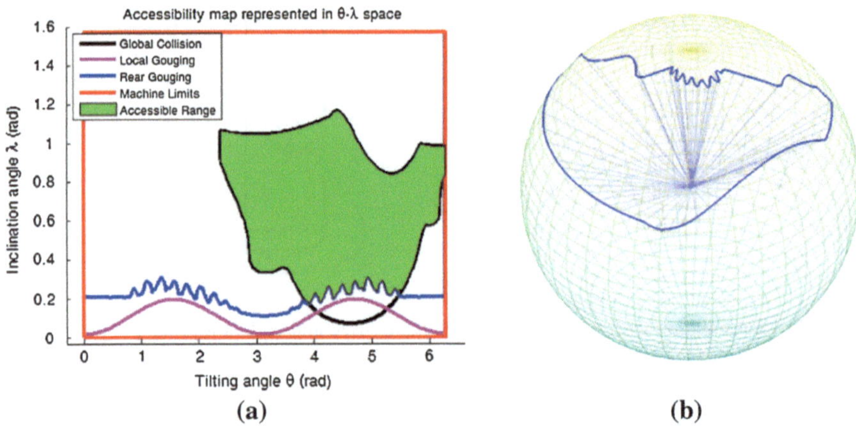


**Fig. 6.8** The A-map at ($u = 0.2$, $v = 0.8$). **a** A normal view of the A-map; **b** a sphere-view of the A-map

**Begin**

(1) Uniformly sample ($\theta_{\min}$, $\theta_{\max}$) into $k$ angles, set $i = 0$.

(2) IF $i \leq (k-1)$, $\theta_i = \theta_{\min} + (\theta_{\max} - \theta_{\min})(i/(k-1))$; otherwise, go to (6).

(3) Find the local-gouging free accessible range [$\theta_i$, ($\lambda_{\theta-lg}$, $\lambda_{\max}$)]. If the accessible range is NULL, $i = i + 1$, go to (2).

(4) Find the rear-gouging free accessible ranges from ($\lambda_{\theta-lg}$, $\lambda_{\max}$). The current accessible posture range is [$\theta_i$, ($\lambda_{\theta-rg}$, $\lambda_{\max}$)] ($\lambda_{\theta-rg} \geq \lambda_{\theta-lg}$). If the accessible range is NULL, $i = i + 1$, go to (2).

(5) Find the global-collision free accessible ranges from ($\lambda_{\theta-rg}$, $\lambda_{\max}$) regarding {$\mathbf{P}_k$, $k = 1, 2, \ldots, m$}. The final accessible range is [$\theta_i$, ($\lambda_{\theta-gc1}$, $\lambda_{\theta-gc2}$)] ($\lambda_{\theta-gc1} \geq \lambda_{\theta-rg}$, $\lambda_{\theta-gc2} \leq \lambda_{\max}$. Return NULL if such a range does not exist. $i = i + 1$. Go to (2).

(6) If the A-map is not NULL, output the A-map. Else, point is not accessible.

**End**

For the workpiece shown in Fig. 6.3a, at $\mathbf{P}_C$ ($u = 0.2$, $v = 0.8$), the A-map is obtained as an example here. The cutter is a fillet-end cutter without holder whose parameter is ($R = 5$ mm, $r_f = 0.5$ mm, $L = 60$ mm). The obtained A-map is plotted in Fig. 6.8a. The range for $\theta$ is taken as [0, $2\pi$] and $\theta$ is evenly sampled into 72 discrete values. The A-map is expressed in context of [$\theta$, $\lambda$]. For a more illustrative presentation, the A-map is transformed back to the local frame onto a unit sphere as shown in Fig. 6.8b. It can be seen that the accessible range for the cutter is an irregular cone-shaped region.

In summary, the A-map algorithm takes a numerical approach to construct a cutter's accessible posture range to a surface point in terms of the tiling and rotational angles. The requirements for interference avoidance regarding local gouging, rear gouging and global collision have been effectively addressed. Such a method gives the most important information about a cutter's accessibility and will prove very useful for tasks such as cutter selection, tool-posture determination, etc. For more details on A-map construction, readers can refer to [2].

## 6.3 Cutter Selection for 5-Axis Milling

During 5-axis milling, interference avoidance acts as a constraint. A process planner's true concern is the tool path's performance, such as machining efficiency and tool path smoothness. For cutter selection, the biggest concern is machining efficiency, or to be more specific, the cutter's size. Provided two cutters are both accessible to the machining surface, the cutter with a larger major radius $R$ is generally preferable. This is because intuitively, the cutter with the larger major radius will produce a better material removal rate. On the contrary, with the same major radius, cutters with smaller minor radii are more preferable. This is based on the observation that a flat-end cutter is more efficient than a ball-end cutter of the same major radius. In the remainder of this chapter, a larger cutter refers to the cutter with a larger major radius. If two cutters have the same major radius, the larger cutter refers to the one with a smaller minor radius. In cutter selection, a larger cutter is always preferable, provided that it is accessible to the surface.

The rest of this section introduces the cutter selection algorithm for our process planning system. Two modes of cutter selection will be considered: single cutter selection returns the largest cutter to finish a machining surface while multi-cutter selection returns the optimal cutter combination for a machining surface.

### 6.3.1 Single Cutter Selection

The A-map algorithm for obtaining the accessible posture range of a cutter to a surface point has already been introduced in Sect. 6.2. For a cutter to machine a surface without interference, the cutter has to be accessible to all the surface points. Thus, a point-based approach is proposed to check a cutter's accessibility to a surface, utilizing the high-density point set constructed for accessibility checking. Suppose $m$ points have been sampled from the machining surface, if the cutter is tested accessible to all the sampled points, the cutter is taken as accessible to the surface. Such a method, if carried out for all of the $n$ cutters in the cutter library, will have a computational complexity of O($kmn$), where $k$ is the number of discrete $\theta$ sampled in the A-map algorithm. This is computationally quite expensive. To bring

**Fig. 6.9** Dummy cutter at a *convex* surface point

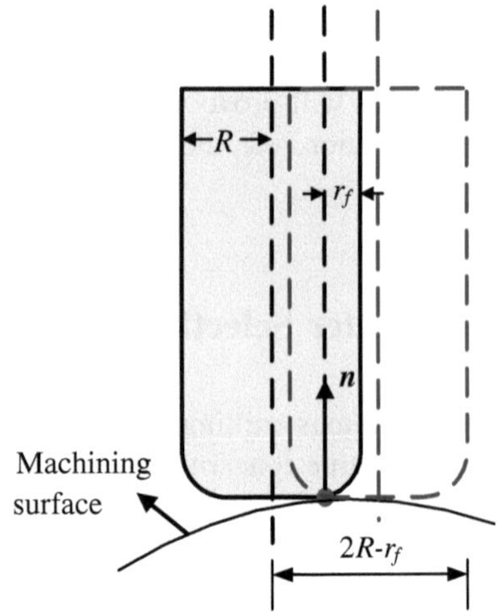


down the computational load, a desirable method is to reduce $n$, i.e., to reduce the number of points checked without affecting the accuracy of the algorithm.

At a surface point, the local surface shape falls into three categories: convex, concave and saddle. Such a classification is based on the maximum and minimum surface curvatures $\kappa_{max}$ and $\kappa_{min}$ at the point [3]. As the surface geometry is smooth, points of a certain category do not appear individually, but will form regions. Using the grid-based method proposed in [4], the sampled points can be grouped into convex, concave and saddle regions with boundaries also made up of sampled points. For a convex point, we have $\kappa_{max} < 0$ and $\kappa_{min} < 0$. Considering the criteria for local gouging avoidance given in Sect. 6.2.1, we can safely conclude that local gouging does not exist for convex surface points. If we eliminate those points with possibility of rear gouging and global collision, we can identify a group of interference-free convex points, for whom the A-map algorithm does not need to be run.

Only one accessible posture is needed to prove a cutter's accessibility to a surface point. Suppose a cutter approaches the point in the direction of the surface normal, the CC point will be in contact with the bottom circle of the cutter. With this posture, there will be an indefinite number of cutter positions (different values of $\theta$) and their enveloping body will be a cylinder of radius $2R-r_f$ (or $2R(z_T)-r_f$ for a holder segment) named as a *dummy cutter* (see Fig. 6.9). As long as no sampled points from the machining/non-machining surfaces fall inside this dummy cutter, the surface normal at the point in question will be an accessible posture for the cutter, meaning that the cutter is accessible to the point. The dummy cutter method has a much smaller computational load than the A-map algorithm. Moreover, it has been proved that if every point on the boundary of a convex region is tested accessible using the dummy cutter method, so are the points inside the region [2]. Based on this, the following algorithm is proposed for finding the interference-free points among the sampled points from a machining surface:

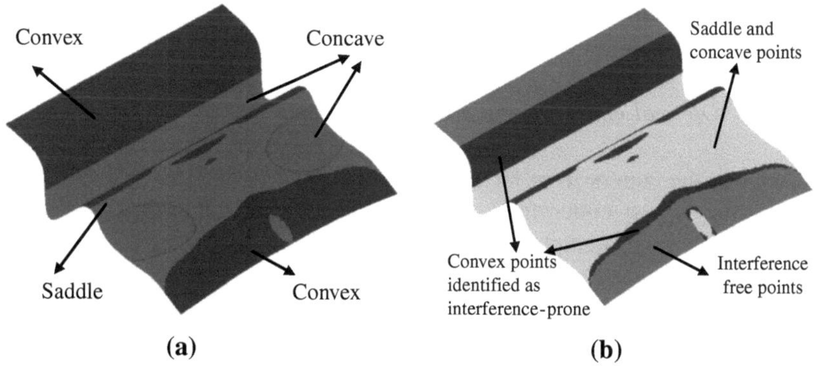


**Fig. 6.10** Identification of interference-free points based on surface division. **a** Surface division result; **b** identification of interference-free points

**Algorithm: Identifying the interference-free regions on a surface**
**Input**: (a) *Sampled points from machining surface* $\{S_i\}$
(b) *A given fillet-end cutter*
**Output**: *Interference-free set* $\{S_{i\text{-}free}\}$ *and interference-prone set* $\{S_{i\text{-}prone}\}$
**Begin**
1. Identify all convex regions and their boundaries. Put the concave and saddle points into $\{S_{i\text{-}prone}\}$.
2 . Pick a convex region *r* and its boundary *e* that hasn't been checked. Set $j = 0$. If all convex regions have been traversed, stop.
3. For $\mathbf{P}_j \in e$, conduct the dummy cutter algorithm. If $\mathbf{P}_j$ is accessible, go to (4). Else, go to (5). If all points on the boundary have been checked, go to (6).
4. Put $\mathbf{P}_j$ into $\{S_{i\text{-}free}\}$, $j = j + 1$. Go to (3).
5. $\mathbf{P}_j$ in $\{S_{i\text{-}prone}\}$. Replace $\mathbf{P}_j$ with the nearest point $\mathbf{P}$ that satisfies $\mathbf{P} \in r$ and $\mathbf{P} \notin e$. Go to (3).
6. Put all points that satisfy $\mathbf{P} \in r$ and $\mathbf{P} \notin e$ into $\{S_{i\text{-}free}\}$. Go to (2).

**End**

The identification of interference-free points can significantly reduce the computational load for checking a cutter's accessibility to a surface. Fig. 6.10 shows an example of the above algorithm applied to a machining surface for a cutter of size ($R = 12$ mm, $r_f = 0.5$ mm, $L = 85$ mm).

With the method for checking a cutter's accessibility to the machining surface, algorithms for single cutter selection become quite straightforward. Accessibility checking will start from the largest cutter in the cutter library and will stop when an accessible cutter is found or when the cutter library is exhausted. The algorithm for single cutter selection is given in the following:

**Algorithm: Selecting a single cutter for a machining surface**
**Input**: (a) Sampled point cloud of machining surface $\{S_i\}$
(b) A library of n fillet end cutters
**Output**: *Optimal cutter for the machining surface*
**Begin**
(1) Arrange the cutters from large to small $\{C_1, C_2, \ldots, C_n\}$.
(2) Pick the largest cutter from the cutter library and set it as *current cutter*. If cutter library is empty, go to (7).
(3) For *current cutter*, use the identification algorithm to get the sets of interference-prone and interference-free points $\{S_{i\text{-}free}\}$ and $\{S_{i\text{-}prone}\}$.
(4) At each point **P** in $\{S_{i\text{-}prone}\}$, construct the cutter's *A-map*. If *A-map* = *NULL*, go to (5). If for all the **P** in $\{S_{i\text{-}prone}\}$, *A-map* ≠*NULL*, go to (6).
(5) Discard *current cutter*, go to (2).
(6) Output *current cutter* as the optimal cutter. Stop.
(7) No feasible cutter in the cutter library. Stop.
**End**

## *6.3.2 Multi-Cutter Selection*

With the A-map algorithm, the selection of a single cutter to machine a free-form surface is quite straightforward. However, such a choice may not be optimal, as the selected cutter is always constrained by the most critical feature on the surface. For surfaces with a large portion of non-critical areas, low machining efficiency will be resulted. Thus, it is desirable to have multiple cutters for machining such a surface. The larger cutters can be used to machine the flat or convex areas while smaller cutters can be used for the critical concave or saddle areas. Considering that most 5-axis machines are equipped with the fast tool change mechanism, the time penalty from tool change can be compensated with the time gain from better machining efficiency.

### 6.3.2.1 Identification of Candidate Cutter Sets

Based on its A-maps, a cutter in the cutter library falls into one of three categories: (1) *accessible*, which has non-empty A-maps at all the sampled points; (2) *partially accessible*, which has non-empty A-maps at some of the sampled points and (3) *inaccessible*, whose A-map is empty at all of the sampled points. Inaccessible cutters are discarded. To achieve interference-free machining, there has to be at least one accessible cutter in the cutter library to machine the surface. However, if there are multiple accessible cutters, only the largest will be chosen to remain in the library for better machining efficiency. So after this initial

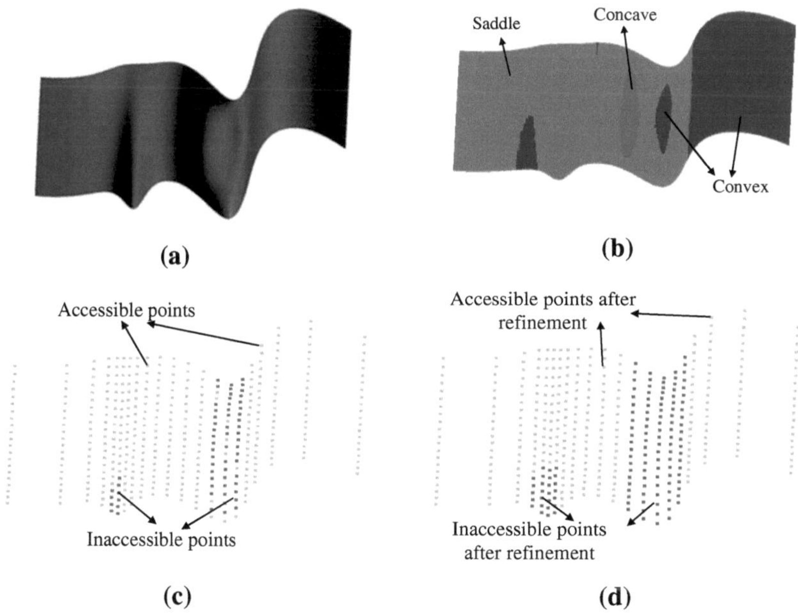


**Fig. 6.11** Accessible/inaccessible regions of a cutter. **a** Machining surface; **b** surface division; **c** the original point classification; **d** point classification after refinement

processing, the updated cutter library for multi-cutter selection contains one accessible cutter $C_a$ and a number of partial accessible cutters $\{C_1, \ldots, C_k\}$.

For a partially accessible cutter, the machining surface can be divided into accessible and inaccessible regions. As the A-map algorithm is a point-based method, these regions can only be approximated by the accessible and inaccessible sampled points as shown in Fig. 6.11c. It can be seen that the obtained accessible region is not continuous. There are isolated accessible points and sometimes, the accessible and inaccessible points are scattered together. This will cause complications for machining region allocation, which makes it necessary to have a refinement process. In this process, all the sampled points within a certain radius of an inaccessible point are labeled inaccessible. Such an approach will classify some accessible points as inaccessible. However, it will put a safety margin for the accessible region of a cutter. From Fig. 6.11d, it can be seen that inaccessible regions become larger after refinement and discontinuities in the surface regions disappear.

A valid cutter set can be formed by any number of cutters from the feasible cutter list as long as $C_a$ is present. For each valid cutter set, some of the accessible regions (*ARs*) of different cutters may overlap. It is preferable to allow the larger cutter to machine the whole of its accessible regions. As a result, the actual region to be machined by a smaller cutter is less than its original ARs. This actual region to be machined by a cutter is named the cutter's *effective accessible regions*

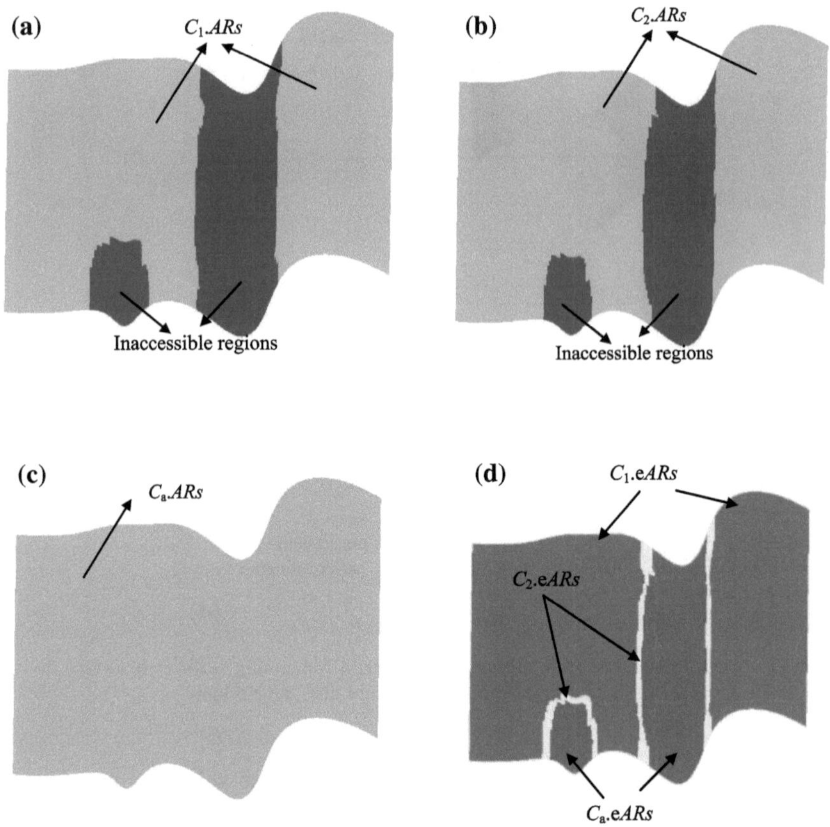


**Fig. 6.12** The construction of eARs. **a** Accessible regions for $C_1$; **b** accessible regions for $C_2$; **c** accessible regions for $C_a$; **d** eARs for $C_1$, $C_2$ and $C_a$

(*eARs*), which is a subset of its original ARs. The eARs of a cutter is used as an important measure to determine whether this cutter should go into the candidate cutter set.

In a valid cutter set $CSet_i=\{C_{i1}, \ldots, C_{ij}, \ldots, C_{im}\}$, in which the cutters are ordered from large to small, the eARs for $C_{ij}$ is the subtraction of the sum of eARs of all the larger cutters from its original ARs, i.e.,

$$C_{ij} \cdot \text{eARs} = C_{ij} \cdot \text{ARs} - C_{ij} \cdot \text{ARs} \cap \left( \sum_{k=1}^{j-1} C_{ik} \cdot \text{eARs} \right) \tag{6.10}$$

According to this definition, the eARs for the largest cutter is its original ARs, i.e., $C_1 \cdot \text{eARs} = C_1 \cdot \text{ARs}$. The eARs of the remaining cutters in the set can then be worked out recursively. An example is provided in Fig. 6.12. The eARs of the three cutters in cutter set $\{C_1, C_2, C_a\}$ are shown in Fig. 6.12d. Proportionally, $C_1 \cdot$eARs, $C_2 \cdot$eARs, and $C_a \cdot$eARs form 70.17, 5.90 and 23.93% of the whole surface respectively.

Although a larger cutter generally results in higher material removal rate in its eARs, putting too many cutters in one set may cause the machining efficiency to drop. This is because much time is wasted in the air travel needed for tool change. Besides, too many cutters may cause discontinuities in the finished surface, resulting in poor surface quality. Here, a heuristic is developed to filter the valid cutter sets using a parameter called the *machining area ratio (MR)*. MR refers to the ratio of the area of a single eAR to the area of the whole surface. By setting a minimum threshold value of MR for all the cutters larger than $C_a$, a smaller eAR will be removed from its corresponding cutter. When all the eARs of a cutter have been removed, the cutter itself will be removed from the set. Such a measure can effectively constrain the number of cutters in the cutter set and the number of the sets. After filtering through all the valid cutter sets, the cutter sets left are called the *candidate cutter sets*. The algorithm for the above process is as follows:

**Algorithm: Selection of the candidate cutter sets**

**Input**: (a) *All the feasible cutter sets* $\{Set_1, Set_2, \ldots, Set_k\}$ *and the eARs of the cutters in each cutter set*

(b) *Threshold value for machining area ratio* $MR_{\min}$ *and surface area A*

**Output**: *The candidate cutter sets*

**Begin**

(1) Pick a valid cutter set as the current set (*CSet*) and set the first cutter in *CSet* as current cutter (*C-cutter*).

(2) If *C-cutter* is not the last cutter, check the area of each *eAR*:

(a) Remove the *eAR* if it is smaller than ($MR_{\min} \times A$).

(b) If there are no remaining *eARs*, remove this cutter from the *CSet*. Recalculate the *eARs* of the remaining cutters and update the *CSet*. Go to (d).

(c) If there are still remaining *eARs*, place *C-Cutter* into the corresponding candidate set, together with its *eARs*.

(d) Set next cutter in *CSet* as *C-cutter*, repeat (2).

(3) If *C-cutter* is the last cutter, put this cutter together with its *eARs* into the candidate cutter sets. Go to (1).

(4) If all the valid cutter sets are processed, combine all the identical candidate cutter sets. Output all the candidate cutter sets. Stop.

**End**

#### 6.3.2.2 Identification of the Optimal Cutter Set

In this section, a heuristic is introduced to select the optimal cutter set from candidate cutter sets. The target is set as maximizing machining efficiency. In terms of multi-cutter selection, this target is interpreted as minimized tool-path length. Another more intuitive interpretation should be minimized machining time. However, the machine joints on a 5-axis machine go through frequent acceleration and deceleration processes due to the complicated angular movement of the cutter. This will cause fluctuations in the feedrate, making prediction of machining time difficult. For a surface of area $A$, the total tool-path length can be approximated as:

$$L = \frac{A}{W} \tag{6.11}$$

where $W$ is the average gap between tool paths. For multi-cutter 5-axis machining, this approximated tool-path length is given the name *cutting time index* (*CTI*) and a detailed form is provided as:

$$\mathrm{CTI} = \sum_{j=1}^{n} \frac{\mathrm{Area}(C_j \cdot \mathrm{eARs})}{W_j} \tag{6.12}$$

where $W_j$ is the average strip width of the cutter $C_j$ in its eARs. As CC points have not been generated at the time of cutter selection, we can only use the average strip width at all sampled points as an approximation.

To calculate $W_j$, we need to revisit some previous concepts. Suppose in the local frame at the point $\mathbf{P}_C$, the cutting direction is given by a unit vector $\boldsymbol{f}$ on the $\mathbf{X}_L$–$\mathbf{Y}_L$ plane with an angle $\omega$ from $\mathbf{X}_L$. The strip width is calculated on the plane normal to $\boldsymbol{f}$. Such a plane is denoted by the surface normal $Z_L$ at $P_C$ and the unit vector $\boldsymbol{f}_p$, which has an angle of $\omega + \pi/2$ between $\mathbf{X}_L$. Based on Eqs. 6.1 and 6.2, we know that the curvatures of the surface on the plane normal to $\boldsymbol{f}$ (plane $\mathbf{Z}_L$–$\boldsymbol{f}_p$) can be given as:

$$\kappa_{s\omega} = \kappa_{\max} \cos^2\left(\omega + \frac{\pi}{2}\right) + \kappa_{\min} \sin^2\left(\omega + \frac{\pi}{2}\right) \tag{6.13}$$

On the same plane, the curvature of the cutter is given as:

$$\kappa_{t\omega} = \frac{\cos^2\left(\omega + \frac{\pi}{2} - \theta\right)(\omega - \theta)}{r_f} + \frac{\sin^2\left(\omega + \frac{\pi}{2} - \theta\right)}{\frac{(R - r_f)}{\sin\lambda} + r_f} \tag{6.14}$$

As $\omega$ is still unknown at this stage, it is assumed the cutter follows the direction of maximum material removal rate at every sampled point. Such a direction refers to the direction of minimum principle curvature, i.e., $\mathbf{Y}_T$. This means $\omega = \pi/2$ and

$$\kappa_{s\omega} = \kappa_{\max} \tag{6.15}$$

It can be seen from Eq. 6.14 that $\kappa_{t\omega}$ is actually determined by $\omega$, $\theta$ and $\lambda$. However, cutter postures are unknown at this stage. According to Li and Zhang [2], the strip width is at the largest when the cutter is aligned with the cutting direction, i.e., when $\theta = \omega$. Under this premise, the titling angle $\lambda$ should also be minimized. As $\lambda$ may not exist for $\theta = \omega$ in the A-map, we take $\lambda$ as the average of

**Table 6.1** The cutter library

| Cutter index | $C_1$ | $C_2$ | $C_3$ | $C_4$ | $C_5$ | $C_6$ | $C_7$ | $C_8$ | $C_9$ |
|---|---|---|---|---|---|---|---|---|---|
| Major radius $R$ (mm) | 12 | 10 | 8 | 6 | 4 | 3 | 2 | 1.5 | 1 |
| Minor radius $r_f$ (mm) | 0.5 | 0.5 | 0.5 | 0.5 | 0.5 | 0.2 | 0.2 | 0.2 | 0.2 |
| Length $L$ (mm) | 85 | 80 | 70 | 60 | 50 | 45 | 45 | 45 | 45 |

the minimum $\lambda$ values on all accessible values of $\theta$ from the A-map, denoted as $\lambda = Ave\ (\lambda_{\min})$. Then, Eq. 6.14 becomes:

$$\kappa_{t\omega} = \frac{1}{\dfrac{(R - r_f)}{\sin(Ave(\lambda_{\min}))} + r_f} \tag{6.16}$$

According to Lee and Chang [5], an approximation of the cutting strip width at a sampled point is given as:

$$W_i = \sqrt{\frac{8\tau}{\kappa_{t\omega} - \kappa_{s\omega}}} \tag{6.17}$$

With Eqs. 6.15–6.17, we have the average strip width for a cutter in its eARs (suppose there are a total number of $m$ sampled points in the eARs):

$$W = \frac{1}{m}\sum_{i=1}^{m} \sqrt{\frac{8\tau}{\dfrac{1}{\frac{(R-r_f)}{\sin\left(Ave(\lambda_{\min})_i\right)} + r_f} - \kappa_{\max,i}}} \tag{6.18}$$

Therefore, the CTI of any particular candidate cutter set can be obtained based on Eqs. 6.12 and 6.18. The cutter set with the smallest CTI is taken as the optimal cutter set for machining the surface.

This heuristic-based method for multi-cutter selection has been implemented in the proposed process planning system. An example is given here to show the performance of this method. The machining surface is shown in Fig. 6.11a. The cutter library used is given in Table 6.1. $MR_{\min}$ is set as 20%. As shown in Fig. 6.12d, cutter set $\{C_1, C_2, C_8(C_a)\}$ does not qualify as a candidate cutter set as the eARs of $C_2$ only take 5.9% of the total surface area. The optimal cutter set that has been finally selected is $\{C_1, C_5, C_8\}$. The $AR$ for $C_5$ is shown in Fig. 6.13a. The eARs of the three cutters are shown in Fig. 6.13b, where $C_1$.eARs, $C_5$.eARs, and $C_8$.eARs form 70.17, 25.36 and 4.45% of the whole surface, respectively (Originally, $C_5$ has two eARs, one of which is removed as it does not reach the 20% threshold for MR).

The tool paths (iso-planar pattern) generated using this cutter set and the largest single accessible cutter ($C_8$) are shown in Fig. 6.14a and b, respectively. A comparison between these two sets of tool paths based on various aspects is given in Table 6.2. It can be seen that using optimal cutter set can produce significantly shorter tool paths (7,650.43 vs 17,477.5 mm), resulting in higher machining efficiency.

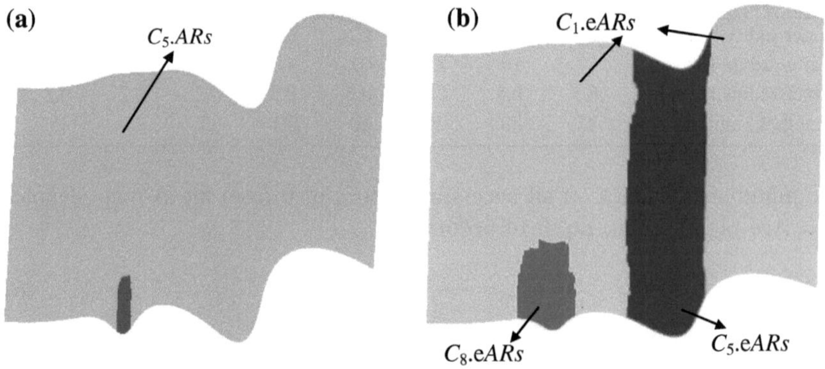


**Fig. 6.13** Machining region allocation for the optimal cutter set. **a** Accessible region for $C_5$; **b** *eARs* of the optimal cutter set

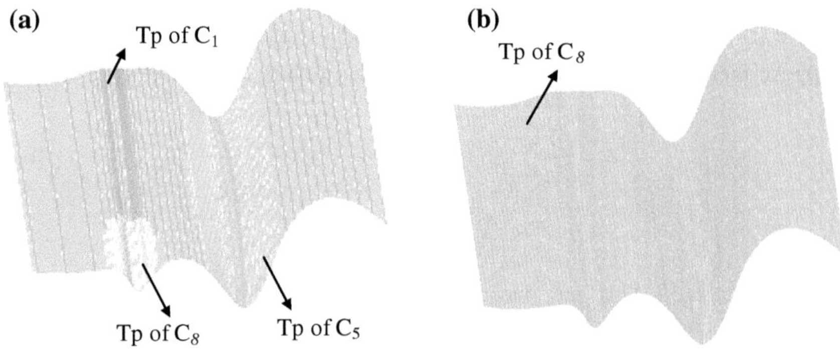


**Fig. 6.14** Tool paths of single/multiple cutter(s). **a** Tool paths using multiple cutters; **b** tool paths using a single cutter

**Table 6.2** Comparison between tool paths of single/multiple cutters

| | Cutter | No. of passes | No. of CC points | Path length (mm) |
|---|---|---|---|---|
| Multi-cutter machining | $C_1$ | 36 | 334 | 3,865.91 |
| | $C_5$ | 26 | 781 | 3,076.26 |
| | $C_8$ | 20 | 482 | 708.265 |
| | ($C_1$, $C_5$, $C_8$) | 82 | 1,597 | 7,650.43 |
| Single-cutter machining | $C_8$ | 134 | 11,710 | 17,477.5 |

## 6.4 Tool-Path Generation for 5-Axis Milling

Over the years, many patterns have been adopted by researchers for the generation of 5-axis tool paths, such as iso-parametric, iso-planar and iso-scallop-height. Among these patterns, the iso-planar pattern is widely used for its simplicity and robustness. Iso-planar tool paths are generated on the intersection curves between the machining surface and a series of parallel cutting planes. The current process planning system employs this pattern for tool-path generation. The generation of iso-planar tool paths can be roughly divided into three tasks: (1) selection of cutting direction, (2) generation of single tool paths, and (3) determination of path intervals. In this section, these three tasks are addressed and the corresponding methods are introduced.

### *6.4.1 Selection of Cutting Direction*

During multi-cutter selection, it is assumed that the cutter follows the direction of maximum material removal rate at each sampled point. However, in practice, there will be only one cutting direction taken for iso-planar tool paths. In general, the selection of a cutting direction will impact the quality of the generated tool paths, e.g., machining efficiency and cutting dynamics. In 5-axis machining of sculptured surfaces, to ensure the quality of the machined surface, the smooth dynamics of cutter motion is a must, i.e., the posture change from one point to the next must be minimized. Extreme change in cutter posture, which is necessary for interference avoidance, is a major cause for the unnatural movement of the cutter and will lead to over and under cutting in 5-axis finish and undesirable irregularity of the surface appearance. Therefore, the tool path smoothness is set as the top priority for cutting direction selection.

To select the optimal cutting direction with respect to the tool path smoothness, we simply take every sampled direction and assign a cutter posture at every sampled surface point. This posture assignment is simply an estimation aiming at maximizing machining efficiency. With the cutter postures at every sampled point, we can check the smoothness of the potential tool paths (to be generated later) by using a proposed measuring factor, called the *posture change rate* (*PCR*). The direction with the best overall PCR is finally chosen.

#### 6.4.1.1 The Cutter Posture Along a Path Direction at a Sampled Surface Point

A-maps contain an indefinite number of feasible cutter postures for each sampled point. The choice of cutter postures affects tool path performance in several aspects. For example, the cutter posture determines the cutting strip width at a CC

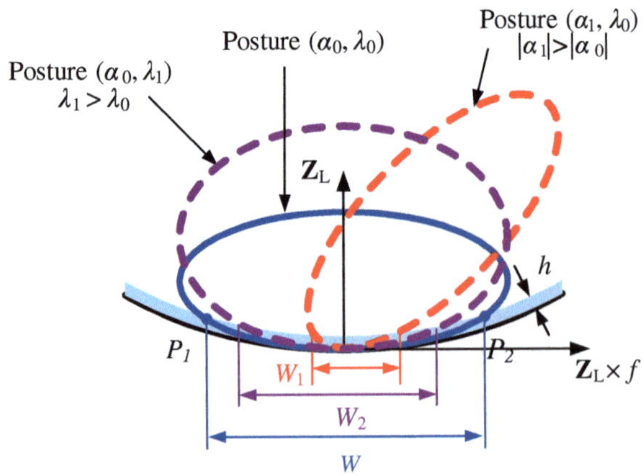


**Fig. 6.15** Cutter posture's effect on machining strip width

point, which in turn will affect the machining efficiency directly. Therefore, it is preferable to have a cutter posture with the largest machining strip width when other requirements are fulfilled.

Figure 6.15 shows the elliptical cross-sections of the cutter on the normal plane perpendicular to the feeding direction $\boldsymbol{f}$ with different postures. The strip of the material that can be effectively removed by the cutter is determined by the elliptical cross-section and the surface with offset $h$ from the part surface (a more detailed discussion is provided in Sect. 6.4.3). The machining strip width decreases with the increment of the tilt angle $\lambda$ since the major axis of the eclipse moves away from the machining surface as shown in Fig. 6.15 [6]. Furthermore, the machining strip width also decreases with the increment of value $|\alpha|$ since the cutter curvature increases at the point touching $\mathbf{P}_C$ with $|\alpha|$, where $\alpha$ is the angle between the cutter axis and feeding direction on $\mathbf{X}_L$–$\mathbf{Y}_L$ plane (see Fig. 6.15).

For better machining efficiency, a heuristic is adopted here to assign the cutter posture such that maximum machining strip width is achieved. From the A-map, we find $\theta$, which is the closest to the feeding rotational angle $\omega$ in the local frame, and then take the minimum feasible inclination angle $\lambda$ to form the posture ($\theta$, $\lambda$) at $\mathbf{P}_C$, as shown in Fig. 6.16a and b. However, a special situation may exist when two postures have the same $|\alpha|$ value (i.e., $\alpha_1$ and $\alpha_2$) in the A-map along clockwise and counterclockwise directions, respectively (see Fig. 6.16b). To address this, another heuristic is proposed. At $\theta_1$ and $\theta_2$, the minimum feasible $\lambda_1$ and $\lambda_2$ are taken to form the postures $\mathbf{T}_1$ and $\mathbf{T}_2$ in global frame. The one (i.e., $\mathbf{T}_1$ in Fig. 6.16b) that is nearer to the cutter orientation $\mathbf{T}_{prev}$ at the previous sampled point is selected as the cutter posture at $\mathbf{P}_C$.

#### 6.4.1.2 Posture Change Rate at a Point

Given a sampled point $\mathbf{P}$ and one of its neighboring sampled point $\mathbf{P}_{next}$, the PCR along the direction of $\mathbf{PP}_{next}$ is defined as the difference between the two

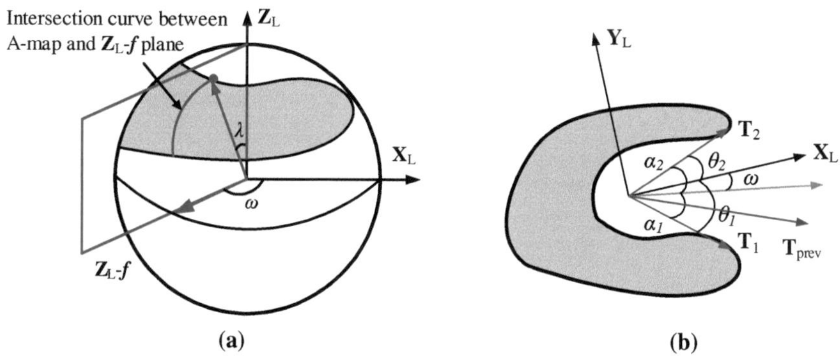


**Fig. 6.16** Selection of posture from the A-map based on cutting direction. **a** Selection of posture based on cutting direction; **b** projection of A-map onto $X_L$–$Y_L$

corresponding postures normalized by the distance between **P** and $\mathbf{P}_{\text{next}}$. If we can obtain the PCRs of all the sampled points along any given cutting direction, the direction that possesses the minimum $\sum PCR_i$ is chosen as the optimal cutting direction.

A cutting direction is represented by the angle $\beta$ between the cutting direction vector and $\mathbf{X}_G$ of the global frame. Based on this definition, the cutting direction determined by **P** and $\mathbf{P}_{\text{next}}$ is given as$\beta = \tan^{-1}\left(\dfrac{\mathbf{P}_{\text{next}} \cdot y_G - \mathbf{P} \cdot y_G}{\mathbf{P}_{\text{next}} \cdot x_G - \mathbf{P} \cdot x_G}\right)$. To obtain the PCR, the cutter postures at **P** and $\mathbf{P}_{\text{next}}$ are obtained based on the heuristic given in Sect. 6.4.1.1. This requires $\beta$ to be transformed into the local coordinate system at **P** and $\mathbf{P}_{\text{next}}$ to find $\omega$ and $\omega_{\text{next}}$. Suppose the postures are selected as $\boldsymbol{p}$ and $\boldsymbol{p}_{\text{next}}$, the PCR between **P** and $\mathbf{P}_{\text{next}}$ can be given as:

$$\text{PCR}_{\mathbf{P}-\mathbf{P}_{\text{next}}} = \frac{\left|\tan^{-1}(\boldsymbol{p} \cdot \boldsymbol{p}_{\text{next}})\right|}{\left|\mathbf{PP}_{\text{next}}\right|} \tag{6.19}$$

For implementation, a local neighborhood of **P** is first defined that contains $n$ sampled points $\{\mathbf{P}_j, j = 1, 2, \ldots, n\}$. As shown in Fig. 6.17a and b, the neighborhood is formed based on the distance between **P** and $\mathbf{P}_j$, i.e. $|\mathbf{PP}_j| \leq d_0$, where $d_0$ is a predefined value according to the curvature at **P**. Intuitively, a small $d_0$ is preferred at the point with a large curvature, while a large $d_0$ at the point with a small curvature. Based on experiments, we take$d_0 = \min\{1/\max(|\kappa_{\max}|, |\kappa_{\min}|), R\}$, where $\kappa_{\max}$ and $\kappa_{\min}$ are the principal curvatures respectively, and $R$ is the cutter major radius. Vector $\mathbf{PP}_j$ can be transformed into a discrete cutting direction $\beta_j$. To obtain the PCR along an arbitrary direction $\beta$ in the global frame, we first find the two closest $\beta_j$ to $\beta$ from $\{\text{PCR}_j, j = 1, 2, \ldots, n\}$. The PCR along $\beta$ can then be obtained by applying linear interpolation of the two corresponding $\text{PCR}_j$ (see Fig. 6.17c). As the point sampling density on the surface is quite high, such a method is expected to produce a close approximation.

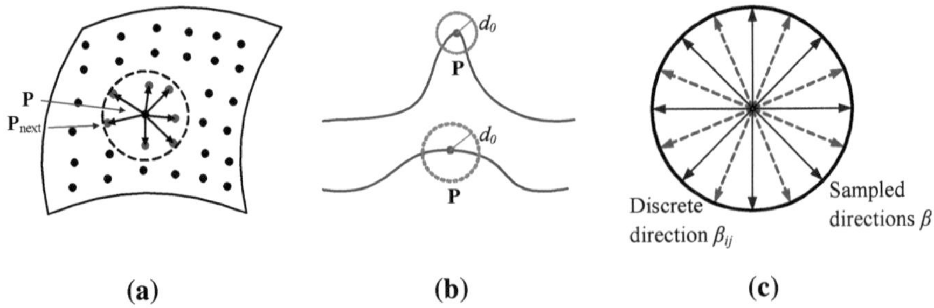


**Fig. 6.17** 17 Calculation of *PCR* in all directions at a sampled point. **a** Neighborhood of **P**; **b** $d_0$ based on local curvature; **c** sample and interpolation

## *6.4.2 Generation of a Single Tool Path*

A single path on the machining surface is produced by intersecting the surface with the current cutting plane (see Fig. 6.18a). In this section, we will focus on how to obtain the discrete CC points on this path to satisfy the predefined profile tolerance. The first CC point can be obtained by intersecting the cutting plane and the surface boundary (region boundary for multi-cutter tool paths). Beginning from the first CC point, the process can be interpreted as "finding the next CC point on the same path such that the deviation of the tool motion trajectory is within the specified tolerance $\tau$ from the part surface". The motion of the cutter between two CC points is determined by the type of interpolator on the 5-axis machine. In this study, it is assumed that a linear interpolator is used and the trajectory of the cutter is a line segment.

As shown in Fig. 6.18b, at a CC point $\mathbf{P}_i(u_i, v_i)$, the next point on the path, $\mathbf{P}_{i+1}(u_{i+1}, v_{i+1})$, can be determined such that the largest deviation $d$ from the line segment $\mathbf{P}_i\mathbf{P}_{i+1}$ to the part surface is very close to but smaller than $\tau$ as:

$$\begin{aligned} S_y(u_{i+1}, v_{i+1}) &= y_i \\ d(\mathbf{P}_i\mathbf{P}_{i+1}, S(u,v)) &\leq \tau \end{aligned} \tag{6.20}$$

Here, we assume that the cutting plane is normal to $\mathbf{Y}_G$ axis as $y = y_i$ (if not, a trivial frame transformation can be applied).

To determine $\mathbf{P}_{i+1}(u_{i+1}, v_{i+1})$ based on Eq. 6.20, a numerical solution is adopted as follows:

(1) Set the initial value for step-forward length between $\mathbf{P}_i\mathbf{P}_{i+1}$ regarding local surface geometry at $\mathbf{P}_i$;
(2) Search for an estimated point $\mathbf{P}_{i+1}$ from $\mathbf{P}_i$ based on the step-forward distance;
(3) Check whether the deviation between the tool trajectory and the machining surface is within range $(1-\delta)\ \tau \leq d \leq \tau$, where $\delta$ is a predefined small value, such as 0.05.

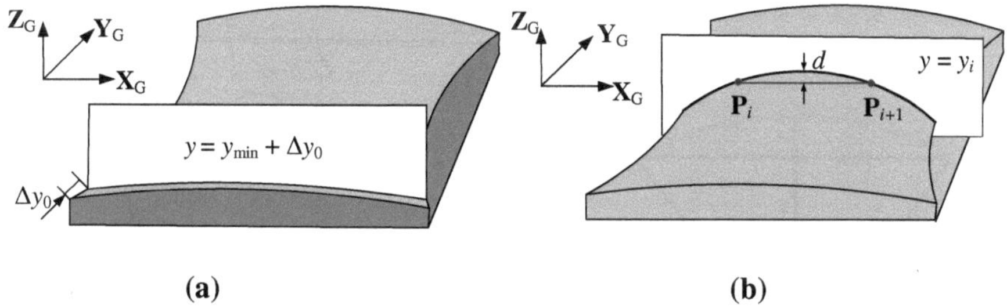


**Fig. 6.18** A single planar tool path. **a** First tool path $y = y_{\min} + \Delta y_0$; **b** next CC point on a single tool path

If the condition is not satisfied, the step-forward length is accordingly changed, and steps (2) and (3) are repeated till the suitable $\mathbf{P}_{i+1}$ is obtained. The following sections will address these three steps separately.

#### 6.4.2.1 Setting the Initial Step-Forward Length

To keep deviation $d$ below a certain level, the density of CC points should grow with the curvature of the tool path. Thus, the initial estimation of step-forward length $L_i$ for $\mathbf{P}_{i+1}$ can be made based on the local surface curvature and $\tau$, as shown in Fig. 6.19a. Here, the geometry of surface curve on the cutting plane in the vicinity of $\mathbf{P}_i$ is approximated by a circular curve, and the curvature $\kappa$ of the circle is that of the curve at $\mathbf{P}_i$ on the plane $y = y_i$. $L_i$ is then given as:

$$L_i = \sqrt{8R\tau - 4\tau^2} = \sqrt{8\tau/\kappa - 4\tau^2} = 2\sqrt{\frac{\tau(2 - \kappa\tau)}{\kappa}} \tag{6.21}$$

According to Meusnier theory in differential geometry, the value of $\kappa$ can be calculated as:

$$\kappa = \kappa_n / \cos\alpha \tag{6.22}$$

where $\kappa_n$ is the normal curvature of the surface curve on plane $\mathbf{Z}_L$–$\boldsymbol{f}$. $\boldsymbol{f}$ is the positive tangent direction of the path curve at $\mathbf{P}_i$ on plane $y = y_i$, and $\alpha$ is the angle between the normal plane $\mathbf{Z}_L$–$\boldsymbol{f}$ and the plane $y = y_i$, as shown in Fig. 6.19b. $\kappa_n$ can be calculated using Eq. 6.1 after $\boldsymbol{f}$ is transformed into the local coordinate system at $\mathbf{P}_i$.

#### 6.4.2.2 Determining the Next Estimated CC Point

Based on the estimated value of $L_i$, we proceed to search for the next CC point $P_{i+1}$ on the path with the distance $L_i$ from $\mathbf{P}_i$ along the cutting direction. The method proposed by Hwang [7] is used here to obtain next point $P_{i+1}$. First, as shown

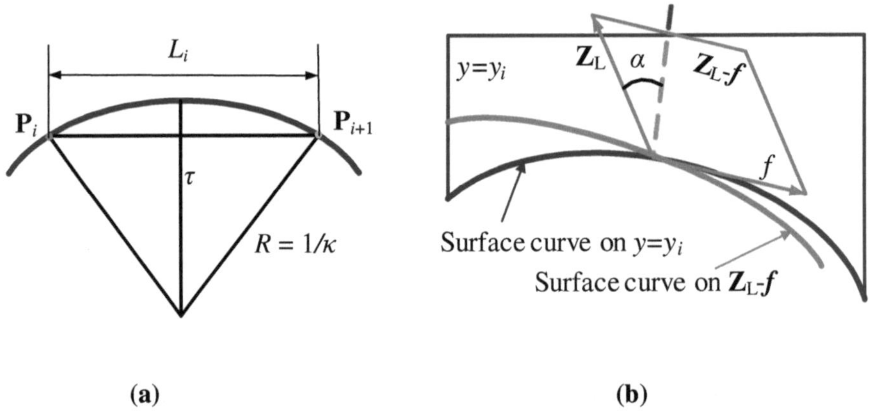


**Fig. 6.19** Initial estimate of the step-forward length at $\mathbf{P}_i$. **a** Step-forward length; **b** surface curve curvature $k_n$ at $\mathbf{P}_i$

in Fig. 6.20a, the tangent distance $s$ is calculated based on the approximated circular curve as:

$$s = (R - \tau)L_i/R \tag{6.23}$$

This gives us $\mathbf{P}' = \mathbf{P}_i + s\boldsymbol{f}_i$. $\mathbf{P}_{i+1}(u_{i+1}, v_{i+1})$ is then obtained to be the intersection point of the part surface with a line through $\mathbf{P}'$ and along $\boldsymbol{n}_i$ direction (see Fig. 6.20b) as:

$$\mathbf{P}_{i+1} = \mathbf{P}' + l\boldsymbol{n}_i \tag{6.24}$$

where $l$ is the distance of $\mathbf{P}_{i+1}$ from $\mathbf{P}'$, which is calculated using the method proposed by Scherrer and Hillberry [8]. It is worth noting that the sign of $l$ can be positive or negative.

#### 6.4.2.3 Calculating Deviation Between Line Segment and Path Curve

To evaluate the maximum deviation between line segment $\mathbf{P}_i\mathbf{P}_{i+1}$ and the path curve, we need to find the corresponding point $\mathbf{P}(u, v)$ at which the deviation reaches its maximum, as shown in Fig. 6.21. First, $\mathbf{P}$ is located on the path curve and therefore on the plane $y = y_i$. Second, the tangent vector of the tool path curve at point $\mathbf{P}$ should be parallel to $\mathbf{P}_i\mathbf{P}_{i+1}$, i.e., the normal $\boldsymbol{n}(u, v)$ at $\mathbf{P}$ is perpendicular to $\mathbf{P}_i\mathbf{P}_{i+1}$. These two conditions can be expressed by:

$$\begin{aligned} \mathbf{S}_y(u, v) &= y_i \\ \boldsymbol{n}(u, v) \cdot \mathbf{P}_i\mathbf{P}_{i+1} &= 0 \end{aligned} \tag{6.25}$$

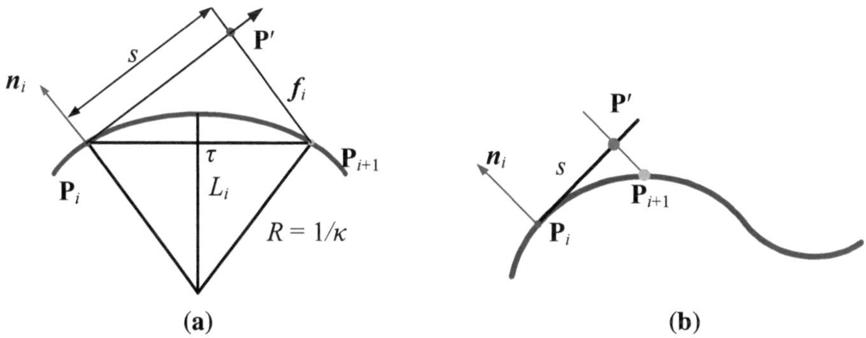


**Fig. 6.20** Determination of next estimated point $\mathbf{P}_{i+1}$. **a** Tangent distance for next CC point; **b** next CC point

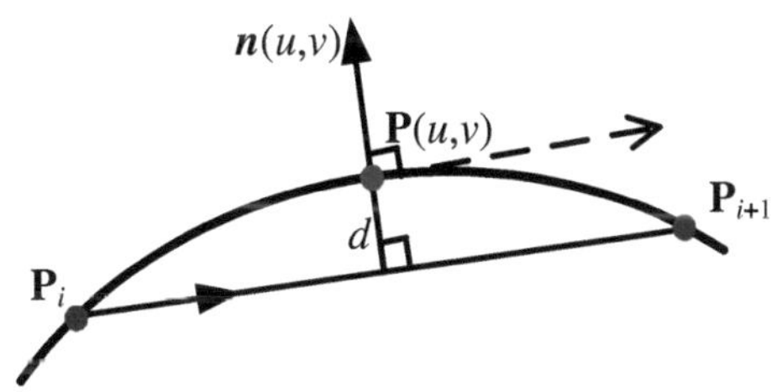


**Fig. 6.21** The deviation of the chord from the path curve on plane $y = y_i$

Here, the secant numerical approach is employed to solve this equation system. After obtaining **P**, the deviation distance $d$ between $\mathbf{P}_i\mathbf{P}_{i+1}$ and path curve is represented as:

$$d = \left|(\mathbf{P} - \mathbf{P}_i) - \left[(\mathbf{P} - \mathbf{P}_i) \cdot \frac{\mathbf{P}_{i+1} - \mathbf{P}_i}{|\mathbf{P}_{i+1} - \mathbf{P}_i|}\right] \frac{\mathbf{P}_{i+1} - \mathbf{P}_i}{|\mathbf{P}_{i+1} - \mathbf{P}_i|}\right| \tag{6.26}$$

#### 6.4.2.4 The Overall Algorithm for Determining CC Point Locations

The algorithm to search for the CC points on a single tool path is as follows:

**Algorithm: Searching for the CC points on a single tool path**
**Input**: (1) *A machining surface or machining surface region* $S_k(u, v)$
(2) *The path cutting plane* $y = y_i$
(3) *The desirable profile tolerance* $\tau$
**Output**: *A set of CC points on the path*
**Begin**
(1) Search for the boundary points on plane $y = y_i$ on $\mathbf{S}_k(u, v)$. Set the first boundary point as the current CC point $\mathbf{P}_i$.

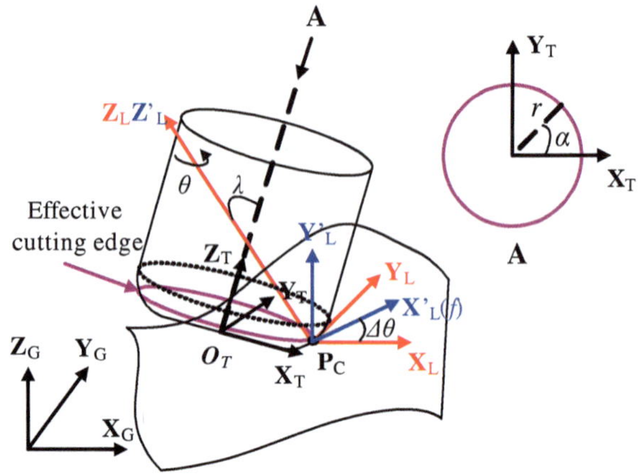


**Fig. 6.22** The new local coordinate system

(2) Estimate the step-forward length $L_i$.
(3) Search for the next CC point $\mathbf{P}_{i+1}$ according to $L_i$.
(4) Calculate the deviation $d$ of $\mathbf{P}_i\mathbf{P}_{i+1}$ from the path curve.
(5) IF $(1-\delta)\ \tau \leq d \leq \tau$, where $\delta$ is a small positive value (e.g., 0.05), accept $\mathbf{P}_{i+1}$. Otherwise, decrease $L_i$ if $d > \tau$ or increase $L_i$ if $d < (1-\delta)\ \tau$, and go to (3).
(6) Save $\mathbf{P}_{i+1}$. If $\mathbf{P}_{i+1}$ does not reach the curve segment boundary, set it as the current one and go back to (2).
(7) Output all CC points on the path. Stop.

**End**

After the locations of the CC points have been obtained, cutter postures will be assigned to them based on the heuristic described in Sect. 6.4.1.1. With CC point locations and cutter postures, the cutter center's locations can be easily calculated. Cutter center locations and cutter postures make up the CL data for the current path.

### *6.4.3 Evaluation of the Path Interval Between Adjacent Paths*

Based on the CL data (including position and posture) on the current path, the method to calculate the maximum allowable path interval $\Delta y$ for setting the next cutting plane is discussed in this section. At each CC point on the current path, a maximum allowable interval $\Delta y_j$ can be calculated that satisfy the given scallop height limit. The minimum of $\Delta y_j$ at all the CC points is taken as $\Delta y$. In the following discussion, the computation of machining strip width is first presented, followed by the detailed algorithm for calculating the path interval.

#### 6.4.3.1 Calculation of Machining Strip Width

A new local frame ($\mathbf{P}_C - \mathbf{X}'_L\mathbf{Y}'_L\mathbf{Z}'_L$) is defined in which $\mathbf{X}'_L$ is set along the feeding direction $\boldsymbol{f}$ and $\mathbf{Z}'_L$ along the surface normal vector (see Fig. 6.22). Therefore, a cutter posture ($\lambda$, $\theta$) expressed in the old local frame is represented as ($\lambda'=\lambda$, $\theta'=\theta + \Delta\theta$) in the new frame, where $\Delta\theta$ is the angle between the maximum principle direction and $\boldsymbol{f}$. Starting from here, we still denote the orientation angles as ($\lambda$, $\theta$) in the new system.

When the cutter is placed at a CC point $\mathbf{P}_C$, its effective cutting edge refers to the circular curve through $\mathbf{P}_C$ on the cutter filleted portion and normal to the cutter axis (see Fig. 6.22). With a cutter posture ($\lambda$, $\theta$), the radius of the cutting edge can be given as $r = (R - r_f) + \sin\lambda \cdot r_f$. Here, we approximate the fillet-end cutter with a flat-end cutter with radius $r$. Such an approximation is on the safe side for the evaluation of machining strip width. The proof is given in the following. As shown in Fig. 6.23a, the cutting portion of the fillet-end cutter can be represented as a set of parallel cutting circles $\{C_i, i = 1, \ldots, m, \ldots, \infty\}$, where $R(C_m) = r$. The cutting shape of the fillet-end cutter can thus be represented as $\boldsymbol{E} = \sum \boldsymbol{E}_i(i = 1, \ldots, m, \ldots, \infty)$, where $\boldsymbol{E}_i$ is the cutting shape of the flat-end cutter with radius as $R(C_i)$. Therefore, $\boldsymbol{E}_m$ must be totally contained inside $\boldsymbol{E}$ on plane $\mathbf{Y}'_L - -\mathbf{Z}'_L$, as illustrated in Fig. 6.23b. When using this flat-end cutter for the evaluation of machining strip width, the resultant result $W$ must be smaller than the actual one $W_{\text{real}}$. Such an approximation will result in a slightly better machining accuracy than required.

In the local frame, the effective cutting edge of a cutter with orientation ($\lambda$, $\theta$) is expressed as [9]:

$$\boldsymbol{P}_L(\alpha) = \begin{pmatrix} r\cos\lambda\cos\theta\cos\alpha - r\sin\theta\sin\alpha - r\cos\lambda\cos\theta \\ r\cos\lambda\sin\theta\cos\alpha + r\cos\theta\sin\alpha - r\cos\lambda\sin\theta \\ -r\sin\lambda\cos\alpha + r\sin\lambda \end{pmatrix} \tag{6.27}$$

where $\alpha \in [0°, 360°]$ gives the angular position of the point on the cutting edge (see Fig. 6.22). The effective cutting shape on $\mathbf{Y}'_L - \mathbf{Z}'_L$ plane (see Fig. 6.23b) is found from Eq. 6.27 as:

$$\boldsymbol{E}_L(\alpha) = \begin{pmatrix} 0 \\ r\cos\lambda\sin\theta\cos\alpha + r\cos\theta\sin\alpha - r\cos\lambda\sin\theta \\ -r\sin\lambda\cos\alpha + r\sin\lambda \end{pmatrix} \tag{6.28}$$

As shown in Fig. 6.23b, the machining strip width is calculated as the distance of $\mathbf{P}_a\mathbf{P}_b$ along the axis $\mathbf{Y}'_L$, where $\mathbf{P}_a$ and $\mathbf{P}_b$ are the intersection points between the effective cutting shape and the offset part surface $\boldsymbol{S}_h$ on $\mathbf{Y}'_L - \mathbf{Z}'_L$ plane. Using the second order Tailor expansion, the surface curve on $\mathbf{Y}'_L - \mathbf{Z}'_L$ plane can be approximated as $Z_L = {}^1\!/_2\kappa_n y_L^2$ [10], where $\kappa_n$ is the curvature of surface curve at the point on the normal plane. On the other hand, a point on the exact offset surface $\boldsymbol{S}_h$ is given as $\mathbf{P}_h = \mathbf{P}_s + hn(\mathbf{P}_s)$. Such a definition, although strict, is inconvenient

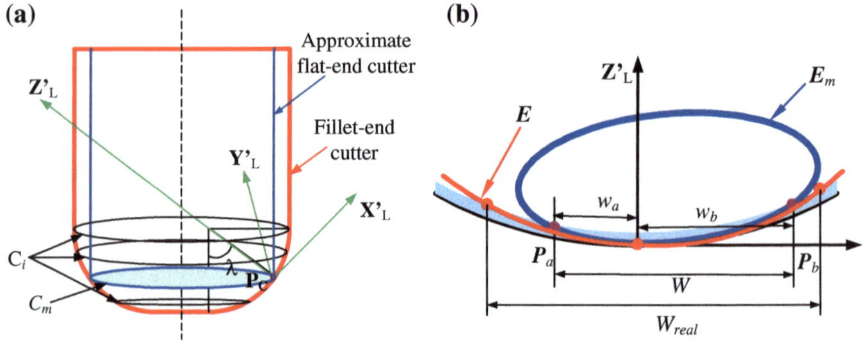


**Fig. 6.23** A fillet-end cutter ant its approximate flat-end cutter. **a** A fillet-end cutter; **b** comparison of machining strip width

for calculating the intersection points $\mathbf{P}_a$ and $\mathbf{P}_b$, as $\boldsymbol{n}(\mathbf{P}_s)$ is always changing on the surface curve. The problem may be substantially simplified if a fixed direction is used, such as the unit normal vector $\boldsymbol{n}(\mathbf{P}_c)$ at $\mathbf{P}_\mathrm{C}$. In this way $\mathbf{S}'_h$ is given as $\mathbf{P}_h = \mathbf{P}_s + h\boldsymbol{n}(\mathbf{P}_c)$. It is suggested in [11] that it is on the safe side working with this approximate version. With this approximation, on plane $\mathbf{Y}'_\mathrm{L} - \mathbf{Z}'_\mathrm{L}$, $\mathbf{S}'_h$ is then given as:

$$z_L = \frac{\kappa_n y_L^2}{2} + h \tag{6.29}$$

Combining Eqs. 6.28 and 6.29, we can obtain the intersection points $\mathbf{P}_a$ and $\mathbf{P}_b$, and then the machining strip width $W$ along $\mathbf{Y}'_\mathrm{L}$ at $\mathbf{P}_\mathrm{C}$. Please note $W$ is divided into two parts $w_a$ and $w_b$ by the CC point.

### 6.4.3.2 Evaluation of the Path Interval

As shown in Fig. 6.24, path interval $\Delta y_j$ at every CC point $\mathbf{P}_j$ is first calculated and the minimum one will be taken as $\Delta y$, i.e., $\Delta y = \min\{\Delta y_j \mid j = 1, \ldots, n_i\}$, where $n_i$ is the number of CC points on the current cutting path. The following discussion will focus on the evaluation of $\Delta y_j$ at a $\mathbf{P}_j$.

When determining $\Delta y_j$, a certain level of overlap should be maintained between the machining strips of the two adjacent paths to make sure that the scallop height is below a given limit $h$, as shown in Fig. 6.25a. On the other hand, the overlapping should be kept as low as possible to maximize machining efficiency. Thus, the following rule is proposed:

$$\Delta y_j \le w_{b,jy} + w_{a,j+1y} \le (1 + \varepsilon)\Delta y_j \tag{6.30}$$

where $w_{b,\,jy}$ and $w_{a,\,j+1y}$ are the projection of machining strip widths at $\mathbf{P}_j$ and $\mathbf{P}_{j+1}$ on the $\mathbf{Y}_\mathrm{G}$-axis respectively, $\mathbf{P}_{j+1}$ is the corresponding CC point on the next path, and $\varepsilon$ is a predefined value such as 0.05.

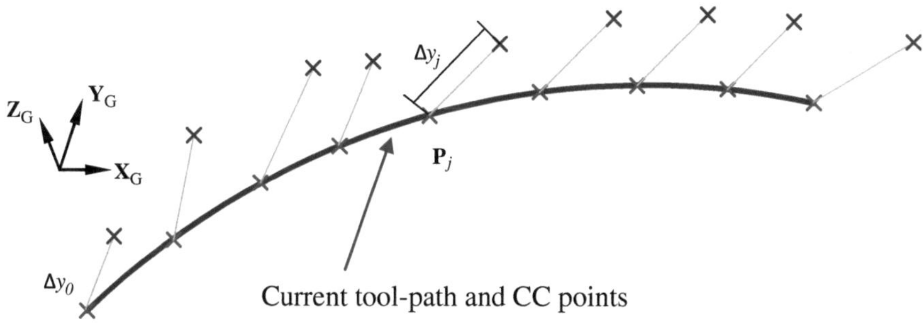


**Fig. 6.24** Evaluation of the path interval

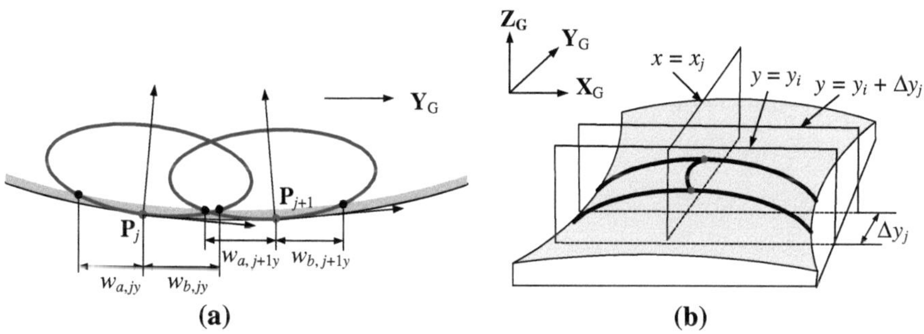


**Fig. 6.25** Calculation of path interval between two tool paths at $\mathbf{P}_j$. **a** CC points on adjacent paths; **b** CC point $\mathbf{P}_j$ and corresponding point $\mathbf{P}_{j+1}$

An iterative checking algorithm is designed to find $\mathbf{P}_{j+1}$ and obtain the largest allowable path interval. First, from $\mathbf{P}_j(u_j, v_j)$, an estimated point $\mathbf{P}_{j+1}(u_{j+1}, v_{j+1})$ is found with an initial path interval$\Delta y_j = w_{b,jy} + w_{a,jy}$. The search method is similar to that used for finding the next CC point in Sect. 6.4.2.2. $\mathbf{P}_{j+1}$ should satisfy (see Fig. 6.25b):

$$\begin{aligned} S_x(u_{j+1}, v_{j+1}) &= x_j \\ S_y(u_{j+1}, v_{j+1}) &= y_i + \Delta y_j \end{aligned} \tag{6.31}$$

where $(x_j, y_i, z_j)$ is the coordinate of point $\mathbf{P}_j$ in the global frame.

Second, the cutter posture is specified at point $\mathbf{P}_{j+1}$ with regard to its feeding direction using the method in Sect. 6.4.1.1 and the machining strip width is calculated using the method in Sect. 6.4.3.1. If the condition given in Eq. 6.30 is not satisfied, the value of $\Delta y_j$ is accordingly changed, and $\mathbf{P}_{j+1}$ will be relocated. This process will be carried out iteratively till convergence. The detailed algorithm is given as follows:

**Algorithm: Calculating the path-interval at a CC point**
**Input**: (1) *A CC point* $\mathbf{P}_C$ *and the corresponding posture*
(2) *A-maps at sampled surface points*
(3) *Desirable scallop height tolerance h*
**Output**: *the path interval at* $\mathbf{P}_C$
**Begin**
(1) Calculate the machining strip width $w_a$ and $w_b$ at $\mathbf{P}_C$, and convert them as $w_{ay}$ and $w_{by}$ along $\mathbf{Y}_G$-axis. Set $\Delta y = w_{ay} + w_{by}$.
(2) Determine the location of $\mathbf{P}_{C1}$ on the next path with $\Delta y$.
(3) Build the A-map at $\mathbf{P}_{C1}$ and determine the cutter posture based on the A-map. Calculate the machining strip width $w_{a1}$ and $w_{b1}$ at $\mathbf{P}_{C1}$ and convert them as $w_{ay1}$ and $w_{by1}$ along $\mathbf{Y}_G$-axis.
(4) Adaptively adjust the value of $\Delta y$ according to the value of $w_{by} + w_{ay1}$ as:
IF $\Delta y \leq w_{by} + w_{ay1} \leq (1 + \varepsilon)\, \Delta y$, go to (5).
IF $\Delta y > w_{by} + w_{ay1}$, decrease $\Delta y$ with a small step and go to (2).
IF $w_{by} + w_{ay1} > (1 + \varepsilon)\, \Delta y$, increase $\Delta y$ with a small step and go to (2).
(5) Output $\Delta y$ as the path interval at $\mathbf{P}_C$. Stop.
**End**

### *6.4.4 Tool Path Generation: The Overall Algorithm and Application Examples*

In the previous sections, the three basic steps for generation of iso-planar 5-axis tool paths, i.e., cutting direction determination, single tool path generation, and path interval determination have been presented. These algorithms can be implemented for the whole machining surface with single-cutter machining, or to a machining region with multi-cutter machining. The overall algorithm is provided in the following:

**Algorithm: Generating iso-planar finish tool-paths**
**Input**: (1) *A NURBS surface or a surface region S(u, v)*
(2) *A fillet-end cutter (R, $r_f$)*
(3) *Machining profile tolerance* $\tau$ *and h*
**Output**: *Tool-paths with a set of CL data*
**Begin**
(1) Cutting direction selection using the *PCR* approach. Selected cutting direction is denoted by $\boldsymbol{f}$.
(2) Set a new workpiece frame with axis $\mathbf{Y}_G$ aligned with $\boldsymbol{f}$. Transform $\boldsymbol{S}(u, v)$ into the new workpiece frame. Find $y_{min}$ and $y_{max}$ for $\boldsymbol{S}(u, v)$. Set $y_i = y_{min} + d$, where $d$ is the preset offset distance between surface edge and first tool path.

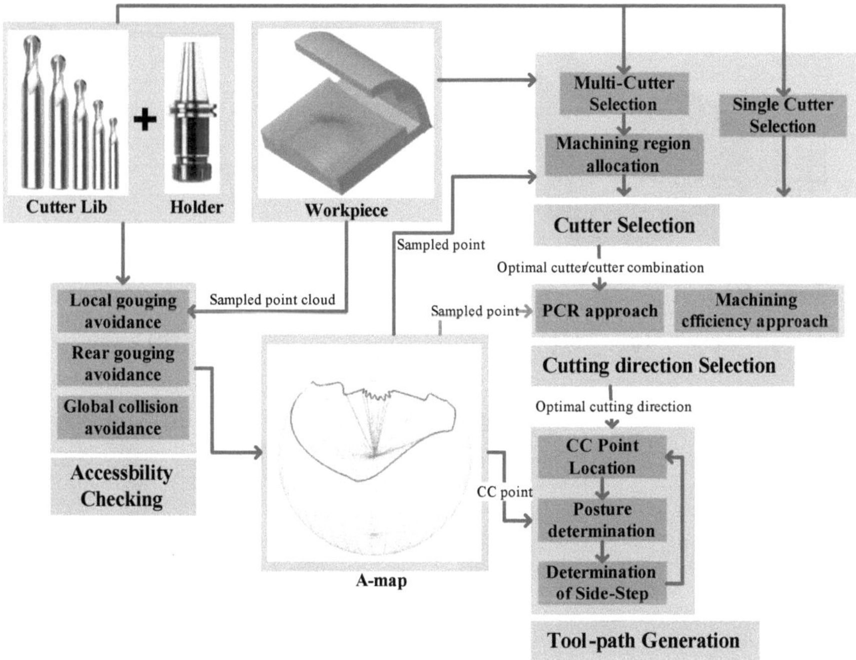


**Fig. 6.26** The automated process planning system for 5-axis milling: CAPP$^{5X}$

(3) Set a new workpiece frame with axis $\mathbf{Y}_G$ aligned with $\boldsymbol{f}$. Transform $\boldsymbol{S}(u, v)$ into the new workpiece frame. Find $y_{\min}$ and $y_{\max}$ for $\boldsymbol{S}(u, v)$. Set $y_i = y_{\min} + d$, where $d$ is the preset offset distance between surface edge and first tool path.
(4) For each CC point on the current path, build A-map for the cutter and select the posture at the point using the strategy given in Sect. 6.4.1.1.
(5) Calculate the path interval $\Delta y_{ij}$, $j = 1, \ldots, n_i$, at each CC point on the current path. Set the path interval $\Delta y_i = \min\{\Delta y_{ij}, j = 1, \ldots, n_i\}$.
(6) Set $y_i = y_i + \Delta y_i$. If $y_i <= y_{\max}$, go back to (3).
(7) Output the CL data.

**End**

The aforementioned algorithms have been implemented using C++ and OpenGL. An automated process planning system (CAPP$^{5X}$) for 5-axis finish milling of sculptured surfaces have been developed (see Fig. 6.26). In the following sections, two application examples are presented to show the efficacy of this system.

In Fig. 6.27, a case study is presented to demonstrate the effectiveness of the system in 5-axis tool path generation for a sculptured surface using a single cutter. Fig. 6.27a shows a part surface to be machined. The total PCR for each $\beta \in (0, 360°)$ has been plotted in Fig. 6.27b. It can be seen that the minimum total PCR is reached when $\beta = 90°$, i.e., when the cutting direction is along $\mathbf{Y}_G$ axis. The iso-

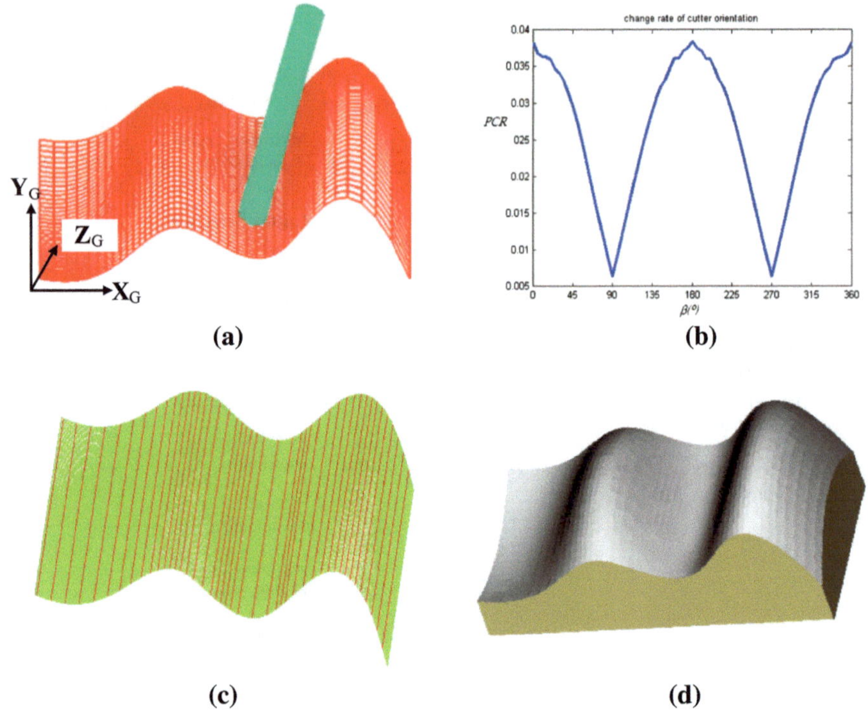


**Fig. 6.27** Case study 1–5-axis machining with a single cutter. **a** Machining surface; **b** *PCR* in all cutting directions; **c** iso-planar tool paths; **d** machining simulation result

planar tool paths generated are plotted in Fig. 6.27c. Machining simulation of this set of tool paths is carried out using VERICUT and the simulation results show that no machining interference happens (see Fig. 6.27d).

The second case study deals with 5-axis machining of a sculptured surface with multiple cutters. The workpiece model is shown in Fig. 6.28a. The workpiece is more complicated with non-machining surfaces acting as obstacles and more critical areas on the surface (see Fig. 6.28b). The cutter library shown in Table 6.1 is used. The optimal cutter set consists of two cutters $C_5$ and $C_9$. The optimal cutting direction is selected to be along the $\mathbf{X}_G$ axis (see Fig. 6.28c). The tool paths for the two cutters are generated separately as shown in Fig. 6.28c. To make a comparison, tool paths are also generated using a single cutter $C_9$ as shown in Fig. 6.28d. The tool paths generated with multiple cutters have a total number of 69 paths with a total length of 6,732.88 mm. Within the cutter set, $C_5$ has 33 paths with a total length of 3,306.16 mm and $C_9$ has 36 paths with a total length of 3,426.72 mm. Meanwhile, tool paths generated with the single cutter $C_9$ have 115 paths with a total length of 11,532.3 mm. Such results demonstrate the superior

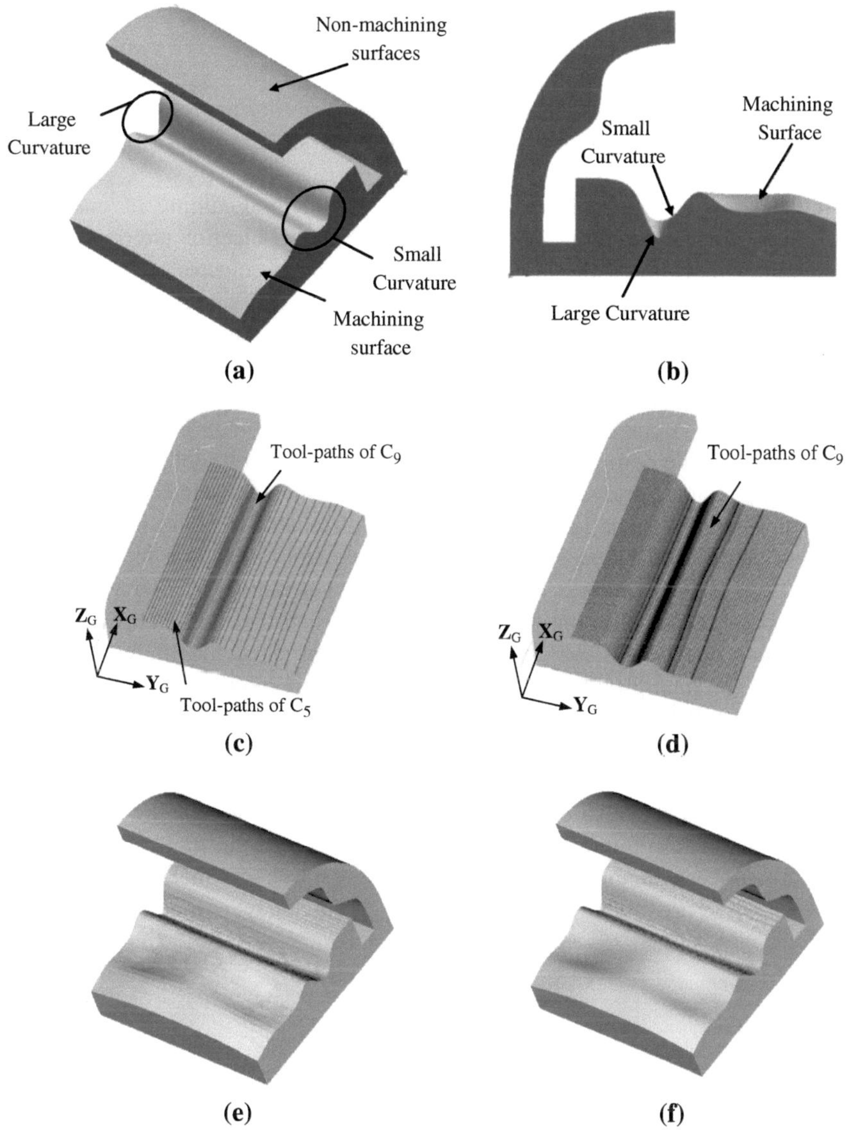


**Fig. 6.28** Case study 2–5-axis machining with multi-cutters. **a** Workpiece model; **b** workpiece model (*side view*); **c** tool-paths with $\{C_5, C_9\}$; **d** tool-paths with $\{C_9\}$; **e** machining simulation with $\{C_5, C_9\}$; **f** machining simulation with $\{C_9\}$

machining efficiency of multi-cutter machining over single-cutter machining. Besides, machining simulations are carried for both sets of tool paths without finding any machining interferences (see Fig. 6.28e, f).

## 6.5 Conclusion

In this chapter, an automated process planning system for 5-axis finish milling is introduced. The whole system is built on an accessibility analysis algorithm that produces a cutter's accessible posture range to a surface point in the form of A-maps.

With A-maps, a point-based cutter selection method aiming at maximized machining efficiency is proposed. Two modes of cutter selection are provided: for single-cutter machining, the largest accessible cutter is selected; for multi-cutter machining, the cutter combination predicted to have the shortest total tool path length is selected. Moreover, the cutting regions for each cutter in the cutter set are also allocated in this step.

The tool path pattern used in this study is iso-planar. Tool paths are generated as the intersection curves between the machining surface and a series of parallel cutting planes. The optimal cutting direction is selected with tool path smoothness as the top priority. For a single tool path, the CC points are selected such that the tool path deviation stays just below tolerance. At each CC point, the cutter postures are determined based on the A-map of the point to produce maximum strip width. Both these strategies aim at maximized machining efficiency. The path interval between neighboring paths is controlled such that the maximum scallop height stays just below the given limit, also to produce maximum machining efficiency. The final product of this system will be the 5-axis finish tool paths for the workpiece in the form of CL data.

In summary, the accessibility analysis plays a fundamental role in all process planning tasks. The A-maps not only provide important information about a cutter's accessibility but also define the solution space for the cutter's posture at a surface point. These important properties make sure that the tool paths generated using this process planning system are interference-free and meet the shape error tolerance while possessing excellent machining efficiency.

## References

1. Jensen CG, Red WE, Pi J (2002) Tool selection for five-axis curvature matched machining. Comput Aided Des 34(3):251–266
2. Li LL, Zhang YF (2006) Cutter selection for 5-axis milling of sculptured surfaces based on accessibility analysis. Int J Prod Res 44(16):3303–3323
3. O'Neil B (1966) Elementary differential geometry. Academic Press, New York
4. Smith TS, Farouki RT (2001) Gauss map computation for free-form surfaces. Comput Aided Geom Des 18(9):831–850
5. Lee Y-S, Chang T-C (1996) Automatic cutter selection for 5-axis sculptured surface machining. Int J Prod Res 34(4):977–998
6. Lee Y-S (1998) Non-isoparametric tool path planning by machining strip evaluation for 5-axis sculptured surface machining. Comput Aided Des 30(7):559–570

7. Hwang JS (1992) Interference-free tool-path generation in the NC machining of parametric compound surfaces. Comput Aided Des 24(12):667–676
8. Scherrer PK, Hillberry BM (1978) Determining distance to a surface represented in piecewise fashion with surface patches. Comput Aided Des 10(5):320–324
9. Sheltami K, Bedi S, Ismail F (1998) Swept volumes of toroidal cutters using generating curves. Int J Mach Tools Manuf 38(7):855–870
10. DoCarmo M (1976) Differential geometry of curves and surfaces. Prentice-Hall, Upper Saddle River
11. Yoon J-H, Pottmann H, Lee Y-S (2003) Locally optimal cutting positions for 5-axis sculptured surface machining. Comput Aided Des 35(1):69–81

# Chapter 7
# Manufacturing of Sculptured Surfaces Using EDM and ECM Processes

**Adam Ruszaj and Wit Grzesik**

This chapter overviews a range of unconventional manufacturing processes in terms of producing macro and micro sculptured surfaces. Special attention is focused on electro-discharge machining (EDM) and electrochemical machining (ECM) processes, as well as hybrid processes due to their wide practical implementations in many advanced manufacturing sectors. Characteristic advantages and disadvantages are discussed in terms of obtainable dimensional accuracy, surface finish and miniaturization. Some typical examples of parts produced by unconventional and hybrid processes along with technological conditions are presented.

## 7.1 Introduction

### *7.1.1 Recent Advances in Unconventional (Physical–Chemical) Machining Technology*

Unconventional processes are manufacturing processes in which a desired part of material is removed not as a result of the action of mechanical forces but due to such physical/chemical processes as melting and vaporization, ablation or electrochemical dissolution. In many cases of machining of advanced materials the combination of at least two processes applied are named as unconventional hybrid

A. Ruszaj
Faculty of Mechanical Engineering, Cracow University of Technology, al. Jana Pawla II 37, 31-864 Cracow, Poland

W. Grzesik (✉)
Department of Manufacturing Engineering and Production Automation, Opole University of Technology, P.O. Box 321, 45-271 Opole, Poland
e-mail: w.grzesik@po.opole.pl

J. P. Davim (ed.), *Machining of Complex Sculptured Surfaces*,
DOI: 10.1007/978-1-4471-2356-9_7, 

processes. In such cases material removal is predominantly performed by mechanical forces but is additionally supported by laser beam or electrochemical dissolution (electrochemical grinding).

In the case of removal by melting and vaporization the material is melted and, in some cases, vaporized by heat generated by various physical phenomena. Material is removed by pressurised gas generated from different sources. Typical examples are: EDM, laser beam machining (LBM) and electron beam machining (EBM).

It is important that mechanical properties of workpiece material do not influence the machining process because there is no mechanical contact and the temperature generated during these processes can easily exceed the boiling point of workpiece materials. On the other hand, thermal properties such as melting point, boiling point, heat conduction efficiency and heat capacitance influence the machining performance.

Among the characteristic features some disadvantages appear including:

- Electrode-tool wear and uncertainty in specifying the workpiece dimensions.
- Discharge gap occurs in EDM.
- The focused spot of beam is not clear in LBM and EBM.
- The formation of the heat affected zone (HAZ) on the machined surface removed by vaporization, skipping the phase of melting.
- Low efficiency in material removal and high cost of equipment.

Material removal by dissolution is based on ionic reaction on the workpiece surface and occurs in the direction perpendicular to the surface. The mechanical properties of workpiece do not influence the removal mechanism, a typical example is ECM. In particular, the mechanical force is equal to zero and, as a consequence, the machined surface is practically free from any damages and residual stresses.

For optimal process parameters the electrode-tool wear tends to be near zero. The workpiece surface is smooth because material is removed in the form of very small portions (atoms). In ECM, machining accuracy is lesser than in EDM because of the delocalization effect. The flow pattern and the temperature of electrolyte also influence machining accuracy.

During recomposition (deposition) metal ions in an electrolyte are deionized to become solid and to form a shape on cathode surface. Typical examples are electroplating or electroforming. It is very easy to create concave microshapes (otherwise it is difficult to fabricate convex microshapes) and they are suitable for mass production. Machining accuracy depends on the accuracy of the mould and materials suitable to recompose from solution are limited.

The new group of manufacturing processes are processes in which the entire product is built by addition of material usually layer by layer or drop by drop. These processes are known as rapid prototyping (RP), rapid tooling (RT) or rapid manufacturing (RM).

The dynamic development of additive manufacturing processes results from the new needs of market clients who continuously expect a wider variety of products. They are used to change products such as domestic equipment and car parts rather often and this is a reason for the shorter lifetime of products and time of product development. Higher expectations for product functionality is a reason for higher and higher product complexity. All the above-mentioned reasons and especially client (market) demands have forced the extremely dynamic development of additive manufacturing processes.

It is worth noticing that the additive processes, including SL, SLS, SLM, FDM, 3-DP and LOM can be applied to RP, RT and RM manufacturing methods. These processes have been optimised and machine-tool systems can work round the clock in a fully automated way. The accuracy of the product can be even 0.01 mm and obtainable surface roughness reaches Rz ~ 0.001 mm.

All the above-mentioned unconventional removal or additive processes can also be used for manufacturing microparts, in which at least one dimension is smaller than 1 mm. It is also important that in the additive processes mentioned each layer can be treated as a microelement because the layer thickness is significantly smaller than 1 mm (in majority cases usually in the range of 20–100 μm). The principles and methods of micromachining, for both classical and hybrid processes, applied in production are [1–3]:

- Mechanical forces—ultrasonic machining (USM).
- Melting/vaporization—EDM, LBM, EB.
- Ablation—LBM (excimer and femtosecond lasers).
- Dissolution—ECM, photoetching.
- Plastic deformation—punching, pressing.
- Solidification—moulding, casting.
- Lamination—SL, SLS.
- Recomposition—electroplating.

The removal or additive processes named above which applied in manufacturing of macro or micro details are being intensively developed. However, it is necessary to mention a few of the most promising directions of development. First of all, the development of additive methods, which closely fulfil the idea of "Computer Manufacturing" should be indicated. The second direction is focussed on the very intensive and dynamic development of micromachining technologies, essential for micro-systems technologies, i.e. Micro-Electro-Mechanical-Systems (MEMS). The development and scale of MEMS productions can be a measure of country civilization level. The third significant direction of this development is hybrid processes.

### *7.1.2 Characterization of Sculptured Surfaces Produced by Machining Processes*

In the production of sculptured surfaces one can apply three basic strategies. The first one can be the process of reproduction of tool dimensions and its shape in the machined material. This can take place in electrical discharge or electrochemical sinking processes. This is because the sculpting electrode-tool is displaced in the direction of machined material and as a result of material melting and vaporization

(EDM), or electrochemical dissolution (ECM), the reproduction of electrode-tool shape in the machined material takes place. Often, when metal cutting cannot machine a whole product shape some local geometries such as a sharp bottom edge of a cavity are finished by EDM. In order to enhance productivity of a product consisting of sculptured surfaces a process planning system is coupled with real-time machining [4, 5].

The second group includes the processes in which the sculptured surface of material is obtained as a reproduction of the trajectory of relative movement between the tool and the workpiece. Here we have two cases. In the first one the tool is a piece of material with very simple shape. In ED or EC milling it can be a cylindrical tool with diameter of a few millimetres. In ED cutting a wire with diameter of about 0.05–0.2 mm is applied. In the second one the tool is a light or electron beam which usually scans the machined surface in such shaping operations as cutting, milling, turning or sinking.

The above-mentioned strategies have significant limitations in the shape and dimensions of a sculptured surface formed resulting from tool properties. The quality and micro-geometry of machined surfaces result from the principle of material excess removal [6].

The third strategy is based on the application of additive manufacturing processes. As explained above, in additive processes the product is built layer by layer or even drop by drop. It gives very high flexibility in fabrication of details/special features with sculptured surfaces. This strategy seems to be the most promising way of machining sculptured surfaces.

### *7.1.3 Possibilities of Unconventional Machining Processes in Generation of Sculptured Surfaces*

Possibilities of unconventional machining processes in creation of sculptured surfaces are in very close relation with manufacturing strategies presented in Sect. 7.1.2. The most flexible and prospective in the creation of sculptured surfaces are additive methods (processes). However, their accuracy is still lower in comparison to cutting, or EDM and ECM processes. Below, a few examples of construction and technological possibilities given by unconventional machining processes are presented (Figs. 7.1–7.3).

## 7.2 Electro-Discharge Machining

### *7.2.1 Principles of Electro-Discharge Machining*

In EDM or spark-erosion machining process material excess is removed from workpiece as a result of phenomena which accompany electrical discharges in the space between electrode-tool and machined material, as shown in Fig. 7.4a. These

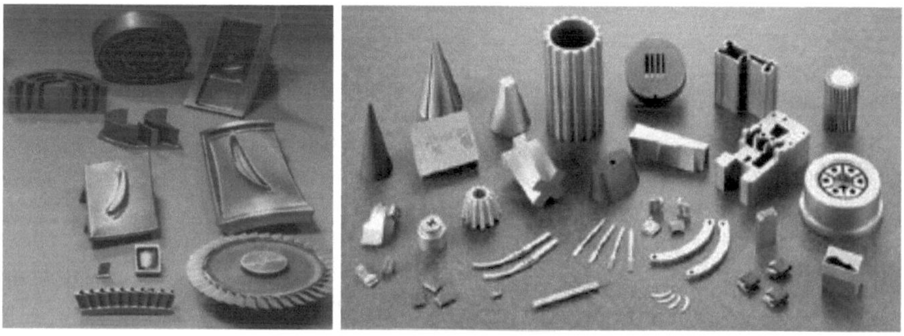

**Fig. 7.1** Examples of parts with complex features and channels produced by EDM machining [1, 7]

**Fig. 7.2** Examples of parts produced by ECM machining (turbine blade-*left* and punches and dies-*right*) [1]

**Fig. 7.3** Insert made of P/M steel manufactured by DLMS (Direct Metal Laser Sintering) process. Very complicated cooling channels inside this part (*invisible*) cannot be manufactured by other methods

fundamental phenomena are based on evaporating and melting of the machined material at a high temperature. The electrode-tool and workpiece are connected to a generator producing interelectrode pulses.

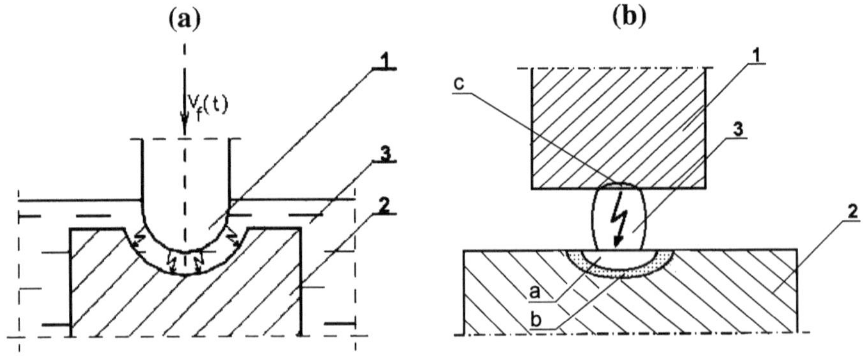


**Fig. 7.4** **a** Scheme of EDM: 1—electrode-tool (usually cathode), 2—workpiece, 3—container with dielectric liquid. $v_f$ (t)—velocity of electrode-feed rate; **b** scheme of material removal during one electric discharge: 1-electrode-tool, 2-workpiece, 3-discharge channel, b—workpiece material removed as a result of melting, a—amount of workpiece material removed as a result of evaporating, c—volume of material removed as a result of melting and evaporating from the electrode-tool [1, 8, 9]

Electric discharges occur at points where intensity of electrical field is about $10^5$–$10^7$ V/cm. This is electrical field intensity at which the electrode-tool begins emitting electrons. Electrons are accelerated in the electrical field and as a result of collisions with particles of the dielectric which is supplied into interelectrode gap, the ionization begins. After some time the ionization stream is created. Electrons are present in the head part and cations in the back part. After the ionization wave reaches the anode surface (usually the workpiece) the first stage of discharge is finished, i.e. the creation of discharge channel and plasma. Plasma in the channel has good electrical conductivity and it allows electrical discharge to develop. During electric discharge the mean temperature of plasma can reach 6,000–12,000 K [1, 8]. The heat energy of plasma is transferred to electrodes (workpiece and electrode-tool), and as a result a certain amount of material of workpiece and electrode-tool is evaporated or melted (Fig. 7.4b). The process parameters are chosen in order to transfer as much energy as possible to the machined material (workpiece) so as to decrease the electrode-tool wear.

EDM can be used for all conducting metallic materials regardless of hardness. In particular, fragile workpieces can be machined without breakage typical for classical machining. At present, EDM is one of the most accurate manufacturing processes available for creating complex or simple shapes and geometries within parts or assemblies. The cutting pattern is usually CNC controlled. In addition, EDM machine electrodes can rotate about two–three axes allowing for cutting of complex internal cavities.

The above described material removal process is applied to the following specific technological operations:

- Electro-discharge (probe) sinking.
- Electro-discharge wire cutting (WEDM).
- Electro-discharge milling.

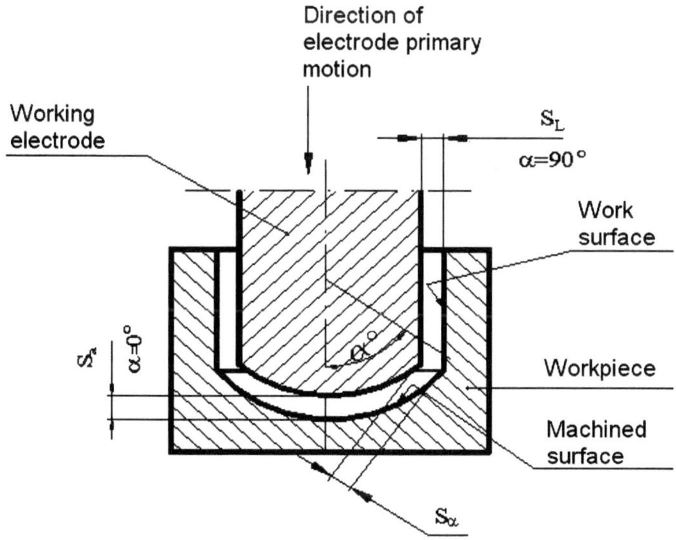


**Fig. 7.5** Scheme of electro-discharge sinking; $S_l$, $S_a$, $S_\alpha$—interelectrode gap thickness for different value of profile angle $\alpha$ [1, 8]

## *7.2.2 Electro-Discharge Sinking*

In the electro-discharge sinking process (Fig. 7.5) electrode-tool is displaced in the direction of workpiece (machined material). Material excess is removed as a result of electrical discharges. As a result, the reproduction of electrode-tool shape and dimensions in workpiece take place. The initial shape and dimensions of electrode-tool are designed as the negative of workpiece shape and dimensions corrected by two factors, i.e. interelectrode gap thickness and electrode-tool wear.

Electro-discharge sinking is applied in machining metal forming tools, shape holes and other parts for space, car and domestic industries. The EDM process finds the greatest application in tool-making including press tools, extrusion dies, forging dies and moulds. Copper or graphite electrodes produced by copy milling are often employed as tools.

## *7.2.3 Electro-Discharge Wire Cutting*

In WEDM or travelling-wire EDM process, shown schematically in Fig. 7.6, the electrode-tool is a wire with diameter 0.05–0.2 mm. During the cutting the wire moves from one reel to another. At that time the workpiece mounted on the table is simultaneously moved in x- and y-axes. Sometimes the wire is inclined in two axes.

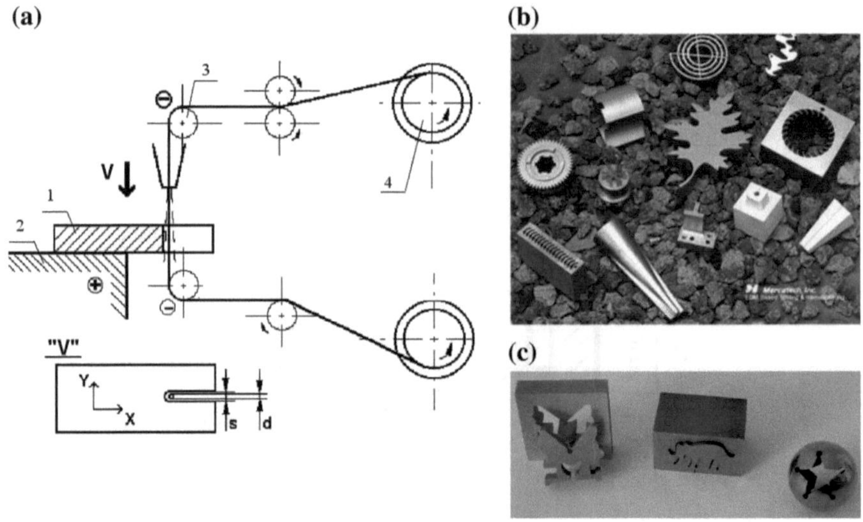


**Fig. 7.6** Scheme of WEDM process **a** 1—workpiece, 2—working table displaced in X- and Y-axes, 3—electrical contact between wire and generator, 4—unit for wire winding, s—thickness of the cut, d—wire diameter; and examples of parts produced (**b**) [1, 10–12] and intricate parts cut in tool and bearing steel workpieces (**c**)

WEDM is widely applied because of very good technological indicators. WEDM metal removal rate can be about 300–400 $mm^2$/min keeping the accuracy of ±(1–2) μm [13]. The process is carried out in deionised water and is friendly for the environment and safe because it excludes the possibility of fire. In WEDM process electrode-tool in the form of a wire (often made of brass) with diameter 0.1–0.3 mm is applied. Of course other materials are also applied. So, the electrode-tool is rather cheap and its wear is not very significant because it moves through machining area only one time. The only limitation for process parameters is to avoid wire breakage. The surface roughness of cutting surface can be even Ra = 0.1 μm [8, 13]. The obtainable dimensional accuracy is ±0.0005 in./in. and the feature profile accuracy of 0.0003 in. is obtainable with cutting path [14]. One direction of WEDM development is the tendency for application of thinner and thinner wires.

Microcracks in the surface layer are eliminated by using special electrical pulses of constant energy. The depth and number of cracks are increased in cases when corrosion occurs. It is possible to decrease this negative phenomenon by bipolar pulses when the mean voltage is zero.

### *7.2.4 Electro-Discharge Milling*

One of the very important directions of EDM process application is electro-discharge milling process (EDMM) shown in Fig. 7.7. In this process the

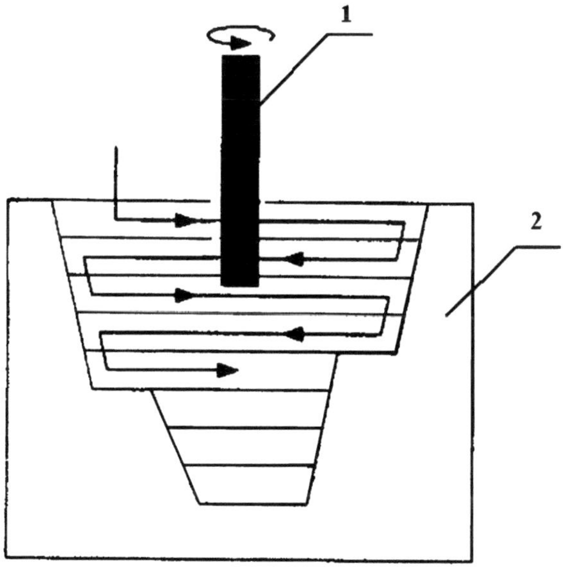


**Fig. 7.7** Scheme of EDM milling process: 1—rotating cylindrical electrode, 2—workpiece [1, 7]

electrode-tool has a simple shape like an end milling cutter, usually cylindrical with the pin diameter of 1–12 mm. This electrode-tool is moved along three-dimensional (3D) trajectory and machined surface shape and dimensions are obtained as a result of the reproduction of this trajectory in the workpiece [7].

Electrode-tool rotation can be 20–600 rot/min and the process is carried out in such a way so as to obtain electrode wear on its flat surface. A control programme automatically corrects electrode trajectory taking into account its wear. So, in comparison to EDM sinking process the problems of electrode-tool manufacturing and influence of its wear on the process technological indicators was rationally solved. EDMM can be efficiently applied for manufacturing sculptured surfaces. Very specialized and very efficient application of EDMM is the manufacturing of micro details.

### *7.2.5 Micro-Electro-Discharge Machining*

For manufacturing micro-parts one can apply EDM sinking, wire cutting and milling processes. From the definition, the micro-part is a part whose at least one dimension is smaller than 1 mm [2, 3, 15]. Because of this fact some new problems of the machining process arise, namely:

- Problem of electrode manufacturing when electrode dimensions can be reduced even to a few micrometers.

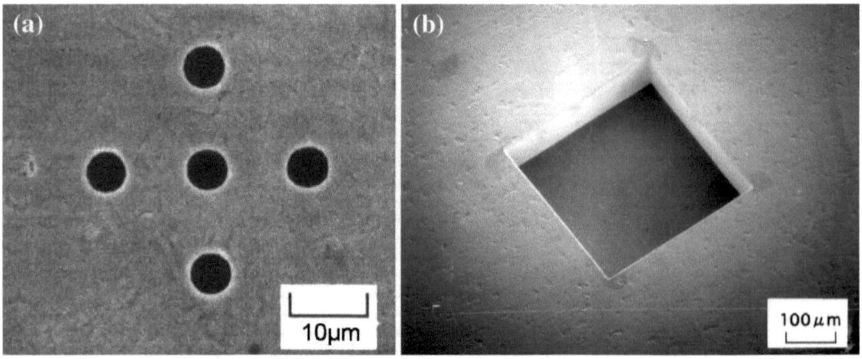


**Fig. 7.8** Holes machined in EDM sinking process: **a** cylindrical, **b** rectangular [20]

- Problem of electrode and workpiece assembly and clamping; there is a tendency for manufacturing electrode-tools on the same machine-tool which will be applied for manufacturing.
- Problem of electrode wear: electrode wear can be even 10–100% and due to this fact electrode-tool is usually made of tungsten or sintering carbides.

The general requirements for micro-EDM are: small pulse energy (less than 1 μJ) for which usually RC generators are applied, high precision of movement and high accuracy of repeated positioning, rapid response of control mechanism. Below some examples of micro details manufactured using EDM process are presented in Figs. 7.8–7.10 using EDM sinking milling and micro-wire EDM processes respectively.

In micro-wire EDM the working principle is the same as that of conventional wire EDM but the wire of diameter 20–30 μm made of tungsten or copper is applied. Special technological possibilities of micro-EDM let it deal with narrow slots producing high aspect ratio features as shown in Fig. 7.10.

Medical shops use both wire and sinker EDMs for micromachining. Working with electrode wire four to five times smaller than a human hair, holding parts barely visible with the naked eye, and accurately producing holes as tiny as 0.0005 in. in diameter is not new in micromachining medical parts with EDM. Obviously, holding tiny medical parts for micro-EDMing is difficult and special equipment is recommended and tiny parts secured using chewing gum. Typically, the smallest tungsten wire used in micro-EDMing is 0.0008 in. in diameter, while common sizes are 0.001 and 0.002 in. The challenge is creating start holes for such tiny wires. Once start holes are burnt in, the wire routes through a machine's tensioning system and into its automatic wire threader. Unlike with conventional EDMs, these two systems must handle the tiny wire diameters for micro-EDMing.

Micro-EDM is a high accuracy method of miniaturization, low speed and shape flexible. Micro-EDM can create the wide range of shapes and cover the wide choice of workpiece materials. It is suitable for the preparation of micro die/moulds, prototype products and small batch production economically and efficiently.

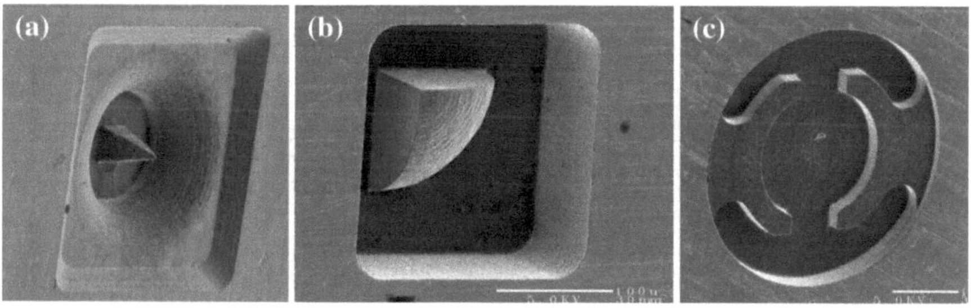


**Fig. 7.9** Micro-parts manufactured by EDM milling process: **a** and **b** rectangular pockets with internal protrusions, **c** cylindrical pocket with side-formed elements [20]

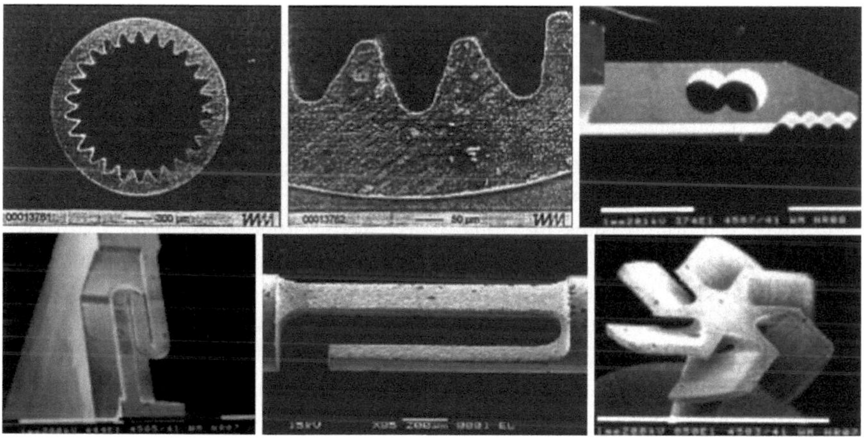

**Fig. 7.10** Examples of micro-parts machined using micro-WEDM [3, 20]

# 7.3 Electrochemical Machining

## *7.3.1 Principles of Electrochemical Machining*

ECM is a method of creating metal shapes by removing metal using an electrochemical process. Similar to EDM, ECM also uses electrical energy to remove material and is the opposite of electrochemical or galvanic coating or deposition process. In other words, ECM is essentially a deplating process that utilizes the principles of electrolysis. In this aspect, ECM is a case of a controlled anodic dissolution at atomic level of the workpiece that is electrically conductive by a shaped tool due to the flow of high current at relatively low potential difference through an electrolyte which is quite often water-based neutral salt solution. The basic principle of ECM, with the tool as the cathode and the workpiece as the anode, is schematically shown in Fig. 7.11. A direct high amperage, low voltage

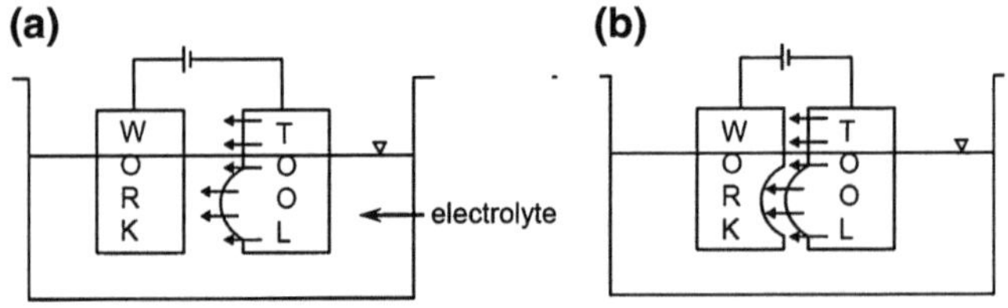


**Fig. 7.11** Schematic principle of electrochemical machining (ECM) [8, 11]. **a** Initial stage of ECM process. **b** Steady-state of ECM process

current is used to dissolve the metal and to remove it from the workpiece which must be electrically conductive. Machining takes place due to anodic dissolution at atomic level of the work material due to electrochemical reaction. A gap between the tool and the workpiece is required to be maintained to proceed with steady-state machining.

The ECM tool is positioned very close to the workpiece (Fig. 7.11b) and a low voltage, high amperage DC current is passed between the two via an electrolyte. The tool is shaped and the shape of the tool is transferred to the workpiece as illustrated in Fig. 7.11. Material is removed from the workpiece and the flowing electrolyte solution washes the ions away. These ions form metal hydroxides which are removed from the electrolyte solution by centrifugal separation. Both the electrolyte and the metal sludge are then recycled.

ECM is important method of removing metal by anodic dissolution and offers a number of advantages over other machining methods. Metal removal is effected by a suitably shaped tool electrode, and the parts thus produced have the specified shape, dimensions and surface finish. ECM forming is carried out so that the shape of the tool electrode is transferred onto, or duplicated in, the workpiece. For high accuracy in shape duplication and high rates of metal removal, the process is effected at very high current densities of the order 10–100 A/cm$^2$, at relative low voltage usually from 8 to 30 V, while maintaining a very narrow machining gap (of the order of 0.1 mm) by feeding the tool electrode in the direction of metal removal from the work surface, with feed rate from 0.1 to 20 mm/min. Dissolved material, gas and heat are removed from the narrow machining gap by the flow of electrolyte pumped through the gap at a high velocity (5–50 m/s).

Being a non-mechanical metal removal process, ECM is capable of machining any electrically conductive material with high stock removal rates regardless of their mechanical properties. In particular, removal rate in ECM is independent of the hardness, toughness and other properties of the material being machined. The use of ECM is most warranted in the manufacture of complex-shaped parts from materials that lend themselves poorly to machining by other methods, mechanical in particular. There is no need to use a tool made of a harder material than the workpiece, and there is practically no tool wear.

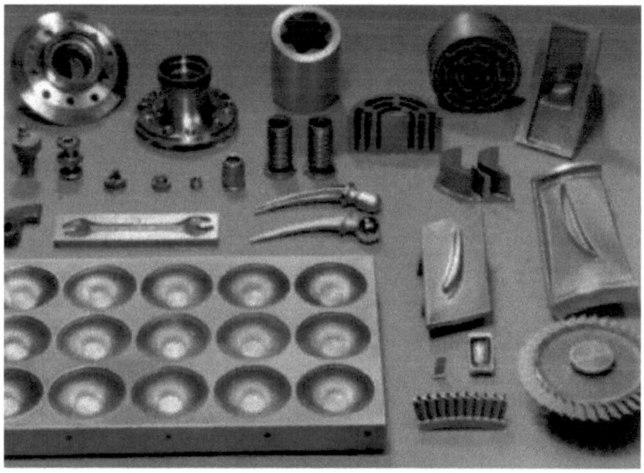

**Fig. 7.12** A variety of parts with internal and external surfaces machined by ECM [15]

Since there is no contact between the tool and the work, ECM is the machining method preferred in the case of thin-walled, easily deformable components and also brittle materials likely to develop cracks in the surface layer. ECM can create normal and delicate 3D shapes but it cannot produce sharp corners and edges. Some typical ECM applications are: copying of complex internal and external surfaces, cutting of curvilinear slots, machining of intricate patterns, production of long curved profiles, machining of gears, etc.

As mentioned above, in most modifications of ECM, the shape of the tool electrode is duplicated over the entire surface of the workpiece connected as the anode. Therefore, complex-shaped parts can be produced by simply moving the tool translationally. For this reason and also because ECM leaves no burrs, one ECM operation can replace several operations of mechanical machining. ECM removes the defective layer of the material and eliminates the flaws inherited by the surface layer from a previous treatment and usually does not generate residual stress in the workpiece. All this enhances the service qualities of the parts manufactured by ECM.

Some examples of parts that are made using ECM include dies, moulds, turbine and compressor blades, cavities, holes, slots, etc., as presented in Figs. 7.12 and 7.13. ECM can be applied to most types of conducting materials and alloys.

Small or odd-shaped angles, intricate contours or cavities in hard and exotic metals, such as titanium aluminides, Inconel, Waspaloy, and high nickel, cobalt and rhenium alloys can be created. In addition, both external and internal geometries can be machined.

Various industrial techniques have been developed on the basis of the process including electrochemical cutting, shaping, broaching, drilling and deburring.

In dissolution process material is removed "atom" by "atom" and as a result the surface quality is higher in comparison to other machining processes. For majority materials (metallic: steels or alloys) there is no problem with receiving roughness value of Ra = 0.32–2.5 μm, however sometimes with optimal process parameters and metallographic material structure it is possible to reach even

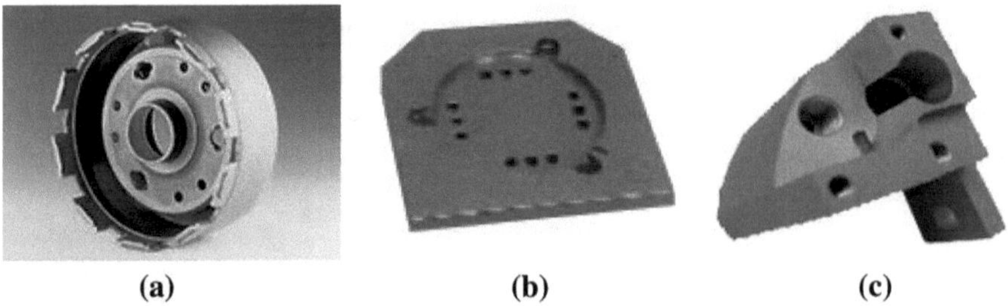

Fig. 7.13 Examples of different parts formed by ECM: a rotor, b sensor, c bracket [16]

Ra = 0.02–0.16 μm. It is important to keep the temperature into machining area significantly lower than 100°C which eliminates any metallographic changes in the surface layer.

## 7.3.2 Electrochemical Sinking

In electrochemical sinking operations the electrode-tool is displaced in the direction of machined material as shown in Fig. 7.14. Electrolyte is flowing into machined area through the hole in electrode-tool.

Hydrodynamic conditions of electrolyte flow are very significant because it is essential for the results of machining to remove from interelectrode area products of electrochemical reactions and heat generated as a result of the flow of current. Material is removed from workpiece and as a result the electrode-tool shape and dimensions are reproduced in machined surface (Fig. 7.14). In order to obtain the desired shape and dimensions of the workpiece, the electrode-tool dimesions should be corrected by interelectrode gap thickness [1, 8].

Classical applications of electrochemical sinking include manufacturing of forming tools, aircraft engine turbine blades. Accuracy of ECM depends on interelectrode gap thickness t∼(0.10–0.15) S, and it usually increases when interelectrode gap thickness decreases. However, when the gap is too small the electrolyte flow is unsteady and shaping accuracy tends to decrease.

## 7.3.3 Electrochemical Milling

In electrochemical sinking some significant problems, such as evaluation of interelectrode gap distribution, keeping optimal conditions of electrolyte flow during machining of large surfaces arise. In order to overcome these difficulties the conception of electrochemical milling using universal electrode is proposed as presented in Fig. 7.15.

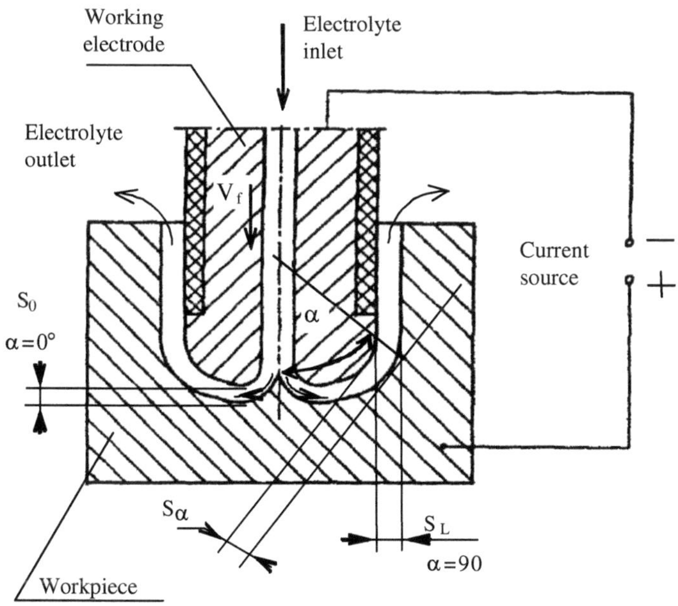


**Fig. 7.14** Scheme of electrochemical sinking (for example forming tool) [1, 8]

**Fig. 7.15** Scheme of electrochemical milling using universal electrode-tool [1]

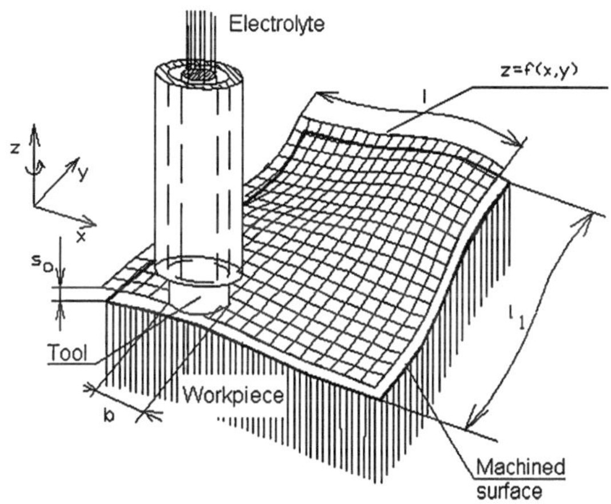


Universal electrode applied in electrochemical process is not exposed to wear and its additional rotation is applied only for improving hydrodynamic conditions. Unfortunately, this process offers low metal removal rate and, in consequence, electrochemical milling is applied mainly in finishing operations (smoothing) or in the machining of miniature parts.

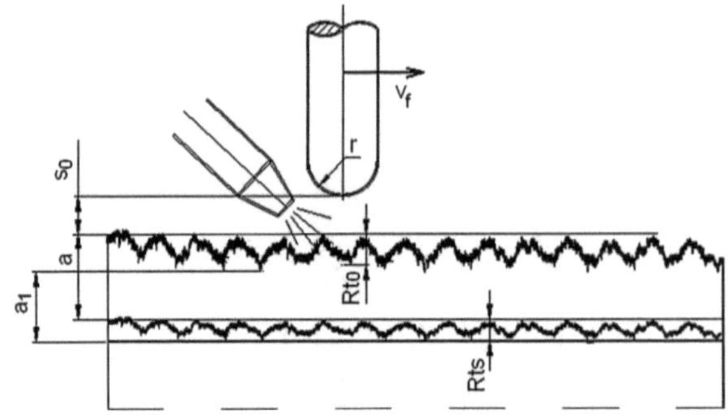


**Fig. 7.16** Scheme of electrochemical smoothing in milling operations; 1—electrode-tool, 2—smoothing surface, a—allowance for removing during smoothing process, $v_f$—electrode-tool feed rate over machined surface, $R_o$—surface roughness parameter before smoothing, $R_k$—surface roughness parameter after smoothing [1, 8, 17]

### *7.3.4 Electrochemical Smoothing*

Electrochemical smoothing can be carried out in a similar way as sinking, milling or sometimes turning. The aim of this operation is to improve surface quality by decreasing surface roughness parameters or removing damages from the subsurface layer which can be induced after rough electrical discharge machining, mechanical milling, etc. Because electrochemical dissolution processes do not change surface layer properties and it is easy to obtain extremely low surface roughness parameters, in this aspect electrochemical smoothing is a very efficient operation. For instance, Fig. 7.16 shows the principle of surface smoothing during electrochemical milling.

A very important advantage of electrochemical smoothing in sinking is very short time of $t \approx 15 \div 90$ s, and the basic disadvantages are low flexibility and high cost of electrode-tool manufacturing. Advantage of smoothing in EC milling is very high flexibility and low cost of electrode-tool. On the other hand, the main disadvantage is the longer time of smoothing.

### *7.3.5 Micro-Electrochemical Machining*

At present, ECM becomes one of the leading methods widely applied in micromachining operations due to many advantages, including [1, 8, 18, 19]:

- Material is removed as a result of electrochemical dissolution process without mechanical forces in low temperature and by very small portions ("atom" by "atom").
- Electrode tool wear does not occur.

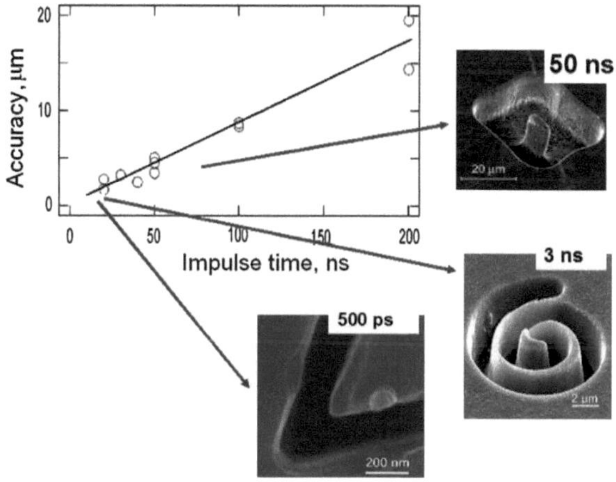


**Fig. 7.17** Influence of voltage pulse time on machining accuracy in µECM

- Metal removal rate and surface quality are significantly higher than in EDM independently of mechanical properties of machined material.

The disadvantages of this process are:

- Lower than in EDM process localization (accuracy),
- Sensitivity for changes of metallographic structure and chemical constitution of machined material; the best surface quality is received for uniform materials with as small as possible grains.

The main electrochemical micromachining process drawback (low process localization) was overcome by using very short voltage pulses (pico and nanoseconds) which is documented in Fig. 7.17. On the other hand, Fig. 7.18 compares qualities of micro-holes of 100 µm in diameter made by micro-ECM (µECM) and micro-EDM (µEDM) processes respectively.

# 7.4 Practical Applications of EDM and ECM

## *7.4.1 Industrial Equipment for EDM and ECM Machining*

Electrical discharge and ECM processes have been applied in the industry for about 50 years. During this time theoretical and technological knowledge databases were significantly developed. Moreover, the advanced systems of computer aided design (CAD) computer aided manufacturing (CAM) were successfully implemented. These advanced systems are usually offered together with advanced machine tools as for example advanced machining centres, which enable manufacturing details with high accuracy at a high level of automation.

Unfortunately, now these systems are very costly and because of this fact they are installed in advanced big industrial companies. In case of small and medium

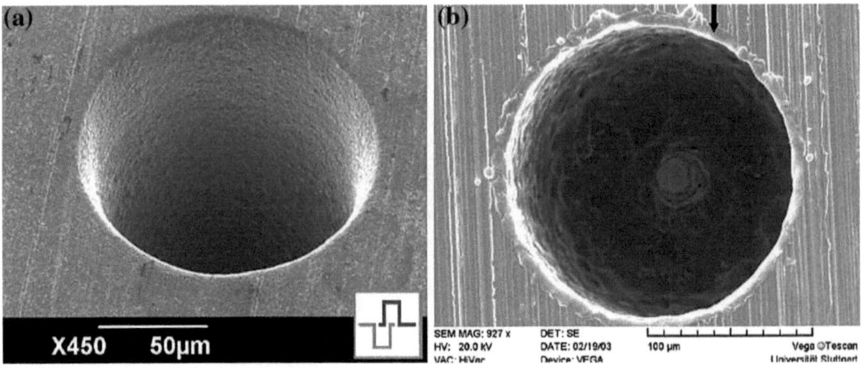


**Fig. 7.18** SEM images of micro-holes machined by ECM (**a**) and by EDM (**b**) [21]

shops less advanced equipment is used so they can also efficiently apply unconventional technologies, mainly EDM and ECM processes. In other words, each company can find machine tools appropriate for their needs and CAD/CAM systems, which make it possible to manufacture sculptured surfaces on parts of advanced materials with accuracy of about a few micrometers and with high quality of the surface roughness of Ra $\sim$ 0.5 μm.

It is only worth noticing that industrial applications of EDM technology are significantly wider than ECM technology. ECM has a few classical applications, predominantly in space, aircraft, car and domestic industries, for instance to produce aircraft engines turbine blades. However, during the past years its industrial applications have increased, mainly for special tasks using dedicated equipment.

Electrical discharge machine tools (centres) for sinking and milling operations as well as machine tools for wire cutting are produced by many companies all over the world. In the range of EDM machine tools there are simple, hand controlled machine tools and very advanced machining centres equipped with sophisticated CAD/CAM systems. Some EDM machine tools for sinking and wire cutting are presented in Fig. 7.19 (exemplary micro-parts produced on the EDM machine shown in Fig. 7.19b are presented in Fig. 7.20).

As mentioned above the equipment for ECM technology is much more complicated than this for EDM and producers usually dedicate it for solving special tasks. Two different designs dedicated to large structures and micro-parts are presented in Fig. 7.21.

### *7.4.2 Details of Manufacturing Using EDM and ECM Processes*

Electrical discharge sinking or milling can be applied first of all for machining working cavities of dies, moulds, shaped or cylindrical holes and many other

(a) (b)

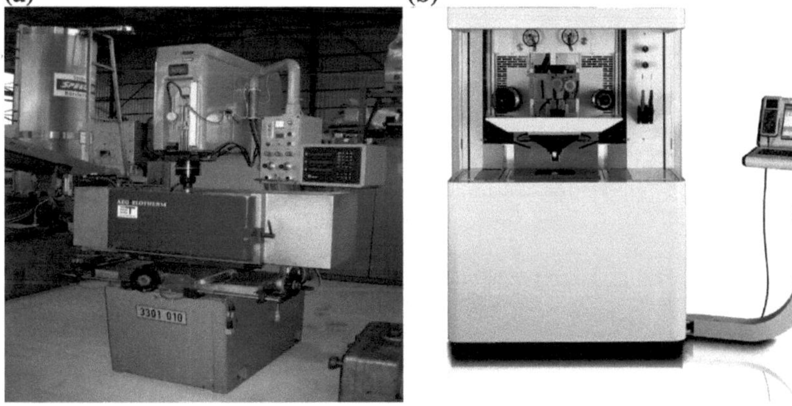

**Fig. 7.19** Examples of EDM machine tools: **a** multifunctional electrical discharge sinking/ milling machine tool, **b** wire cutting machine for ultra-accurate applications [7]

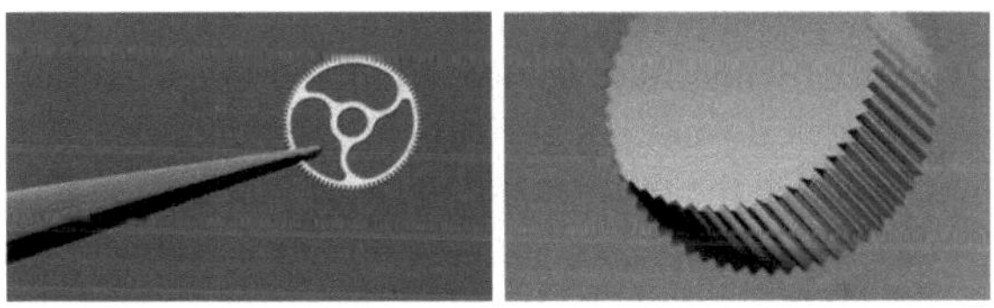

**Fig. 7.20** Micro-parts machined on precision wire cutting machine shown in Fig. 7.19b [7]

sculptured surfaces in details made of materials difficult for cutting. For many years there has been dynamic competitions between high speed milling and EDM sinking or milling processes. Nowadays it obvious that these two technologies can support each other. Because of this fact in advanced space, aircraft and medical industry companies apply HSM and EDM for advanced production of dies and moulds. These parts of dies or moulds which need to remove a big amount of material are machined using HSM technology and parts with precise sculptured 3D shape are machined by EDM. The technical and economical effects are astonishing and result from synergy effect.

Electrical discharge wire cutting process is also very precise and efficient in production of blanking and forming dies and many other details for space, aircraft, car or medical industries. Nowadays it is simply impossible to imagine production of sculptured surfaces without EDM technology.

ECM has also a very important (as mentioned previously) field of application, usually thanks to its basic assets: lack of electrode-tool wear, metal removal rate significantly higher than in EDM and in some cases (also in HSM) and surface layer with quality of the core material. These advantages are essential for production of aircraft turbine blades.

For both EDM and ECM technologies it is essential that during machining there is no mechanical contact between electrode-tool and workpiece and a very

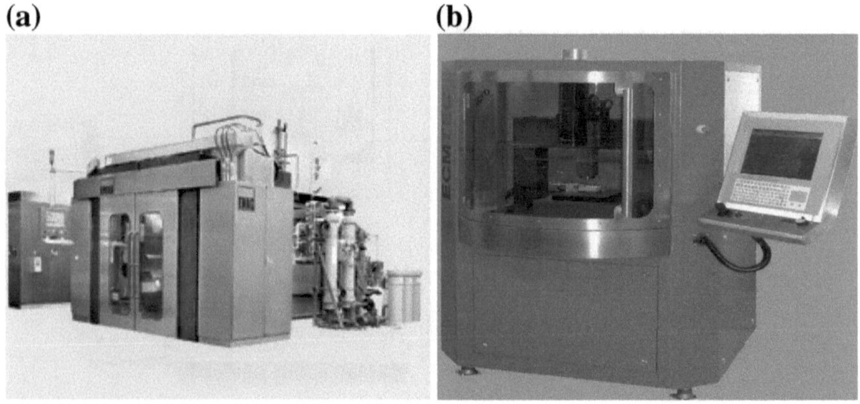


**Fig. 7.21** Examples of EDC machine tools: **a** machine tool with very high metal removal rate dedicated to manufacture of advanced 2D/3D structures, **b** machine tool for micromachining dedicated to manufacture of advanced 2D/3D microstructures [13]

important technological indicator is: thickness of interelectrode gap (S). In EDM processes when interelectrode gap is higher than critical value (Sk), the electrical discharges do not occur. This phenomenon is a limitation for interelectrode gap changes. On the other hand, in ECM process the interelectrode gap thickness S can be increased even to over $\sim 1$ mm as a result of changes of electrolyte electrical conductivity along electrolyte flow or on detail side walls. Because of this fact the problem of electrode correction is much more complicated in comparison to EDM sinking. To simplify this problem the pulse ECM process has been worked out and applied successfully in the industry. In this case, the heat and electrochemical reaction products are removed out of the machining space between voltage pulses or pack of pulses. Thanks to this, the distribution of interelectrode gap thickness has become more uniform. The disadvantage of this solution is decreasing metal removal rate in comparison to machining with constant current.

When there is no mechanical contact between electrode-tool and workpiece one can assume that forces acting on the workpiece are zero. This assumption can be made in EDM sinking or milling macro details. In wire cutting or in micro-EDM the electrode-tool is very small so the electrodynamic forces can be significant for machining process. In WEDM the wire electrode-tool can vibrate and in micro-EDM electrode can be distorted.

In ECM process of large surfaces the forces from electrolyte pressure can be very high and this fact should be taken into account when designing technological equipment. The essential problem which decides about the results of ECM machining is: optimal electrolyte flow through interelectrode area. When electrolyte flow is not optimal the accuracy and surface quality significantly decrease and in many cases the machining process is stopped by electrical discharges.

The diameter of electrode-tools in EDM or ECM micro-milling is usually about 10–100 μm. Because of this fact the very important problems of clamping and positioning these tiny tools occur. These problems were partly solved by applying special units for electrode-tools manufacturing in machine tools.

The producers of EDM and ECM machine tools usually equip them with CAD/CAM systems, however when using simple, hand controlled machine tools the technological process designing becomes more difficult. In this case the results depend on the technologists' experience.

### *7.4.3 Advantages and Disadvantages of EDM and ECM Processes*

Basic advantages of EDM process are [1, 8, 12]:

- Possibility of machining metals, their alloys and electrically conductive composites which are difficult for cutting.
- Possibility of machining fragile materials which can be damaged as a result of cutting forces.
- Possibility of machining elements with thin walls which could be distorted as a result of cutting forces.
- Possibility of machining sculptured surface with accuracy $T = 0.01 \div 0.1$ mm which is satisfying for manufacturing majority of die cavities, metal moulds etc.
- Possibility of machining with high precision microelements.
- High reliability of material removal and possibility of complex automation. EDM machine tools can work round the clock.

Disadvantages of the EDM process are [1, 8, 12]:

- Electrode tool wear which is a reason for machining cost increase.
- Material excess is removed as a result of melting and evaporating and because of this the surface layer properties are changed in comparison to core material (Lower mechanical properties, sometimes cracks, etc.).
- In order to receive high quality of surface layer (low surface roughness) the finishing operations are necessary.
- Relatively low metal removal rate in comparison to HSM or ECM processes.
- Some environment problems (radiation fog, harmful dielectric fluids), however smaller in comparison to ECM.

Basic advantages of ECM process are [1, 8, 12]:

- Possibility of machining metal, their alloys and conductive composite materials with high metal removal rate in comparison to EDM and in some cases HSM.
- Possibility of machining fragile materials which can be broken in cutting operations.
- Possibility of machining details with thin walls which can be distorted during cutting.
- Possibility of machining sculptured surfaces with dimensional accuracy of 0.05–0.20 mm, satisfying in majority of cases.

- Possibility of receiving surfaces with high quality—without internal stresses or cracks and low surface roughness parameters (temperature in machining area $T \ll 100°C$).
- Lack of electrode-tool wear when machining with optimal process parameters—especially electrolyte flow (machining without electrical discharges which are the only reason for electrode-tool wear and damage of machined surface).
- Very high surface quality and very short time of finishing operations (ECM smoothing) in comparison to when finishing operations are applied HSM or EDM.
- In the majority of cases the higher the metal removal rate the higher the accuracy of machining (because of smaller interelectrode gap) and machined surface quality—the higher the current density, the smaller the surface roughness created in ECM process.

The reasons for rather low range of ECM industrial application are its disadvantages [1, 8, 12]:

- Very high costs of machine tools and tooling (in comparison to HSM or EDM equipment), it is necessary to apply special materials because of corrosion problems.
- Environmental problems with electrolytes and products of electrochemical dissolution process (for instance it is necessary to reduce $Cr^{+6}$ to $Cr^{+3}$).
- It is necessary to wash and protect the machined details against corrosion.
- Lower accuracy of machining (low localization of electrochemical dissolution process) in comparison to EDM or conventional HSM processes.
- Lesser possibilities of process automation in comparison to EDM or HSM—it results from lower reliability in material removal because of random disturbances in electrolyte flow or local passivation.

## References


1. Ruszaj A (1999) Unconventional methods of producing machine and tool elements (in Polish). Institute of Advanced Manufacturing Technologies (IZTW), Cracow
2. Masuzawa T (2000) State of the art of micromachining. Ann CIRP 49(2):473–488
3. Grzesik W (2008) Advanced machining processes of metallic materials. Elsevier, Amsterdam
4. Yamazaki K, Kawahara Y, Jeng J-Ch, Aoyama H (1995) Autonomous process planning with real-time machining for productive sculptured surface machining based on automatic recognition of geometric features. Ann CIRP 44(1):439–444
5. Davim JP (ed) (2008) Machining. Fundamentals and recent advances, Chapters 8 and 11. Springer, London
6. Davim JP (ed) (2010) Surface integrity in machining. Springer, London
7. ED die-sinking and wire-cut machines. http://www.gfac.com/agiecharmilles
8. Jain VK (2005) Advanced machining processes. Allied Publishers PVT Ltd, New Delhi
9. Boothroyd G, Knight WA (2006) Fundamentals of machining and machine-tools. CRC Press, Boca Raton
10. Groover MP (2011) Principles of modern manufacturing, 4th edn. Wiley, New Delhi

11. Tlusty G (2000) Manufacturing processes and equipment. Prentice-Hall Inc, Upper Saddle River
12. Nontraditional machining processes. http://www.mechse.illinois.edu
13. Elektrochemische Metallbearbeitung (ECM). http://www.gfac.com/EMAG
14. MicroEDMing medical parts (2005) American Machinist
15. Electrochemical machining. http://www.home-machine-shop.com
16. Electrochemical machining. http://www.emachineshop.com
17. Ruszaj A, Czekaj J, Miller T, Skoczypiec S (2005) Electrochemical finishing surfaces after rough milling. Int J Manuf Sci Technol 7(2):21
18. Jackson MJ (ed) (2006) Microfabrication and nanomanufacturing. CRC Press, Boca Raton
19. El-Hofy HA-G (2005) Advanced machining processes–nontraditional and hybrid machining processes. The McGraw-Hill Companies, New York
20. Rajurkar KP, Levy G, Malshe A, Sundaram MM, McGeough J, et al (2006) Micro and nano machining by electro-physical and chemical processes. Annals of the CIRP 55/2:643–666
21. Electrochemical micro-milling. http://www.ecmtec.eu

# Index





DOI: 10.1007/978-1-4471-2356-9, © Springer-Verlag London Limited 2012

MIX
Papier aus verantwortungsvollen Quellen
Paper from responsible sources
FSC® C105338

If you have any concerns about our products,
you can contact us on
**ProductSafety@springernature.com**

In case Publisher is established outside the EU,
the EU authorized representative is:
**Springer Nature Customer Service Center GmbH**
**Europaplatz 3, 69115 Heidelberg, Germany**

Printed by Libri Plureos GmbH
in Hamburg, Germany